PRIMATE SOCIAL RELATIONSHIPS
AN INTEGRATED APPROACH

 SINAUER ASSOCIATES,INC · PUBLISHERS

SUNDERLAND, MASSACHUSETTS 01375

TO: Please accept this complimentary copy for review. We would appreciate receiving two tearsheets of any notice or review that you publish

Thank you.

Pub Date: January 1984 $21.00 paper
$40.00 cloth

Primate Social Relationships

AN INTEGRATED APPROACH

EDITED BY

ROBERT A. HINDE
ScD, FRS
MRC Unit on the
Development and Integration of Behaviour
Madingley, Cambridge

WITH CONTRIBUTIONS BY

Carol M. Berman	Cynthia J. Moss
Bernard Chapais	Joyce H. Poole
Dorothy L. Cheney	Kathlyn L. R. Rasmussen
John Colvin	Steven R. Schulman
Saroj B. Datta	Robert M. Seyfarth
Robin I. M. Dunbar	Michael J. A. Simpson
Alexander H. Harcourt	Barbara B. Smuts
Robert A. Hinde	Joan Stevenson-Hinde
Jill M. Hooley	Kelly J. Stewart
Phyllis C. Lee	Richard W. Wrangham

CHAPTER HEAD-PIECES BY
Priscilla Barrett

SINAUER ASSOCIATES, INC. · PUBLISHERS
SUNDERLAND, MASSACHUSETTS
BOSTON MELBOURNE

First published 1983

Photoset by Enset Ltd, Midsomer
Norton, Bath, Avon and printed and
bound in Great Britain at the
Camelot Press Ltd, Southampton

Distributed in the USA by
Sinauer Associates, Inc.,
Publishers, Sunderland
Massachusetts

Library of Congress
Cataloging in Publication Data

Primate social relationships.
    Bibliography: p.
    Includes indexes.
    1. Primates—Behavior.   2. Social
behavior in animals.   I. Hinde,
Robert A.   II. Berman, Carol M.
[DNLM: 1. Primates.   2. Social behavior.
3. Socialization.   4. Behavior, Animal.
QL 737.P9 P9532]   QL 737.P9P674
1983   599.8'0451   83-12023
ISBN 0-87893-275-5
ISBN 0-87893-276-3 (pbk.)

Contents

Contributors, ix

Preface, xi

Acknowledgements, xv

1   **A Conceptual Framework,** 1
ROBERT A. HINDE

2   **Species, Study Sites and Methods,** 8
PHYLLIS C. LEE

3   **Description of Social Behaviour**
3.1 General issues in describing social behaviour, 17
ROBERT A. HINDE
3.2 Description of sibling and peer relationships among immature male rhesus monkeys, 20
JOHN COLVIN

4   **Individual Characteristics and the Social Situation**
4.1 Individual characteristics: a statement of the problem, 28
JOAN STEVENSON-HINDE
4.2 Consistency over time, 30
JOAN STEVENSON-HINDE
4.3 Predictability across situations, 34
JOAN STEVENSON-HINDE
4.4 Context-specific unpredictability in dominance interactions, 35
PHYLLIS C. LEE

5   **Influence of Individual Characteristics Upon Relationships**
5.1 Individual characteristics and relationships, 45
ROBERT A. HINDE
5.2 Age-related variation in the interactions of adult females with adult males in yellow baboons, 47
KATHLYN L. R. RASMUSSEN
5.3 Effect of the sex of an infant on the mother–infant relationship and the mother's subsequent reproduction, 53
MICHAEL J. A. SIMPSON
5.4 Rank influences rhesus male peer relationships, 57
JOHN COLVIN

6   **Development and Dynamics of Relationships**

6.1   Development and dynamics, 65
ROBERT A. HINDE

6.2   Feedback in the mother–infant relationship, 70
ROBERT A. HINDE

6.3   Effects of the loss of the mother on social development, 73
PHYLLIS C. LEE

6.4   Effects of being orphaned: a detailed case study of an infant rhesus, 79
CAROL M. BERMAN

6.5   Play as a means for developing relationships, 81
PHYLLIS C. LEE

6.6   Differentiation of relationships among rhesus monkey infants, 89
CAROL M. BERMAN

6.7   Relative power and the acquisition of rank, 93
SAROJ B. DATTA

6.8   Relative power and the maintenance of dominance, 103
SAROJ B. DATTA

6.9   Dynamics of 'special relationships' between adult male and female olive baboons, 112
BARBARA B. SMUTS

6.10 Influence of affiliative preferences upon the behaviour of male and female baboons during sexual consortships, 116
KATHLYN L. R. RASMUSSEN

7   **Effects of Interactions and Relationships Upon the Individual**

7.1   Social experience and individual differences, 121
ROBERT A. HINDE

7.2   Individual characteristics of mothers and their infants, 122
JOAN STEVENSON-HINDE

8   **Influence of the Social Situation on Relationships**

8.1   Dyads embedded in a group, 128
ROBERT A. HINDE

8.2   Differences in the mother–infant relationships between socially living and segregated mother–infant dyads, 130
ROBERT A. HINDE

8.3   Matriline differences and infant development, 132
CAROL M. BERMAN

8.4   Effects of parturition on the mother's relationship with older offspring, 134
PHYLLIS C. LEE

8.5   Influence of siblings on the infant's relationship with the mother and others, 139
JILL M. HOOLEY and MICHAEL J.A. SIMPSON

8.6   Primiparous and multiparous mothers and their infants, 142
JILL M. HOOLEY

8.7   Caretaking of infants and mother–infant relationships, 146
PHYLLIS C. LEE

9   **Triadic Interactions and Relationships**

9.1 Triadic interactions and social sophistication, 152
ROBERT A. HINDE

9.2 Early differences in relationships between infants and other group members based on the mother's status: their possible relationship to peer–peer rank acquisition, 154
CAROL M. BERMAN

9.3 Influence of close female relatives on peer–peer rank acquisition, 157
CAROL M. BERMAN

9.4 Influences of the social situation on male emigration, 160
JOHN COLVIN

9.5 Matriline membership and male rhesus reaching high ranks in their natal troops, 171
BERNARD CHAPAIS

10   **Description of and Proximate Factors Influencing Social Structure**

10.1 Description, 176
ROBERT A. HINDE

10.2 Grooming and social competition in primates, 182
ROBERT M. SEYFARTH

10.3 Familiarity, rank and the structure of rhesus male peer networks, 190
JOHN COLVIN

10.4 Structure of the birth season relationship among adult male and female rhesus monkeys, 200
BERNARD CHAPAIS

10.5 Dominance, relatedness and the structure of female relationships in rhesus monkeys, 209
BERNARD CHAPAIS

10.6 Autonomous, bisexual subgroups in a troop of rhesus monkeys, 220
BERNARD CHAPAIS

10.7 Analysis of social structure, 222
STEVEN R. SCHULMAN

10.8 Ecological influences on relationships and social structures, 225
PHYLLIS C. LEE

11   **Intergroup Relationships**

11.1 Home range, territory and intergroup encounters, 231
PHYLLIS C. LEE

11.2 Intergroup encounters among Old World monkeys, 233
DOROTHY L. CHENEY

11.3 Proximate and ultimate factors related to the distribution of male migration, 241
DOROTHY L. CHENEY

12   **Ultimate Factors Determining Individual Strategies, Relationships and Social Structure**

12.1 General, 250
ROBERT A. HINDE

12.2 Ultimate factors determining social structure, 255
RICHARD W. WRANGHAM

12.3 Special relationships between adult male and female olive baboons: selective advantages, 262
BARBARA B. SMUTS

12.4 Male dominance and reproductive activity in rhesus monkeys, 267
BERNARD CHAPAIS

12.5 Fitness and female dominance relationships, 271
BERNARD CHAPAIS and STEVEN R. SCHULMAN

12.6 Extrafamilial alliances among vervet monkeys, 278
DOROTHY L. CHENEY

12.7 Adaptive aspects of social relationships among adult rhesus monkeys, 286
BERNARD CHAPAIS

12.8 Patterns of agonistic interference, 289
SAROJ B. DATTA

13 **Generality of the Approach to Other Species**

13.1 Applicability of the approach to other species, 298
ROBERT A. HINDE

13.2 Relationships and social structure in gelada and hamadryas baboons, 299
ROBIN I.M. DUNBAR

13.3 Interactions, relationships and social structure: the great apes, 307
ALEXANDER H. HARCOURT and KELLY J. STEWART

13.4 Relationships and social structure of African elephants, 314
CYNTHIA J. MOSS and JOYCE H. POOLE

13.5 Social relationships in comparative perspective, 325
RICHARD W. WRANGHAM

13.6 The human species, 334
ROBERT A. HINDE

References, 340

Author Index, 373

Subject Index, 382

Contributors

CAROL M. BERMAN, Department of Anthropology, State University of New York at Buffalo, 581–L Bldg 5, Spaulding, Buffalo, New York 14261, USA.

BERNARD CHAPAIS, Department d'Anthropologie, Université de Montréal, CP 6218, Succ. A, Montréal H3C 3J7, PO, Canada.

DOROTHY L. CHENEY, Department of Anthropology, University of California, Los Angeles, California 90024, USA.

JOHN COLVIN, Department of Psychology, 8–10 Berkeley Square, Bristol, BS8 1HH.

SAROJ B. DATTA, Sub-Department of Animal Behaviour, University of Cambridge, High Street, Madingley, Cambridge, CB3 8AA.

ROBIN I. M. DUNBAR, Sub-Department of Animal Behaviour, University of Cambridge, High Street, Madingley, Cambridge, CB3 8AA.

ALEXANDER H. HARCOURT, Department of Applied Biology, University of Cambridge, Pembroke Street, Cambridge, CB2 3DX.

ROBERT A. HINDE, MRC Unit on the Development and Integration of Behaviour, High Street, Madingley, Cambridge, CB3 8AA.

JILL M. HOOLEY, Department of Psychiatry, The Warneford Hospital, The University of Oxford, Oxford, OX3 7JX.

PHYLLIS C. LEE, AWLF, Box 48177, Nairobi, Kenya.

CYNTHIA J. MOSS, AWLF, Box 48177, Nairobi, Kenya.

JOYCE H. POOLE, AWLF, Box 48177, Nairobi, Kenya.

KATHLYN L. R. RASMUSSEN, University of Colorado Medical Center, Box C268, 4200 East Ninth Avenue, Denver, Colorado 80262, USA.

STEVEN R. SCHULMAN, Department of Biology, Princeton University, Princeton, New Jersey 08544, USA.

ROBERT M. SEYFARTH, Department of Anthropology, University of California, Los Angeles, California 90024, USA.

MICHAEL J. A. SIMPSON, MRC Unit on the Development and Integration of Behaviour, High Street, Madingley, Cambridge, CB3 8AA.

BARBARA B. SMUTS, Department of Anthropology, Peabody Museum, Harvard University, Cambridge, Massachusetts 02138, USA.

JOAN STEVENSON-HINDE, MRC Unit on the Development and Integration of Behaviour, High Street, Madingley, Cambridge, CB3 8AA.

KELLY J. STEWART, Sub-Department of Animal Behaviour, University of Cambridge, High Street, Madingley, Cambridge, CB3 8AA.

RICHARD W. WRANGHAM, Department of Anthropology, University of Michigan, Ann Arbor, Michigan 48109, USA.

Preface

Human social behaviour displays such great flexibility and diversity that many have sought to understand it in terms of cultural influences alone. However, such influences depend upon, and operate through, capacities and propensities that themselves have multiple determinants. These constraints and predispositions are not easy to study: pan-cultural aspects are elusive because they lie not in the behaviour and cultural forms observed but in propensities which give rise to different forms according to the cultural soil on which they are nourished. Studies of non-human primates may be a powerful tool here. Although cultural influences are at least minimal, the social behaviour of monkeys and apes is of considerable sophistication and great diversity. If we attempt not to draw superficial parallels between the behaviour of monkey and human but to explore the principles on which the diversity of the behaviour of non-human primates depends, we may be able to contribute to a new perspective on the role of cultural influences in the human case.

While their close evolutionary relationship to human beings gives their study a special interest, the complexity of the social behaviour of monkeys and apes provides material from which explanatory principles applicable to many other species may be drawn. In addition, the study of non-human primates, in its own right, presents some of the most exciting problems facing students of behaviour. Practically every study reveals new complexities of social behaviour and structure. One is forced to ask how the behaviour of individuals contributes to and is determined by the social nexus in which they live. One wonders how the behavioural propensities of individuals are shaped by their social and physical environment during development. Can the ways in which that shaping occurs be understood in terms of the action of natural selection? What is the functional significance of the differences between individuals and between species? Why, and also how, has diversity evolved? While the biologist has learned to distinguish clearly between problems of causation, development, function and evolution, it soon becomes apparent that in studying social behaviour he cannot limit his concern to any one of them in isolation from the others. Problems of causation pose and merge with developmental issues, and the facts of causation and development demand functional explanations as a more comprehensive understanding is achieved.

If primatologists are to come to terms with these issues, if they are to use studies of non-human primates to generate principles of wider applicability, and if they are to contribute, in however small a way, to an understanding of the human case, they need more than additional facts about monkeys and apes. They need a conceptual framework within which to arrange the data so that they have an ordered body of knowledge—a science of social behaviour. Such a conceptual framework is in fact used by most research workers but it is seldom made explicit and the research findings remain only partially integrated. Our aim here is to specify and exemplify some of the steps involved in building such a framework. This involves a descriptive base which takes account of the several levels of complexity involved in social behaviour and provides a starting point for a search for principles of explanation. This approach leads to an emphasis not just on descriptions of the social behaviour shown by the different age/sex classes of individuals, such as were common in early field studies, but on the relationships between individuals and on the manner in which each relationship is affected by others.

The book is organized into 13 chapters, most of which contain an introductory survey followed by one or more individual contributions. The latter are varied in nature and include empirical data, reviews and theoretical models, but concern primarily three reasonably well-known species—rhesus macaques, baboons and vervet monkeys.

Limiting primary data to three species has involved sacrificing much of the richness that would have come from a broader comparative survey. The reader not familiar with primate behaviour must remember that the three species discussed all live in multimale troops: there are many other primate species which live in single-male troops, monogamously or in variants of one of these. However, it is our hope that the course followed here will have several advantages. Concentration on a few species will obviate the need for the non-primatologist reader to become familiar with the natural history of a wide range of species. It also facilitates the pursuit of principles in depth, principles which will be applicable, at least in some degree, to species with different life-styles. While reviews of the literature can easily obscure the nature of, and the difficulties inherent in, the original material, it is hoped that the presentation of specific studies, many containing previously unpublished data, will better illustrate the methods, facts and types of argument used for the understanding of social behaviour. For those wishing to find their way into a wider literature, references have been included in the introductions to each chapter and in the individual contributions: furthermore the generality of the approach to other species, including that of our own, is assessed in some more wide-ranging contributions in Chapter 13.

The book has been written by a group of colleagues who, though they never met as a group, have been linked by a network of relationships which permitted the sharing of ideas and a fairly uniform approach. It is a collaborative venture and some of the contributions (including my own) have been improved by comments from other authors. Over some issues, different authors have reached similar conclusions from different starting points, while over others, a degree of disagreement remains. I have not attempted to impress an undue uniformity of style on the contributions, preferring that they should convey not only scientific data but also something of the several interests and personalities of the authors.

Madingley, Cambridge   Robert A. Hinde

Acknowledgements

Nearly all the contributors to this book were at some time members of, or closely associated with, the sub-Department of Animal Behaviour or the Medical Research Council Unit on the Development and Integration of Behaviour, Madingley, Cambridge. We are grateful to many colleagues for their interest, advice and practical help, especially Dr. Patrick Bateson, the Director of the sub-Department, and Richard White, Brian Styles, Rozanne McNamee and Douglas Easton, who have given help with recording and statistical methods.

Much of the field work was carried out in the National Parks of Kenya, Tanzania or South Africa and we gratefully acknowledge their help and co-operation. Through the years the National Parks of Africa have served two roles—that of protecting the animals and habitats within their boundaries for posterity and that of providing facilities for their study. The research presented in this book would not have been possible without the cooperation, interest and support of the various government ministries concerned with wildlife research and conservation. It is our earnest hope that our work will provide information that will facilitate their endeavours.

For the research carried out on Cayo Santiago (Puerto Rico) we are grateful to the Carribean Primate Research Centre and the National Institutes of Health (USA). This colony of rhesus macaques with precisely known genealogical histories offers unique oppotunities for research and we hope our work will go some way towards justifying the efforts that have been put into maintaining this remarkable facility over the years. Several contributors would like to express their special thanks to Donald Sade and to Richard Rawlins, who were in charge of the Cayo Santiago colony while they were there.

Dorothy Cheney, Robert Seyfarth, Phyllis Lee, Cynthia Moss and Joyce Poole would like to thank J. Kioko and B. Oguya, the Wardens of Amboseli during their research, for their help. Dorothy Cheney and Robert Seyfarth are also grateful for demographic data obtained from the long-term records of the Amboseli vervet groups maintained by S. Andelman and P. Lee.

Barbara Smuts would like to thank Drs R. S. O. Harding and S. C. Strum for allowing her to study the Kekopey baboons and Dr James G. Else, Director of the Institute of Primate Research, Kenya, for help with field research in Kenya.

None of this research would have been possible without financial support from funding bodies. We would like especially to thank the: Royal Society, Medical Research Council, Science and Engineering Research Council, and L.S.B. Leakey Foundation in the UK; the National Institutes of Health, National

Science Foundation, Wenner Gren Foundation for Anthropological Research, L.S.B. Leakey Foundation, H. F. Guggenheim Foundation, American Association of University Women, African Wildlife Leadership Foundation, Grant Foundation and New York Zoological Society in the USA; and the Canada Council and le Ministère de l'Education de la Province de Québec in Canada.

Several authors wish to make special acknowledgements for comments on their papers: Barbara Smuts to R. Bailey, J. Watanabe and R. Smuts; Richard Wrangham to Warren Holmes; Robert Hinde to Patrick Bateson and Nick Humphrey; Robert Seyfarth to L. Fairbanks; and John Colvin to Gerry Tissier and Robert Seyfarth.

Chris Reed has typed and retyped many manuscripts with great care. Pat Naylor has been indefatigable in helping to organize the manuscripts and the references and making many of the final corrections. Les Barden has drawn figures and photographed drawings for many of the papers.

Finally, Phyllis Lee, in addition to her role as author, took charge of the editorial work in Robert Hinde's absence for a month and did a great deal of work on the manuscripts. She contributed greatly to the quality of the volume.

Chapter 1
A Conceptual Framework

ROBERT A. HINDE

The approach presented here starts from the belief that the study of behaviour must begin with a descriptive base. However, the description of social behaviour poses a special difficulty because it involves phenomena at a number of levels of complexity: at each of these levels properties simply not relevant to those below become apparent. It is first necessary to make this problem explicit.

One could start with individual behaviour. To describe an individual's behaviour, one must specify what he is doing (e.g. walking) and add a number of qualifiers to describe how he is doing it (e.g. fast, north-west). However, most of our data consist of observations of two (or sometimes more) individuals in *interactions*. An interaction implies that A does X to B and B does Y back again. It may involve a number of repetitions, perhaps with some change in the nature of the behaviour shown, but the precise constraints to be set on the category of 'interactions' need not concern us. Interactions can similarly be described in terms of content and qualifiers: A may be grooming (content) B vigorously or perfunctorily (qualifiers). Both content and qualifier may imply properties just not relevant to the more elementary level of individual behaviour: when alone, one can talk but one cannot converse and one certainly cannot converse competitively.

When two individuals have a series of interactions over a period of time, what happens in any one interaction may affect what happens in subsequent ones. The

1

two individuals can then be said to have a *relationship* with each other (Hinde 1979a). Their relationship can be described in terms of what they do together (the content of their interactions), how they do it (the qualifiers) and the relative frequency and patterning of those interactions in time. The last issue introduces properties which are not relevant to interactions in isolation. The interactions may be consistent, unpredictable or improving in quality and their relative frequencies may provide properties, such as permissiveness or warmth, that apply only in a much more impoverished sense to individual interactions (see Chapter 3). These properties may be crucial for the dynamics of the relationship:

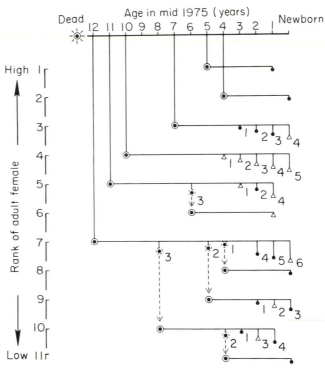

Fig. 1.1. Social structure of rhesus macaques: the rank relationships within one lineage or matriline. Each group may contain several lineages. Age reads horizontally, the eldest individuals being to the left of the figure. Rank reads vertically. The female from whom the matriline originated is shown at the top left hand corner but was dead at the time to which the figure refers. The numbers indicate the rank relationships between siblings within families. The broken arrows lead from the ranking of a particular adult female among her maternal siblings to her ranking among the adult females. The solid and vertical lines indicate descent. ● = non-adult female; △ = non-adult male; ◉ = adult female; ◈ = the position of a female, now adult, among her maternal sibs (the same individual female is connected by a broken arrow to her position as an adult female with respect to other adult females). (After Datta 1981.)

for instance the participants may evaluate the relationship as it has developed so far partially in terms of such properties, and act accordingly.

Each participant in a relationship is involved also in others, so that the relationship forms part of a network of relationships or *social structure*. This structure can be described in terms of the properties of the constituent (dyadic or higher-order) relationships and how those relationships are patterned: the structure could be linear (A has a relationship with B, who has a relationship with C, who has a relationship with D, etc.) or involve ramifying networks of varying density and extent, entailing properties simply not relevant to particular dyadic relationships. Knowledge of the social structures of monkey groups is still far from complete, but it indicates very considerable complexity (Hinde 1979b) (see Figs 1.1, 9.6, 10.3 and 10.4 and Chapters 10 and 13).

Description, of course, is but a first step. We must next seek to understand the behaviour we have described. This can be seen as a search for principles of explanation, though it will seldom yet be profitable to attempt to state those principles formally. In discussing them, we may move away from the data language used for describing social behaviour at its several levels of complexity and employ new concepts not present in the data, just as we explain an apple's fall by the 'force of gravity' (Table 1.1).

'Explanations' of biological characteristics, whether of structure or behaviour, can be of four logically distinct types, concerned with causation, development, function or evolution (Tinbergen 1963). For example the fact that our thumb moves differently from the other digits may be explained in terms of the structures of the muscles and bones which move it (causation), in terms of the way in which embryonic cells differentiate and grow into muscles, bones and nerves (development), by reference to its usefulness in gripping branches or other objects (function) or by comparison with the monkeys and apes, believed to be closely related, which also have opposable thumbs (evolution). These types of explanation, though logically distinct, are both complementary to each other and inter-related. Full understanding requires examination of all four and comprehension of the relations between them.

A major task is to consider how these distinct yet inter-related types of explanation bear on the distinct yet inter-related levels of individual behaviour, interactions, relationships and social structure. Causal questions may be asked first but it is evident that these soon merge with those of development.

The behaviour of individuals appears to have some consistency over time. We may feel ourselves to be consistently ill at ease in social groups or perceive particular monkeys to be consistently subservient to others. At the same time, there are some situations in which we surprise ourselves with our own self-confidence and even a subordinate monkey may boss a stranger. In spite of the apparent consistencies in individuals, their social behaviour may yet change with

Table 1.1. Relations between interactions, relationships and social structure. Some of the proximate factors influencing them are shown in the first column, while the ultimate factors are shown in the last two columns. No critical significance attaches to the ordering of items in the first, third and fourth columns but those at the top are perhaps more relevant to social structure and those at the bottom to interactions.

Proximate factors	Description	Ultimate factors	
Consanguinity and familiarity	STRUCTURE (Nature and patterning of relationships)	Alliances—aiding relatives, grooming high-status individuals	KIN SELECTION
Status		Mutualism, reciprocal altruism, altruism	
Exchange	RELATIONSHIPS (Content, quality and patterning of interactions)	Mate choice	INDIVIDUAL SELECTION
Age and sex			
Personality		Parent—offspring conflict	
Learning	INTERACTIONS (Content and quality)	Competition—for food, mates, space	SEXUAL SELECTION
Need satisfaction			
‿‿‿ (Causal) Theory language	‿‿‿ Data language	‿‿‿ (Function) Theory language	

the company they keep. The behaviour an individual shows thus depends in part on his own characteristics and in part on whom he is with. It will be evident that this raises a problem about what is meant by an 'individual characteristic' of behaviour (section 4.1). Nevertheless, a first requirement for understanding the causal bases of social behaviour is an understanding of how individuals differ in their propensities to behave and in their behaviour with particular others (Chapter 4). Primatologists have been primarily concerned with contrasting individuals differing in age or sex (Chapter 5), or with the influence upon interactions of the kinship relations of the individuals concerned (Chapters 9–12). However, aspects of 'personality' (a term which, with the possible reservations in mind, one may apply to monkey as well as human) may also be important (Chapter 4).

While the properties of each interaction depend on the natures of the participants, each interaction within a relationship may affect subsequent interactions. In other words, the behaviour of each participant towards the other in future encounters may be affected (not to mention their attitudes to or beliefs about each other, in so far as such terms may be applicable to non-human primates): for example from now on B will always yield to, or expect to obtain food from, A. Such consequences constitute effects upon the relationship between A and B. A second requirement for understanding the causal bases of social behaviour is thus a set of principles concerned with how interactions affect subsequent interactions within the same relationship and, more generally, with the development of relationships.

The problems of the dynamics of relationships and of their development in fact merge. Since each interaction within a relationship affects subsequent ones and since the participants are themselves likely to change with time, every relationship is likely to change and any stability it may have must be dynamic in nature. The way relationships change with time and the study of their development and decay, is thus inseparable from that of their dynamics (Chapter 6). While laws of learning will clearly be basic here, work on human interpersonal relationships has suggested other principles, e.g. those related to theories of attribution, exchange, interdependence, balance and systems, whose applicability to non-human species remain largely unexplored. Yet many of the concepts on which these principles depend, such as expectations, goal-seeking, emotions and relative values, may well be applicable to non-human species (Hinde & Stevenson-Hinde 1976).

Furthermore, not only are relationships affected by the nature of the participants but individuals are affected by their relationships. As a consequence of their continued interactions, B may yield not only to A but may become subservient to all he meets. The effects of a particular relationship may be far-reaching, affecting the participants' attitudes towards, beliefs about or behaviour with a whole range of other individuals. In so far as they are durable, these constitute changes in the nature of the individual concerned, i.e. changes in personality. While such changes are most marked in young individuals, they continue in some degree throughout life (Chapter 7). There is, in fact, a continuous dialectic between the natures of individuals and the interactions and relationships in which they participate (Fig. 1.2), and questions about the development of social behaviour in individuals may be closely related to questions about the dynamics (and development) of relationships. There is thus a need to know how individuals are shaped by their social and other experiences.

The participants in a relationship will also have relationships with others and the latter will affect their relationship with each other. In general, each relationship is set in a nexus of other relationships, which mutually affect each other (Chapters 8 and 9). A fourth set of principles necessary for understanding the causal bases of relationships (i.e. the nature and patterning of the interactions within them) will concern the social influences on relationships. In the human case, this will include also the cultural norms which affect the expectations of the individuals in the relationship in question.

However, as seen above, the social group not only affects the dyadic and higher-order relationships within it but also is constituted by those relationships. In addition, therefore, to the dialectic between individuals' characteristics and relationships, there is also a dialectic between relationships and social structure (Fig. 1.2). Thus, there is a need also for principles to understand the way the relationships are patterned to compose the structure. For understanding of the patterning of relationships in (i.e. the structure of) groups of non-human primates, principles concerned with the natures of individuals (e.g. the propensities to form relationships within or between age/sex classes), with blood relationship or some correlate such as familiarity and with status have so far proved useful (Chapter 10). In the human case, those concerned with social norms and institutions are of course paramount (section 13.6).

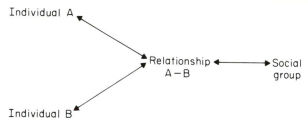

Fig. 1.2. Dialectics between individuals, their relationship and the social nexus in which it is embedded.

The social groups of primates do not exist in isolation from each other. There is increasing evidence to suggest that each troop has a specific relationship with each other troop in its vicinity, the relationship being characterized by varying degrees of tolerance or aggression. There is thus the yet higher order of the structure of troops within a community to be considered (Chapter 11).

So far we have considered principles concerned with the development and causation of the immediate patterning to be observed at each level. Such principles pave the way for causal analysis at yet finer (e.g. endocrine, neurophysiological) levels of analysis but may well not be predictable solely from the latter, since they may be concerned with properties emergent at their own level.

As we have seen, however, two other issues must be met before understanding is complete. One concerns the consequences of social behaviour and in particular its biological functions, i.e. its costs and benefits to the individual concerned measured in terms of its effects on his and his relatives' reproductive success. Of course, any one act has many consequences, some advantageous in terms of natural selection and some not: most behaviour is a compromise to the conflicting demands of a variety of requirements. Thus we must not fall into the trap of seeking a functional explanation for every instance of behaviour: just as structural characteristics may be profitable in some contexts but not in others, so also may behavioural propensities occasionally be used inappropriately and to the individual's detriment. It is the overall consequences of the behavioural propensities that must be considered, not those of isolated instances of their realization.

The functional and the causal explanations of behaviour must of course mesh with each other. Thus, observed data on a conflict between mother and infant at weaning may be (causally) explicable in terms of (for example) differing behavioural propensities in mother and infant to maintain contact with each other, themselves (causally) explicable in terms of previous experience and/or neurophysiological mechanisms, and also functionally in terms of the relative benefits and costs (measured in terms of gene survival) to mother and to infant of continued nursing (see section 12.1). Similarly, if young males and females are observed to differ in their tendencies to continue to associate with their mothers, this may be understandable not only in terms of the causal (and ontogenetic) mechanisms involved but also in terms of the differences between the adaptive consequences for males and for females of making and breaking matrilineal ties (Table 1.1). An issue not yet wholly closed is whether, in addition to seeking explanations of individual propensities in terms of individual (or inclusive, see section 12.1) reproductive advantage, it is appropriate also to seek explanations of group characteristics in terms of group advantage.

The final set of questions, namely those concerned with the course of evolution of social behaviour, is one that will be mentioned only in passing here. The comparative methods pioneered by Lorenz (1935) and Tinbergen (1959) have been used to formulate hypotheses about the evolution of certain aspects of the social behaviour of individuals, most especially the behaviour used in social communication in interactions. However as yet practically nothing is known about the course of evolution of the behavioural propensities that lead to the formation of long-term relationships or to social structures of particular kinds.

Chapter 2
Species, Study Sites and Methods

PHYLLIS C. LEE

The contributions in Chapters 3–12 are concerned primarily with three species or species groups—rhesus macaques, baboons and vervet monkeys. In this chapter, their characteristics, the areas where they were studied and the study methods used are considered.

Study animals

With the growth of research on primates in the 1950s and 1960s, the rhesus macaque, *Macaca mulatta*, was the species most used in the laboratory. As a result, its behaviour and physiology are better known than those of many other species. The rhesus is a medium-sized monkey, the males (5.5–11 kg) being rather larger than the females (4.5–10 kg) They range over much of the Indian subcontinent and numerous studies of animals in their natural habitat provide a comparative basis for interpreting data on behaviour derived from laboratories. Groups kept in the laboratory provide an ideal opportunity for the investigation of social behaviour under conditions that can be, at least partially, controlled.

In their natural habitat of the African savannahs and woodlands, baboons captured the attention of early field primatologists: their large size and terrestrial habits facilitate observation. At one time, baboons were proposed as a model for understanding the evolution of human social groups, since they live in groups similar in size and complexity to those postulated for ancestral hominids (Pilbeam 1972; Washburn & DeVore 1961). The baboon species discussed in the following chapters are collectively known as the savannah baboons (Jolly 1972). They comprise races or subspecies of the yellow baboon, *Papio cynocephalus*, which is found from Kenya to Angola. In parts of its range further to the south, it is known as the chacma baboon, *Papio cynocephalus ursinus*. The olive baboon, *Papio cynocephalus anubis*, is found from Niger to the north-west of eastern Africa and often inhabits more densely wooded areas as well as the open savannah. Males (20–30 kg) are considerably larger than females (10–15 kg) and have consipicuous canines. A vast literature now exists on almost all details of the ecology and behaviour of baboons in many different areas in Africa.

The vervet monkey, *Cercopithecus aethiops*, is widely distributed in Africa and various subspecies are found from Sudan in the north to the Cape in the south and from the east to the west coast. The vervet is a smaller primate than either the rhesus or the baboon, though males (3–4 kg) are still somewhat larger than females (2–3 kg). Vervets are less terrestrial than the baboons or rhesus and live in areas of gallery forest, forest fringe, bush and open grassland. The range of the vervet overlaps extensively with that of the baboons. In comparison with the rhesus and the baboons, studies of vervets have been few since the early work of Struhsaker (1967a) and Gartlan (Gartlan & Brain 1965). Discussions of the taxonomy and morphology of these species can be found in Hill (1966, 1970), Napier and Napier (1967) and Szalay and Delson (1979).

All three primates share characteristics of social organization. The basic social unit is the group or troop composed of two or more adult males and several adult females and their offspring. The sex ratio of adults varies from group to group and from population to population but is often two females to one male (Clutton-Brock & Harvey 1977). Females form the stable social core of the group, with daughters remaining in the group of their birth and sons generally leaving for another group around the time of sexual maturity (reviewed in Harcourt 1978b; Packer 1979a). The movement of females between groups has been observed only infrequently in baboons and never in vervets. Through time mothers and their daughters form distinctive interactive subunits called matrilines (Sade 1972b) and the adult males are usually unrelated to the females in their group.

Both vervets and rhesus are seasonal breeders, copulating and giving birth at specific times each year (see Koford 1965; Rowell & Richards 1979; Southwick *et al.* 1965). Baboons often mate and give birth throughout the year, although there can be clusters of births within a group (Altmann & Altmann 1970; Hall & DeVore 1965). Males compete with each other for access to mates (for baboons see Hall & DeVore 1965; Packer 1979b; Rasmussen 1980; for rhesus see Southwick *et al.* 1965; review in Chapais 1981; for vervets see Struhsaker 1967b). Female baboons have conspicuous perineal swellings that indicate their sexual state (Hendricks & Kraemer 1969), while rhesus females have marked changes in the colour of the sexual skin during their oestrus cycle (Czaja *et al.* 1975). Females in all three species almost invariably give birth to a single infant at a time and the period of infant-dependence can last for a year or more (Altmann 1980; Berman 1978a; Lee 1981). The period of immaturity is extended through the first three or four years of life, though the timing of sexual maturity may depend on early growth rates and differs between areas (e.g. see Altmann *et al.* 1977).

Some form of dominance relations exists between the individuals in each group and these have generally been described as linear hierarchies (Bernstein 1976a; Deag 1977). In most cases, males are dominant to females and both can be ranked in separate hierarchies. Immature animals, especially females, assume their mother's rank (reviewed in Walters 1980). In rhesus, each female rises in rank above her elder sister (reviewed in Datta 1981) and this also has been occasionally observed in baboons (Moore 1978). The complexities of the social structures to be found in primate groups are yet not fully understood but are illustrated in Fig. 1.1.

In the wild, baboons live in the largest groups and vervets in the smallest, but group size is probably more influenced by qualities of the habitat than by any species-specific characteristic (Clutton-Brock & Harvey 1977). The vervets are unique among the three as the only species that actively and regularly defends a small territory against other vervet groups (Cheney 1981; Struhsaker 1967c). The baboons and rhesus live in larger home ranges whose boundaries overlap with those of other groups and territorial defence is not generally observed (but see Hamilton *et al.* 1976; Hausfater 1972). Intergroup relations range from complete tolerance and mixing of individuals, to groups avoiding each other, to aggressive encounters (see Chapter 11).

Study areas (Table 2.1)

The observations on rhesus macaques presented in the following chapters were made in two areas. At Madingley, the rhesus live in six small groups, each with two to five adult females, their young and one adult male, enclosed in an outdoor pen (5.4×2.4×2.4 m) with access to a heated indoor room (1.8×1.35×2.25 m). The colony was established in 1959 and detailed continuous information on genealogies and individual histories are available. On Cayo Santiago, a 15.5 ha island off the coast of Puerto Rico in the Carribean, the monkeys are free to range over the island. This colony was started in 1938 by Carpenter (see Carpenter 1942) as a breeding group. The subtropical climate and wooded terrain are similar to parts of their range in India. Food and water are provided for the monkeys but they also forage in the natural vegetation. The number of monkeys

Table 2.1. Study areas, species and aspects of the study groups whose behaviour is discussed in this volume.

Species	Study area	Habitat	Study duration	Home range	No. males	No. females	Group size	Researcher
Rhesus monkey	Madingley	Caged	22 years		1	2–4	7–12	Hinde *et al.*
	Cayo Santiago	Provisioned	22 months	15 ha (I)	10–15	17–19	53–77	Berman
			12 months	(F)	16	17–20	88–106	Chapais
			18 months	(I)	16	20–23	91–112	Colvin
			16 months	(J)	17–21	20–24	105	Datta
			12 months	(L)	14–23	19–21	91–104	Schulman
Yellow baboons	Mikumi	Savannah	11 months	43 km²	13–15	36–37	102–120	Rasmussen
Olive baboons	Ruaha	Bushwoods	7 months	61 km²	8	20	72–78	Lee and Oliver
	Gombe	Forest	6 months	1 km²	9	15	45–51	Oliver and Lee
	Gilgil	Grasslands	18 months	20 km²	14	34	115	Smuts
Chacma baboons	Mount Zebra Park	Montaine bush	15 months	3 km²	2	8	24–30	Cheney and Seyfarth
Vervet monkeys	Amboseli	Wooded grasslands	5 months	22 ha (A)	3–7	3–8	19–29	Cheney, Lee, Seyfarth
				18 ha (B)	1–2	7	16–21	Seyfarth and Wrangham
				30 ha (C)	3–4	5–8	27–29	Wrangham

on the island changes as a result of births and removals but the birth rate and the population densities are high. Even on this small island, at a high population density, there are well-defined, separate groups, ranging in size from under 50 to well over 100. Maternal genealogies are available for most individuals born on the island and the continuity of observations again allows for detailed knowledge of individual histories. The Cayo Santiago monkeys are fully habituated to observers and data collection is relatively easy.

The rhesus on Cayo Santiago can be viewed as an intermediate condition between the free-ranging macaques in their natural habitats and those kept in the more controlled environment at Madingley. In particular, valuable comparisons can be made between the behaviour of individuals in large or small groups, from large or small matrilines or at different population densities.

All the baboons were observed in their natural habitats. The studies of Cheney and Seyfarth were carried out in Mountain Zebra National Park, South Africa on the chacma baboons. There, the baboons lived in rocky, mountainous terrain at elevations ranging from 1000–2000 m above sea level. Rainfall was seasonal and low (less than 350 mm year^{-1}), in common with many other areas inhabited by baboons. However, at this elevation and latitude, there were occasional snowfalls and temperatures ranged from a minimum of 0 to 30° C. During the dry season, water was restricted to widely dispersed river courses. The baboons foraged in the dry, open grasslands and interspersed scrub and their diets consisted predominantly of grasses. The groups were generally small and the relatedness of the adult animals was unknown.

The yellow baboons were observed in two areas. Rasmussen's study was carried out in Mikumi National Park, Tanzania. The baboons lived in the short-grass savannah and deciduous woodland typical of the habitat of the yellow baboon in eastern Africa. Rainfall was seasonal, falling in the two periods of November–December and February–May. Permanent water during the dry season was located in widely spaced water-holes. The baboons had a large home range which included the seasonally flooded short-grass plains and mixed woodlands and extended into the surrounding hilly areas covered with shrubs and tall grass. The groups here were large and numerous and relatedness between the adults again was unknown. The other brief study on yellow baboons was done in Ruaha National Park, in south-central Tanzania. This was a more hilly region of mixed bush and woodland, with the same pattern of seasonal rainfall. Water was continuously available from the Great Ruaha River running through the large home range of the study troop. The study troop was large and genealogies were known only for infants and a few young juveniles.

Olive baboons were observed at Gombe Stream National Park, located on the rift escarpment on the edge of Lake Tanganyika, Tanzania. The terrain was steep and the vegetation consisted of closed deciduous forest in valleys and open grasslands on the tops of ridges. The study group had been observed for several years by many different observers (e.g. Nash 1976; Owens 1975a; Packer 1979a) and genealogies were available for all adolescent and immature animals. Rainfall at Gombe was seasonal but higher than that of the other study areas and the vegetation comparatively denser. Water was permanently available from the lake.

The olive baboons were also observed in less forested areas on Kekopey Ranch near Gilgil in Kenya. Here, a large troop ranged through grasslands and scattered trees, sleeping on tall cliffs. The study troop, called Eburru Cliffs, was part of the same population as the Pumphouse Gang troop studied by Harding (1973a,b, 1976), Strum (1975, 1982, in press a,b) and others. The baboons shared their range with cattle and small farmers cultivating grain crops. Rainfall was seasonal and higher, averaging over 350 mm year^{-1}. The baboons primarily ate grasses (wet season) and corms (dry season). Water was available year round from cattle troughs located throughout their range. Genealogies were known for immature individuals only.

The vervets were studied in Amboseli National Park, Kenya, located on the drainage basin for Kilimanjaro. The vervets were found throughout the park in areas where there was at least some *Acacia* (fever tree) woodland. They also ranged onto the alkaline short-grass plains and into the interspersed bushlands. Rainfall was again highly seasonal, in the same pattern as that at Mikumi. Three separate groups were observed at the same time and genealogies were known for all immature animals. Water was continuously available from spring-fed water-holes in swampy areas for two of the groups (B and C) but the third group (A) had no access to standing water throughout the dry season.

The baboons and vervets shared their ranges with many other species of grazers, browsers and primates. They were also exposed to the predators and illnesses that would affect any population. In this sense, they were 'typical' of animals in their habitats. However, it is worth noting that the groups studied were probably not chosen at random from the possible population of groups. There is a tendency to ignore the smallest and the largest groups when selecting a study group and to choose a group that is larger than the mean size, especially when the mean group size is small (Sharman & Dunbar in press). No one study group or study area can be considered to represent either the full range or the norm of all types of behaviour for a species. However, when data are drawn from a number of different areas and groups, they are more likely to reveal consistent patterns of behaviour.

Study techniques

In all the studies of the free-ranging primates, observations were made while on foot, often close to the animals. All individuals were recognized and either named or numbered. In most cases they were used to the presence of an observer. Habituation is an important element of a field study, since the observer is attempting to record the interactions of the subjects with other animals, not himself. Individual recognition has become increasingly important in studies of behaviour since individual differences in behaviour are important for the understanding of relationships and the adaptive significance of different types of behaviour. In the studies where complete genealogical information was available, questions about the influence of kinship on behaviour could be investigated in detail. The majority of the studies lacking this information used a variety of associations and interactions, such as grooming and alliance formation, to infer kinship. Such interactions have been reliably demonstrated to correlate with kinship (see Walters 1981).

Observations were recorded in most cases on predesigned check-sheets or spoken into tape-recorders. Check-sheets attained popularity because observations can be coded directly into categories, superfluous data can be disregarded and the transcription and analysis of data are greatly facilitated (Hinde 1973). Observations spoken directly into a tape-recorder require laborious transcription but are advantageous in allowing behaviour which does not fit easily into a predetermined category to be recorded and the sequences of interactions to be preserved.

An 'event-recorder' was used to collect data on mother–infant pairs at Madingley. This machine allows for a sophisticated combination of coded categories of behaviour with all sequences and time intervals preserved. The event-recorder transmits signals to a tape which can then be fed directly into a computer for transcription and analysis (Simpson 1979b; White 1971). The advantages of preserving sequences and time intervals in great detail and the ease and rapidity of the analysis are considerable, but there are still disadvantages for field-work. When event-recorders become small enough to be completely portable without sacrificing flexibility in the number of interactions that can be coded and when they become less subject to mechanical failure under extremes of weather, dust and damage in isolated field conditions, they will probably become the most efficient and accurate recording technique.

Sampling techniques

The majority of the data described in this book was compiled using a sampling technique called 'focal animal sampling' (Altmann 1974) and supplemented by several other techniques. In a focal sample, one individual is observed for a predetermined time period and the observations are repeated at intervals. Within a focal sample, behaviour is recorded either as an event, an instantaneous occurrence, or as a state, a behaviour with a measurable duration. Focal samples generally combine instantaneous records with records of duration and the sequences in which the behaviour occurs. Focal sampling is considered to be the most reliable technique and free from most biases such as those resulting from differences in visibility and unmatched sampling intervals (see Altmann 1974; Dunbar 1976).

Within a focal sample, behaviour can be recorded each time it occurs. However, this is often not possible and two common alternatives are to record whether or not it occurs within a specific time interval (one-zero sampling) or at preselected moments in time (instantaneous sampling). One-zero sampling has the advantage of being a quick method of scoring activities and is especially useful when behaviour occurs in bouts with indefinite beginnings and endings. However, one-zero records introduce problems for the analysis of rates and durations of behaviour, the potential errors varying with the relation between the lengths of the recording interval and the durations and frequencies of the behavioural events (Kraemer 1979; Rhine & Flanigon 1978; Simpson & Simpson 1977). Instantaneous sampling is more accurate except for rare events.

A closely related technique, 'scan sampling', differs from instantaneous sampling only in that the behaviour of a number of individuals at closely spaced points in time are sampled. This technique is widely used to census groups or for recording individuals' nearest neighbours.

A quite different type of record is provided by *ad libitum* sampling. In this case, all observations of certain behaviours are recorded. This technique is particularly valuable for recording rare but important events that might not be picked up in focal samples. Several assumptions must be met before ad lib samples can be used in any analysis. First, all animals must be equally visible and have the same chance of being observed when engaged in the rare behaviour. If this condition is not met, then ad lib samples provide unreliable data on rates or frequencies of occurrence. Second, the categories of behaviour that are sampled in this way must be equally likely to be observed before comparisons between categories can be made (see Altmann 1974). In most studies, this method is used to supplement data on rare interactions between specific individuals and is particularly useful in assessing interactions with directional components or those which are summed into a matrix.

Focal animal, and indeed all, sampling techniques assume that the questions to which the observer is addressing himself have been clearly defined and that the events or states which are recorded are in some way meaningful categories or elements for that question (Sackett *et al.* 1978). The problems of choosing which behaviour to sample have been discussed by Dunbar (1976), who points out that a careful choice must be made if the external index (i.e. the behaviour measured) is to be appropriate for answering a specific question relating to the motivational state of the animal. This is discussed further in section 3.1.

Analytical procedures

However the data are recorded, they are liable to consist of frequencies or durations of events. These may be used in three ways:

1   *Directly.* The behavioural counts or durations are expressed as a rate per unit observation time. The majority of the measures used in this book are of this type.

2   *As a ratio.* For many purposes it is important to know, not just how often an event occurs but how often it occurs relative to the frequency of some other event, e.g. the frequency with which an infant monkey is rejected by its mother may be of less or different developmental significance than the frequency with which it is rejected relative to the frequency with which it attempts to make contact (Hinde & Herrmann 1977).

3   *To derive an index.* Behavioural counts or durations may be combined to produce indices which assess aspects of behaviour or relationships not susceptible to immediate measurements. One such index is used in a number of studies that follow to assess the relative roles of two partners in the maintenance of proximity in a relationship. This index has been widely used in the study of mother−infant pairs and is defined as the 'percentage of approaches' (%Ap) due to an infant minus the 'percentage of leavings (%L) due to the infant, where approaches and leavings are defined in terms of changes in distance between mothers and infants across a certain threshold. If the index is positive, the infant is primarily responsible for the maintenance of proximity but if it is negative, the mother is (see also Hinde & Atkinson 1970). The use of such an index is not restricted to mother−infant pairs but can also be applied to partners in other relationships.

Chapter 3
Description of Social Behaviour

3.1 General Issues in Describing Social Behaviour

ROBERT A. HINDE

Many of the issues involved in the description of individual behaviour have been considered previously by ethologists and we have already seen that social behaviour, involving phenomena at different levels of analysis (Chapter 2), poses further problems. In this section, some further matters of special importance for the study of social behaviour are emphasized.

Description in physical terms versus description by consequence

In studying the behaviour of individuals, it is sometimes convenient to describe behaviour in terms that are ultimately reducible to patterns of muscular contraction (e.g. the resting posture implys particular positions of the limbs with respect to the body) and sometimes to refer to the consequence of the behaviour without specifying the precise movements (e.g. climbing the tree; opening the nut) (Hinde 1970). Descriptions of social behaviour usually contain mixtures of both, with communicatory gestures receiving physical description (e.g. threat posture) within descriptions by consequence of longer episodes (e.g. repelling an intruder). It is, of course, important to remember that even where physical description is most convenient, the behaviour may be governed by its consequences (e.g. the threat posture and its sequelae may be attuned to the moment-

17

to-moment responses of the opponent). Neglect of this issue has in fact distorted the analysis of social signalling (Hinde 1981a).

Importance of context

While good description demands selectivity in what is recorded, social behaviour can only be meaningfully interpreted with respect to its context, which must not be neglected. Three categories of context may be recognized. The first is the physical context: behaviour may be affected by whether an individual is on or off his territory, on the ground or in a tree. The second is the social context: in seeking for precise recording methods, it must not be forgotten that what a monkey does may depend on its hopes for assistance from neighbouring animals or that behaviour directed towards one animal may affect a third (Chapters 6, 8 and 9). Third, the temporal context may be crucial, e.g. descriptions of mother–infant interaction can be enormously enhanced if periods of rest and activity are distinguished (Harcourt 1978a; Simpson 1979a).

Description of interactions

Descriptions of social behaviour must focus first on what the individuals are doing together, for instance are they fighting, grooming or playing? As seen above, however, a mere statement of what they are doing conveys too impoverished a picture of the interaction: it is also necessary to know how they are doing it. For instance are some females grooming assiduously or perfunctorily, or did a mother pick up her baby roughly or gently? Data on the qualities of interactions are not easy to obtain but there is no reason in principle why they should not be quantified.

A second issue concerns the initiation of activities. It is often much easier to record what two animals are doing than who started it or who broke off first, but the latter often has greater prognostic value. In predicting the consequences of temporary separation between mother and infant rhesus, the relative role of the infant in maintaining proximity to the mother during the preseparation period is more valuable than the actual amount of time he spent near her (Hinde & Spencer-Booth 1970) and the direction as well as the amount of grooming between chimpanzees is related to aspects of social structure (Simpson 1973).

Description of interindividual relationships

A relationship between two individuals extends over time and involves a series of interactions each one of which may influence subsequent interactions. Description must therefore concern first the content and quality of the component interactions. Their diversity may also be important: the more different things two individuals do together, the more diverse the ways in which interactions may influence other interactions.

In addition, the absolute and relative frequencies of interactions and the manner in which they are patterned in time may present new and emergent properties. Some properties of relationships depend on the correlated presence of interactions of different types: for instance a mother–infant relationship is more likely to be labelled as involving 'maternal warmth' if the mother is solicitous of the infant's welfare in diverse contexts than if she is merely meticulous about grooming it or merely holds it a lot. Dominance–subordinance is the more valuable as a label for a dyadic relationship, the more aspects of behaviour can be subsumed by it (see below).

Other properties of relationships depend not merely on the extent of correlations between interactions of different types but on their relative frequencies. A mother who often picked up her infant and never rejected it might be described as possessive (in the context of that relationship) and one who did the opposite as rejecting. However, a mother who often did both, or seldom did either, might be described respectively as controlling or permissive: these properties depend on interactions of different types (e.g. Altmann 1980; Hinde 1979a; Simpson *et al.* 1981).

The interactions within a relationship can be characterized according to whether the participants do the same thing (either simultaneously or in turn, i.e. *reciprocal* interactions) or different but *complementary* things. Thus, when young monkeys take it in turn to chase each other, the interaction is reciprocal, but when one adult bosses and another is bossed, or a male mounts and a female is mounted, the interaction is complementary (cf. Wade 1977).

Thus, the content, diversity, qualities, relative frequency and patterning and reciprocity versus complementarity of interactions may all constitute important characteristics of a relationship. Colvin (section 3.2) seeks a general description of relationships through these parameters, discussing also the category of interpersonal perception. We shall see later that yet other categories of properties are important in the interpersonal relationships of our own species (section 13.6).

Comparable issues arise in the description of social structure. These are discussed in section 10.1.

Data and theory language

If there is to be a science of social behaviour, it is essential that a clear distinction between data and explanation is maintained. Terms like 'competitive' as applied to interactions, 'permissive' as applied to relationships or 'hierarchical' as applied to social structure are clearly mere short-hand terms referring to properties of the original data and need involve no departure from a language of description. By contrast, terms such as 'sex drive', 'learning' and 'familiarity' do not refer to directly observable entities but are explanatory concepts which are used to account for the phenomena observed. It is crucial to maintain a clear distinction between the language used for recording data and the language of explanation. Yet the boundary between description and explanation is not always easy to draw, and some terms need special care. Consider 'dominance'. If dominance is defined in terms of who bosses whom, it does not add anything to the observation that A bosses B to say that A is dominant to B. However, 'dominance' becomes more useful descriptively and also begins to become explanatory if it is relevant to more than one aspect of behaviour. If we find in a number of dyads the pattern of A bosses B, B avoids A, A has priority of access to resources over B, B grooms A more than A grooms B, etc, then the statement that A is dominant to B acquires an explanatory value: it implies a difference between the characteristics of A and B which determines the direction of complementarity in interactions of a number of types. 'Dominant' here is taking on the properties of an intervening variable and belongs properly in the 'theory language' (see Hinde 1978; Hinde & Datta 1981). Its explanatory value depends in part on the diversity of contexts to which it applies. Thus, if A bosses B in the context of food disputes but B has priority of access to females, the explanatory value of dominance is limited to particular contexts (see Chapters 4, 6, 9 and 10). In the same way, 'maternal warmth' is a descriptive term but can acquire explanatory value when a number of aspects of behaviour are correlated with each other. In all such cases, however, care must be taken of slipping inadvertently from data language to theory language, for that would be to fall into the instinct fallacy, e.g. 'the bird builds a nest: the bird has an instinct for nest building: the nest-building instinct makes the bird build the nest'. 'A bosses B: therefore A is dominant to B: A bosses B because A is dominant to B' is an argument of the same type. However, if it is observed that the several types of interaction mentioned above go together, e.g. that Z grooms Y and Y has access to food, it may be deduced that Y is dominant to Z and predicted that if the occasion arises Y may boss Z (see also Chapter 10).

Finally, we must maintain not only a clear distinction between description and explanation but also one between different types of explanation. As has been seen, causal and functional explanations must be compatible but are not identical.

3.2 Description of Sibling and Peer Relationships among Immature Male Rhesus Monkeys

JOHN COLVIN

Relationships are usually differentiated according to the age/sex classes of the individuals concerned. However, we can also distinguish between relationships in terms of the behaviour involved, an approach which has already met with some success in studies of the effects of separating infants from their mothers (Hinde & Spencer-Booth 1971b) and in describing the adult male–female relationships of baboons (Seyfarth 1978a). In this section, those criteria for describing social relationships outlined by Hinde (section 3.1) are used to distinguish between relationships of immature male rhesus monkeys. Two types of peer–peer relationships, distinguished on the basis of proximity time, are contrasted with each other and with sibling relationships.

In a study of immature males in two troops on Cayo Santiago (Colvin & Tissier in prep.) the author focused on the relationships of 20 three-year-old males with their male peers and their younger brothers (13 males had younger brothers, in 12 cases aged two years and in the remaining case aged one year). The 14 males of Group I were studied in both spring and summer; the six males of Group L in summer only. Those few relationships that appeared to be the most important to the males themselves were concentrated on and for each male those three peer relationships in which the individual was most involved (in terms of time spent with the partner) were selected.

Among the three most preferred relationships of each male a distinction could be drawn between 'strong' and 'weak' relationships. A strong relationship was defined as one in which each partner was involved for more than 15% of his total peer proximity time. A weak relationship was one in which at least one partner (and normally both) was involved for less than 15% of his total proximity time. There are two corollaries of this distinction. First, strong relationships reflected reciprocated preferences, whereas weak relationships reflected mainly

Table 3.1. Differences between strong and weak peer relationships in the reciprocation of partner preferences. A partner preference was defined as the rank given by a male to a particular partner, in terms of the time spent in proximity with that partner compared with that spent with other peer partners. A 'reciprocated preference' was one in which two males ranked each other within their top three peer relationships, in such a way that the differences between their rankings of each other was one or zero.

	Spring		Summer	
	Strong	Weak	Strong	Weak
Dyads in which preferences reciprocated	13	2	16	3
Dyads in which preferences unreciprocated	0	10	1	9
	$P < 0.005$		$P < 0.001$	
	(Fisher test, two-tailed)			

unreciprocated preferences (Table 3.1). Second, strong relationships involved more proximity time, including more close proximity time, than weak peer relationships.

Beyond the definitive differences between sibling, strong and weak peer relationships, there were several further clear differences and, in addition, some similarities. These are summarized briefly below. Characteristics of these relationships are then considered in greater detail in terms of Hinde's criteria.

Table 3.2. Characteristics of strong peer relationships in contrast to weak.

---

1. More reciprocal partner preferences.*
2. More proximity and close proximity.*
3. Proportionately more feeding together but proportionately less play.
4. Differences in the patterning of play.
5. Greater likelihood of grooming and friendly gestures.
6. Greater likelihood of alliance formation, including aiding.
7. Lower rates of agonistic interaction, particularly in 'aggressive' relationships.

---

*Definitive characteristics.

While strong and weak peer relationships showed certain similarities, e.g. in overall rates of play and in the content and reciprocity of play components, these were outweighed by many differences, as summarized in Table 3.2. In turn, these two types of relationship contrasted with that found between brothers. In general, sibling relationships could be described as more reciprocal and affiliative than strong peer relationships and these in turn as more reciprocal and affiliative than weak peer relationships. All sibling relationships were characterized by grooming and many by alliance formation, especially aiding*, while agonistic interaction was rare. In contrast, in weak peer relationships grooming and alliance formation were rare and aiding was absent, while rates of agonistic interaction were high. Strong peer relationships were intermediate between sibling and weak peer relationships on these measures (Table 3.3 and Fig. 3.1a).

The differences between sibling, strong peer and weak peer relationships were apparent in several of Hinde's categories of dimensions, namely the content, diversity, qualities and relative frequency and patterning of interactions, reciprocity versus complementarity and interpersonal perception (see also Hinde 1979a, 1981b).

*Following Cheney (1977), in this section and section 5.4 a distinction is recognized between two types of alliance formation: aids and coalitions. In aids, the interfering individual helps the victim of an agonistic interaction, whereas in coalitions, the interferer allies itself to the aggressor. Clearly, aiding carries greater risks than coalition formation. (Some other authors use coalitions for short-term liaisons, alliances for longer term ones.) (Editor's footnote.)

Table 3.3. Differences between sibling, strong and weak peer relationships in friendly gestures, grooming and alliance formation.

	Spring			Summer		
	Sib	Strong	Weak	Sib	Strong	Weak
A. Friendly gestures*						
No. dyads showing friendly gestures	8	11	2	9	14	5
No. dyads not showing friendly gestures	1	2	10	2	3	7
	(Strong versus weak, $P < 0.001$, Fisher test)			(Strong versus weak, $P < 0.05$, Fisher test)		
% friendly gesture dyads in which friendly gestures reciprocated	38	55	*50*	56	29	40
B. Grooming†						
No. dyads showing grooming	9	8	3	11	13	2
No. dyads not showing grooming	0	5	9	0	4	10
				(Strong versus weak, $P < 0.002$, Fisher test)		
% grooming dyads in which grooming reciprocated	67	63	0	73	54	*50*
C. Alliance formation†						
No. dyads forming alliances (No. forming coalitions/aids)	4(4/4)	6(6/1)	2(2/0)	8(6/5)	9(9/4)	1(0/1)
No. dyads not forming alliances	5	7	10	3	8	11
% alliances reciprocated	25	33	*0*	13	67	*0*

*Based on focal data only.
†Based on focal and *ad libitum* data.
Percentages based on small *n*s are italicized.

All three types of relationship shared a fundamental set of contents: agonistic interactions, playful interactions and mutual resting and feeding. Moreover, similar contents within play, namely rough-and-tumble play, play hits and bites and approach–withdrawal play, were also found in all three types of relationship. However, additional contents (friendly gestures, grooming and alliances) were found only in strong peer and sibling relationships, with aiding, a particular type

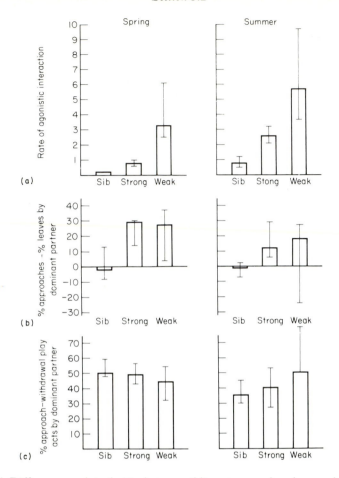

Fig. 3.1. Differences and similarities between sibling, strong and weak peer relationships. All three types of relationship differ in their patterning regarding rates of agonistic interaction (a) and this may also reflect differences in the quality of agonistic interaction. Further differences in patterning of the close proximity index (b) distinguish the sibling relationship from the other two types. Thus, responsibility for close proximity is more reciprocal in the former than in the latter. On the other hand, all three types of relationship share a similarity in the reciprocity of approach–withdrawal play (c).

of alliance formation, more common in the latter (Table 3.3). As a result, sibling and strong peer relationships contained a greater diversity of interactions than weak peer relationships.

Qualities of interactions are less easily described. While no specific attempt was made in this study to assess the qualities of each particular type of interaction, indications of qualities pertaining to a relationship may sometimes be obtained by studying how the interactions of that relationship are patterned (Hinde 1979b), e.g. agonistic interactions between individuals who are rarely in proximity may cause the victim to be more fearful of its aggressor than agonistic interactions occurring at a similar frequency between individuals who spend greater amounts of time in proximity. As a result, these interactions may qualify as more 'serious' in the former case than the latter, even though they may occur at similar intensities in both cases. The rate of agonistic interaction (min^{-1} in proximity) was highest in weak peer relationships and lowest in sibling relationships (Fig. 3.1a). Thus, agonistic interactions in weak peer relationships would be regarded as the most serious and those in sibling relationships as the least. Some support for this came from data on the patterning of approaches and leavings (see below).

The three types of relationship differed also in the relations between the different types of interaction. Sibling dyads whose play showed high rates of approach–withdrawal were also the dyads whose play contained frequent play hits and bites. In contrast, in weak peer relationships the strong positive correlation between the rates of these components was absent and in strong peer relationships the correlation was strongly negative.

Reciprocity was a key feature of the sibling relationship and of some strong peer relationships. However, other strong peer relationships and all weak peer relationships showed a greater complementarity across their component interactions. Thus, while responsibility for close proximity (see Chapter 2) was more reciprocal in sibling relationships than in strong peer relationships (Fig. 3.1b), partner preference was more reciprocal in strong than in weak peer relationships. Furthermore, grooming was reciprocated in many sibling and strong peer dyads but rarely in weak peer dyads. On the other hand, in all three types of relationship there was considerable reciprocity of play components, especially approach–withdrawal play (Fig. 3.1c), while agonistic interactions were in almost all relationships highly complementary.

In non-human primates the category of interpersonal perception is probably restricted to the issue of each partner's 'view' of the other. Operationally, this is likely to be related to the degree to which each is able to interact effectively with the other through an ability to anticipate and predict the other's actual behaviour and/or propensities for behaviour. Such interpersonal perception may depend both on the degree of familiarity and of similarity between the partners (Colvin 1982). It might therefore be predicted that interpersonal perception is more accurate in sibling relationships (high familiarity) and in strong peer relationships (moderate familiarity and high similarity) than in weak peer relationships (low familiarity and moderate similarity).

Since it is evidently not an easy task to measure this dimension, the above prediction cannot be readily tested. One possible resolution of this difficulty is that we might expect different levels of interpersonal perception to find their correlates in different qualities of relationships, e.g. on the basis of data on the rhesus mother–infant relationship, Hinde and Simpson (1975) have suggested that behavioural 'meshing', with each partner adjusting his behaviour to harmonize with the ongoing goals of the other, may be the behavioural correlate of some aspects of intersubjectivity. In the present instance data are not yet available.

Thus, at least five of Hinde's categories or dimensions permit distinctions between different types of relationships in monkeys. Not only do they permit an analysis of the differences between relationships differentiated in terms of the kinship relations of the participants but they also permit differentiation within such a category.

In the current instance, the differentiating characteristics can be subsumed under the more global concept of 'affiliation'. With increasing affiliation, the diversity and reciprocity of interactions also increase. Of the three types of relationship between immature males described, the weakest type, the weak peer relationships, had only one affiliative component—play—and even this was different in patterning, although not in content, from play in stronger relationships. On the other hand, there were many unaffiliative and imbalanced components in these relationships, including low proximity times, low reciprocity of partner preference, high rates of agonistic interaction and a stronger role of the dominant partner in maintaining proximity.

In the more affiliative strong peer relationships, there were, in comparison to the weak peer relationships, more proximity and close proximity time, accompanied by more mutual feeding and, in addition, grooming and coalition formation, although generally at low rates. At the same time, rates of agonistic interaction were lower and partner preference more reciprocal.

In the most affiliative relationship, the sibling relationship, proximity times were very high and in some dyads greater than 50% of the total sampling time. All dyads showed grooming, which was usually reciprocated, and in some dyads this activity accounted for a substantial proportion of close proximity time. The proportion of alliances that were aids and therefore contained a higher element of risk was high. On the other hand, rates of agonistic interaction between siblings were very low and in many dyads the roles of each partner in maintaining close proximity was highly reciprocal.

The nature of these strongly affiliative relationships in young, male rhesus monkeys fulfills many of the criteria used to describe friendships in very young children. While the description of friendships in children depends a great deal on the levels of interpersonal awareness involved, in one and two year olds friendships can probably be adequately characterized, first, in terms of a greater frequency of proximity seeking, sharing, positive affect and play and less hurting than in other dyads and, second, in terms of mutual preference (Vandell & Mueller 1980).

According to these criteria, in young, male rhesus monkeys most sibling relationships and possibly some strong peer relationships could be described as 'friendships'. Mutual preference and proximity seeking are evident, while proximity during feeding and the reciprocal exchange of 'social resources' (such as grooming and alliance formation) might be described as a form of sharing. In terms of positive affect, these relationships are clearly highly affiliative, while levels of agonistic interaction are low. Play, on the other hand, would appear not to be a distinguishing characteristic of immature rhesus friendships, except possibly in terms of its patterning.

In order that it retain the greatest descriptive value, the limited application of the term 'friendship' to sibling relationships and to the strongest peer relationships is proposed for this study. Only six of the 22 strong peer dyads of the current study fulfilled the majority of the criteria defining friendship.

It is suggested that in non-human primates certain adult relationships may also qualify as friendships. The 'persistent high frequency' social bonds between certain dyads of adult male and female baboons (Seyfarth 1978 a,b) may be a case in point. These dyads exhibited frequent proximity and grooming and the preferences of the dyad's members for each other as grooming partners were reciprocal. Agonistic interactions, while frequent, were most likely to be followed by proximity, and males in these dyads aided their partners more frequently than they aided other females (see also section 6.10).

Friendship in children also implies the idea that greater expectations are possible with friends than with others. By two years of age children appear to have an initial concept of 'friend' as a familiar peer from whom one expects particular responses (Rubin 1980), although it is not until much later that a clear conception of an enduring, reciprocal relationship develops (Selman & Jaquette 1978). While care must be taken how much one imputes to young rhesus monkeys expectations of each other's behaviour, one can tentatively suggest that the presence in a relationship of reciprocal grooming and alliance formation (especially aiding), both interactions with potential or real costs, implies, if not actual expectations of reciprocity, some mechanism governing the stability of these components which is sensitive to their reciprocation.

Chapter 4
Individual Characteristics and the Social Situation

4.1 Individual Characteristics: a Statement of the Problem

JOAN STEVENSON-HINDE

Terms such as 'personality', 'traits' or 'individual characteristics' are usefully employed to refer to aspects of behaviour that differentiate one individual from another. While such terms may carry conceptual or explanatory implications when used within specified theoretical frameworks, they do not carry any extra meaning on their own. In particular, they do not imply anything concrete that an individual carries about with himself. Neither do they imply anything permanent, valid for all time and all situations. Indeed, it has proved fruitless to search for generalized traits as the main determinants of individual differences in behaviour (Mischel 1973). Conversely, the search for entirely situational determinants has been equally fruitless. In that sense, there is a parallel between the study of personality and that of behavioural development. Just as the innate versus learned dichotomy was resolved by adopting an interactionist view (e.g. Hinde 1968; Lehrman 1953), so the trait versus situation controversy is being resolved in a similar way. Neither an individual himself nor the situation he is in is to be seen as the main determinant of his behaviour; rather behaviour is 'determined by a continuous process of interaction between the individual and the situation he encounters' (Endler & Magnusson 1976, p.12). Since an individual never develops or behaves in a vacuum, the search for his 'true characteristics', like the

28

search for 'innate behaviour', was misguided.

Such an interactionist view implies that one cannot always separate aspects of an individual from aspects of his relationships with other individuals. This is especially true for those characteristics which involve behaviour with others, e.g. characteristics such as Confident or Sociable (which shall be considered in section 4.2) could be affected if relationships with mother or siblings should change.

Thus, any assessment of personality must incorporate effects of the situation, especially its social aspects. By taking situations into account, it should be possible to predict more individuals more of the time, at least for particular characteristics (Bem & Funder 1978; Kenrick & Stringfield 1980). Individuals and situations of course change with time. However, it is likely that some characteristics and some individuals are more consistent across time than others. In the human being, consistency is generally greater from infancy to pre-school age and from adolescence to adulthood than from middle childhood to adolescence (Yarrow & Yarrow 1964).

For humans, various means of assessing personality have been presented and critically discussed by Block (1977). For animals, there are at least three sources of data for assessing individual characteristics. First, behaviour may be objectively recorded in a free situation. While this can provide useful data, there are some difficulties. The situation is uncontrolled and there is no easy way to take a changing context, including the social one, into account. With developing individuals, the meaning of behavioural items may change with age, though this is a problem for any method.

Second, behavioural tests may be used. This has the advantage that the situation is controlled and objectively similar for all individuals. Of course, it is not subjectively similar: the degree of stress imposed might be high for one animal but low for another. Even with a period of 'settling in' to the test situation, most tests must be fairly novel to ensure performance. There is thus a danger that the situation has a relatively high value in the 'individual–situation' interaction. This has the consequence that generalization to another test, let alone another situation, may be low (see e.g. Stevenson-Hinde *et al.* 1980b). Last, although responses may be objectively recorded, it is often difficult to know which of the many possible measures to use (e.g. Spencer-Booth & Hinde 1969).

The third method involves the use of rating scales. These have been surprisingly little used with non-humans. However, ratings based on Plutchik's theory of human emotions have been made on baboons (Buirski *et al.* 1973) and chimpanzees (Buirski *et al.* 1978). The studies described in section 4.2 used items developed in discussion with those who knew the monkeys well: they were thus based on behaviour rather than theory (Stevenson-Hinde & Zunz 1978). Of course, there are dangers inherent in subjective evaluations but the human literature shows that there are also counterbalancing advantages. Block (1977) contrasts an observer's role in collecting behavioural data with that in making subjective assessments. In the former case, if an observer is necessary he is merely another kind of recording instrument, recording preselected, readily identified behaviour patterns. With subjective assessment, the observer is an active instrument, filtering, cumulating, weighting and integrating the data. While this carries the danger that the observer may introduce his own bias, especially in the form of an 'implicit personality theory' (Bruner & Tagiuri 1954), the active role of the observer gives flexibility, e.g. he may incorporate either a rare but crucially important behaviour pattern or a changing context into a subjective assessment more readily than into a recording of preselected behaviour patterns.

Whatever source of data is used, the separate items may be brought together by intuitive, theoretical and/or statistical means (see section 4.2) to obtain dimensions with supposedly greater reliability and generality than the individual items on their own. In the sections 4.2 and 4.3 the issues of temporal and cross-situational consistency are illustrated with data from the Madingley colony of rhesus monkeys. In section 4.4 the issue of relative dominance being context specific is discussed.

4.2 Consistency over Time
JOAN STEVENSON-HINDE

One approach to consistency has involved applying a common set of ratings to all the monkeys in the Madingley colony over four successive years. Ratings have the advantages mentioned in section 4.1, but since they have been under-used for non-human primates, the general issues of reliability, validity and data reduction must first be considered. With any rating instrument, the items, their definitions or even the rating scale must depend on the situation and purpose of the study. Two observers (who were not always the same in different years) rated the monkeys over a list of 25 behaviourally defined adjectives, using a seven-point scale. For the last year of rating, interobserver reliability reached the $P < 0.001$ level (Pearson product-moment correlations, one-tailed) for 21 items.

One of the advantages of observers' ratings is that they can capture aspects of behaviour at a more global level than can direct observations. Validation of ratings against direct observation is therefore not always possible. However, in the last year of rating it was possible to compare the observers' ratings on six of the 21 items with behavioural records of each member of the colony. As shown in Table 4.1, all six Spearman correlation coefficients were significantly positive ($P < 0.001$, one-tailed).

Table 4.1. Validation of six questionnaire items against behavioural observations.

Item	Definition	Observation	Validity (r_s)
Effective	Gets own way; can control others	Displaces others; is avoided by others	0.73
Aggressive	Causes harm or potential harm	Hits, threatens, chases others	0.49
Fearful	Fear-grins; retreats from others or from outside disturbances	Fear-grins; avoidance of others	0.57
Excitable	Over-reacts to change	Displays; threats directed outside pen	0.45
Sociable	Seeks companionship of others	No. of others in contact with focal animal (excluding mother's offspring of less than four years old)	0.46
Playful	Initiates play; joins in when play is solicited	Play	0.71
			$n = 48$
			$P < 0.001$, one-tailed

Having established interobserver reliability and validity where possible, it is usually necessary to achieve some reduction in the number of items to be considered. As a first step, observers' ratings were standardized, to set the mean rating for each item at zero and the standard deviation at one. Combining the different observers' ratings gave a standardized mean rating for each item for each monkey. From these, an item×item correlation matrix showed those items which clustered together. If it makes good intuitive or theoretical sense to sum items within a cluster, data reduction can be achieved in this way, for some of the items at least. However, more sophisticated ways of data reduction are also available. A principal components analysis, for example, can show how much each of the items contributes to a number of main components or dimensions. Each component can then provide a score to which the items contribute according to their loadings (see Nie *et al.* 1975). However, the loadings provided by such an analysis are subject to sampling error and therefore will vary from sample to sample, even if in theory the population of animals remains constant. Indeed, in our study, although similar components did emerge in each year, the loadings did not remain exactly the same. Therefore, rather than relying on any particular set of loadings, we used the principal component analysis as a guide for summing. It provided three main components (accounting for over 65% of the variance each year): C1—Confident to Fearful; C2—Excitable to Slow; and C3—Sociable to Solitary. Thirteen items loaded highly positively (+) or negatively (−) on these components and included five of those that had been validated against direct observations. These items were as follows:

C1 (+)   *Confident* (Behaves in a positive, assured manner; is not restrained or tentative)

    (+)   *Effective* (Gets its own way; can control others)

    (+)   *Aggressive* (Causes harm or potential harm)

    (−)   *Apprehensive* (Is anxious about everything; fears and avoids any kind of risk)

    (−)   *Subordinate* (Gives in readily to others; submits easily)

    (−)   *Fearful* (Makes fear-grins; retreats from others or from outside disturbances)

C2 (+)   *Excitable* (Over-reacts to change)

    (+)   *Active* (Moves about a lot)

    (−)   *Equable* (Reacts to others in an even, calm way; is not easily disturbed)

    (−)   *Slow* (Moves and sits in a relaxed manner; moves slowly and deliberately; is not easily hurried)

C3 (+)   *Sociable* (Seeks companionship of others)

    (+)   *Opportunistic* (Seizes a chance as soon as it arises)

    (−)   *Solitary* (Spends time alone)

        (Solitary was in fact not used here, since it was introduced for only the last two years of rating.)

A score on each component for each monkey was calculated by multiplying the standardized rating by +1 if the item loaded positively and −1 if it loaded negatively, summing over items and dividing by the number of items (e.g. 6 for a Confident score, 4 for an Excitable score, etc.). Thus, each score had a mean of zero over all the individuals rated that year and a standard deviation of slightly less than one. Further details are given by Stevenson-Hinde *et al.* (1980a).

Consistency from one year to the next was assessed by asking if individuals maintained the same rank-ordering as they grew older. Table 4.2 presents Spearman correlation coefficients over successive years. Confident scores were significantly correlated from one to two years of age, two to three, three to four and into adulthood. However, consistency does not imply stability, in the sense of remaining at a constant level, e.g. even though the rank-order correlation from two to three years was high (0.88) for Confident scores, they did increase significantly ($P < 0.05$, Wilcoxon matched-pairs test, two-tailed). For adults, the Confident, Excitable and Sociable scores were consistent and they did not increase or decrease over four years of rating (Stevenson-Hinde *et al.* 1980a).

However, when scores were considered over a longer time period than four years, Excitable scores did decrease with age. For 15 different mothers ranging in age from five to 20 years, age was significantly correlated with Excitable scores ($r_s = -0.67$, $P < 0.01$, two-tailed) but not with Confident or Sociable scores.

Consistency and/or stability may be upset by life events. For example, the Sociable scores of a 16-year-old female (Eliane) dropped severely when another female aged 12 years and a male aged 16 years that had always been with her had to be removed. Eliane's Sociable score, which had been 0.87 and 0.84 over the

Table 4.2. Correlation coefficients from one year to the next (r_s) or over four years (r_s average), for Confident, Excitable and Sociable scores. M = male; F = female.

	Confident	Excitable	Sociable
1–2 years (n = 11M + 10F)	0.65‡	0.32	0.04
2–3 years (n = 5M + 4F)	0.88†	0.44	0.49
3–4 years (n = 3M + 5F)	0.67*	0.20	0.70*
Adult females (excluding primips) over four years of assessment (n = 11, aged 6–20 years at first assessment)	0.90‡	0.80‡	0.59†
Adult males over four years of assessment (n = 5, aged 8–15 years at first assessment)	0.65‡	0.48*	0.72†

$*P < 0.05$.
$†P < 0.01$.
$‡P < 0.001$, Spearman (one-tailed) or Kendall tests.

two preceding years, fell to -0.30 after removal of these two peers, even though other peers were introduced. With young males there is some evidence that adverse early experience increases Excitable scores (Stevenson-Hinde *et al.* 1980a).

In summary, consistency appears earlier for some scores than others, consistency over time need not imply stability and scores may be affected by life events. This is in harmony with an interactionist position, with an individual's characteristics reflecting aspects of itself and its situation, including its social aspects.

4.3 Predictability across Situations

JOAN STEVENSON-HINDE

The Confident, Excitable and Sociable scores referred to in section 4.2 may also be used to address the issue of consistency across situations. However, with monkeys any situational change is liable to be more drastic than those normally used with humans. The ratings had been done in a social context and one cannot simply transfer a monkey to another social context and ask meaningful questions about consistency, without allowing for a long period of adaptation to the new group. This period may in turn create some real changes in his characteristics. The issue has therefore been rephrased, as implied by the section heading.

To assess behaviour in another situation, infants (at ages one and two-and-a-half years) were removed from their groups in the main colony to a neighbouring building with no other monkeys in it (except the mother of each one year old) for one to two weeks. Each infant was given behavioural tests and was also observed for 'baseline' periods without any test objects. The tests involved presentation of novel objects, such as a ball, banana or mirror, and the two-and-a-half year olds received operant tests as well. We are not concerned here with particular test measures but with correlations between measures from the colony to the test situation. At each age, there were more significant correlations for males than for females. Between the Confident, Excitable and Sociable scores based on behaviour in the colony and eight behavioural measures taken during the baseline periods in the test situation, there were seven significant ($P < 0.05$, two-tailed) correlations out of 24 for one-year-old males but none for females (Stevenson-Hinde *et al.* 1980b, Fig. III). Between the colony scores and tests at one year there were two significant correlations out of 12 for males but none for females (Stevenson-Hinde *et al.* 1980b, Table 2). Finally, between colony scores and measures in the test situation at two-and-a-half years, there were five significant correlations out of 27 for males, compared with only two for females (Stevenson-Hinde *et al.* 1980b, Table 3). Indeed, the two correlations for females may be regarded as one, since they involved only one test measure with two colony scores, Confident and Sociable, which were themselves highly correlated (0.86) for females but not for males (0.26). Thus, at both ages the males showed some predictability from colony scores to the test situation, while females did not.

The lack of cross-situational predictability for females is surprising since: (1) each measure in the test situation was either reliable (i.e. the test/retest measures) or was taken over a sufficiently long period to be representative (i.e. the baseline measures); (2) the observer ratings were reliable between observers and correlated meaningfully with earlier social behaviour (Stevenson-Hinde *et al.* 1980a); and (3) the use of colony scores based on observers' judgements, rather than isolated behavioural measures, should maximize the possibility of finding significant cross-situational correlations (see e.g. Block 1977).

The fact that higher correlations were not found underlines the different nature of the two situations, one a social group in which the monkey was born and raised and the other a strange situation without any other monkeys (except the mother at one year old). Indeed, the cross-situational correlations that did emerge for males were not straightforward, e.g. it was not the Confident but the Excitable males who could be described as 'confident' in the test situation. At one year, the males with higher Excitable scores spent more time with the strange ball ($r_s = 0.75$, $n = 11$) and distress-called relatively infrequently (-0.80). At two-and-a-half, Excitable scores were significantly correlated with working quickly through a series of operant tests for Smarties and holding down a lever for relatively long periods to view slides (Stevenson-Hinde *et al.* 1980b). To obtain a high Excitable score in the colony, an animal would have been rated highly on the items Excitable (over-reacts to change) and Active (moves about a lot). These characteristics do not describe the above test behaviour. Indeed at neither age was there a significant correlation between males' Excitable scores and activity in the test situation. Interestingly, although males rated as Confident in the colony did not tend to behave confidently in the tests, the males who did behave confidently in the tests tended to have mothers rated as Confident in the colony (Stevenson-Hinde & Simpson 1981). Thus, the search for predictability across situations takes unexpected paths, which in turn appear to be better defined for males than for females.

4.4 Context-specific Unpredictability in Dominance Interactions
PHYLLIS C. LEE

The usefulness of the concept of dominance has been continuously debated (e.g. Bernstein 1970, 1976a, 1981; Hinde 1978; Rowell 1974) but it is perhaps more applicable when viewed as an intervening variable for describing or explaining interactions between two individuals. As such, dominance is valuable for describing the outcome of interactions between individuals when the directionality of a number of dependent variables co-vary, e.g. when the pattern, A supplants B, A has priority of access to water over B, A has priority of access to grooming partners over B, B grooms A more than A grooms B, B yields to A, etc., applies to the majority of dyads. It is even more useful if it can be linked to independent variables such as maternal rank, size or sex (Hinde & Datta 1981; see section 10.1).

Often, the individuals in a group can be arranged in a linear hierarchy, such that A is dominant to B and all others, B is dominant to C and all others, etc. The hierarchy is then based on the transition matrix of the direction of the outcome of a number of interactions between different individuals (e.g. Dawkins 1976) and interactions where a normally subordinate individual wins over a dominant are rare (e.g. Table 9.1). Such linear hierarchies are often based on only one or a few types of interactions (fights or supplants over food) and their predictive value is limited (Datta 1981; Richards 1974). However, the concept has been usefully extended by the models of Seyfarth (1977) and Cheney (1978b), wherein dominance determines the distribution of friendly behaviour among the members of the group and subordinates are often excluded from receiving the benefits of such behaviour.

During immaturity, competitive dominance rank is acquired through a variety of interactions with others in the social group. In many species, the rank of young females, and in some cases that of males, depends on that of their mothers and is a result of the support given during aggression and competition (Cheney 1977; Datta 1981; Kawai 1958; Lee & Oliver 1979; Sade 1967). The dominance rank of an immature may also depend on individual aggressiveness which increases at a particular time during development, such as the first oestrus (Walters 1980) or just before transfer from the natal group (Packer 1979a). Fedigan (1972) found that immature vervets in a captive group had the same dominance ranks as their mothers, with important consequences for the nature and distribution of their social interactions.

Since competitive dominance rank among the young is not fixed at birth but rather gradually acquired through the interactions of the young and the assistance of others, it is necessary to examine the contexts in which competitive interactions take place. For some individuals, dominance may be predictable in certain contexts but not in others, depending on the resources involved in competition or on the presence or absence of supporting individuals. The outcome of a dominance interaction may also depend on the benefits of participating, e.g. an individual's initial probability of success may be low but gradually rise through time if weight determines the outcome (e.g. Packer 1979b): in such a case, a young animal might attempt to initiate competitive interactions against dominant individuals in order to test when it is sufficiently heavy to win consistently. The competitive rank of young animals who are in the process of acquiring rank may thus be unpredictable in some contexts.

In this section, the competitive interactions of immature vervets and the nature of the resources involved in competition will be examined in two ways. The first is to assess the underlying stability of immatures' interactions in the framework of their dominance relationships, particularly with respect to their mothers' rank order. The second is to examine in detail those cases where the outcome of an interaction was reversed from that predicted by the rank orders of the participants. These reversals have been considered as 'wins against the hierarchy' which Dunbar (1983) defines as occasions when a normally subordinate individual is able to defeat a more dominant individual in a particular agonistic encounter.

The assessment of interindividual competition was based on the direction of dyadic approach–retreat interactions, using criteria established by Struhsaker (1967a) and Seyfarth (1980). Approach–retreat interactions were separated into two categories. The first is non-specific competition, where one animal moved away at the approach of another to within a 2 m radius (avoid), cringed or cowered at the approach of another or ceased it's activities when another passed by within contact distance (pass and interrupt). The second is context-specific competition, where one animal was able to take the food resource of another (food supplant), take over the grooming partner of another (groom supplant) or take over the space another occupied, such as a protected or shady resting spot (place supplant). An approach during feeding which did not result in a direct take-over but did produce a cessation of the feeding activity was scored as a feeding interruption and considered as a food supplant.

The identities of both the initiator and the recipient were scored in each type of competitive interaction. If a food resource was contested, the neighbours within 2 m of the competing dyad were recorded. When the resource was a potential grooming partner, its identity was recorded.

Rank-order approach–retreat hierarchies were constructed from focal and *ad libitum* data. The approach–retreat hierarchies were considered to be dominance hierarchies (e.g. Bernstein 1976a; Bernstein & Sharpe 1966; Deag 1977; Kaufmann 1967; Rowell 1974) and the individual's rank order was called its dominance rank. The author used the approach–retreat criterion rather than the aggressive–submissive criterion (Altmann 1980; Sade 1967; Walters 1980) since the outcome of an aggressive interaction could be altered by the probability of the aggression becoming polyadic (Seyfarth 1980; Struhsaker 1967b). Individuals were ranked in an order which minimized the number of 'wins against the hierarchy'. Immatures were ranked in two ways. First, all ages and both sexes were ranked within the entire group, giving their position with respect to adult females as well as to all other immatures. Second, immatures of each sex were ranked only with other members of the same sex but of all different ages.

The presence of related individuals nearby (within 2 m) during an interaction has been examined for its effects on the outcome of a variety of immature competitive interactions, since for the immature vervets in these groups most of the alliances took place between mothers and offspring or between siblings (Lee 1981). In all analyses, competition between related animals has been excluded, since dominance between siblings is often flexible (Lee 1981). Immatures have been considered in two age categories: infants (up to 12 months of age) and juveniles (between 12 and 48 months for females and 60 for males) (Lee 1981).

Resources involved in competition

In all three study groups, competitive interactions were more frequent in the
six-month wet season of high food availability than they were in the dry-season
period of low food availability ($P < 0.02$). In spite of these seasonal differences in
the rates of competitive interactions, there were no significant differences in the
proportion of supplants of each type which took place during the dry or wet
seasons in any of the study groups (Fig. 4.1). Immatures were not competing for
different resource categories in the two seasons and most of the competition for
resources took place over food. During the wet season, the highest frequency of
food supplants took place over grass, while during the dry season, they were over

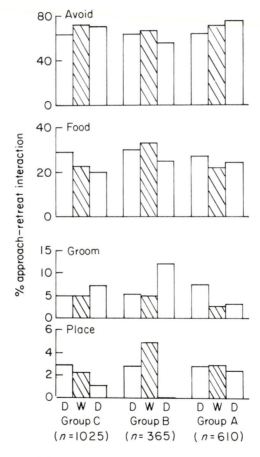

Fig. 4.1. Percentage of all approach–retreat interactions that took place over different
resources. Each group was studied separately in each of three different seasons. D = dry
season; W = wet season.

Acacia xanthophloea exudate. It is interesting to note that exudate was the major dry-season food and of high caloric value, while grass was one of the major wet-season foods and was high in protein (Klein 1978; Lee 1981).

Participants in competitive interactions

All age/sex classes competed with each other for resources and the majority of all competitive interactions were in the direction established by the hierarchies (Lee 1981). In the following analyses, the factors determining the ability of an immature who was generally ranked as subordinate to take resources from a dominant animal will primarily be considered.

Most (61%, $n = 18$) juveniles in all three groups were able successfully to initiate approach–retreat interactions against adult females who were subordinate to their mothers in the first year of the study. Only two of the 18 (11%) infants were able to initiate approach–retreat interactions against adult females. In the second year, 22 of the 31 (71%) juveniles were able to initiate interactions against females subordinate to their mothers, as could three of the four infants.

Increasing age or experience appeared to increase the proportion of juveniles who would attempt to win in interactions with adult females. As males approached adolescence (at about five years old), they began consistently to dominate the adult females at the bottom of the hierarchy, in that those females no longer initiated an interaction against them. Some of the younger juveniles were also able to dominate adult females subordinate to their mothers and this was most conspicuous for females. By the time a female was two years old, she was able to dominate at least some of the adult females subordinate to her mother. By the age of three, juvenile females with high-ranking mothers could dominate most other adult females in context-specific and avoidance interactions. Of the seven three-year-old females, six could dominate at least two-thirds of the females subordinate to their mothers. Thus, three factors initially affecting the outcome of competitive interactions were the age and sex of the participants and the relative rank of the mother in the adult female hierarchy.

Twenty-eight per cent (33 of 119) of all the approach–retreat interactions initiated by immatures against adult females were supplants over access to grooming partners. This proportion was significantly different from that of grooming supplants initiated by adult females against immatures (55 of 513). Reversals from the expected outcome appeared to be common for interaction over grooming partners. The supplants over other resources and avoids did not show this tendency, thus suggesting that the motivation for engaging in grooming may have been higher than that for other resources. Immatures may have been willing to take more risks to get access to grooming partners.

Maternal dominance also affected the dominance relations among the immatures. In hierarchies between same-sex immatures, age and maternal dominance interacted to determine the ranks of the immature males. Significantly more immature males in all three groups were dominant to males who were younger than those who were older, but younger sons of high-ranking mothers were dominant to older males of lower-ranking mothers (Table 4.3). However, maternal dominance alone predicted the rank order for all but two of the immature females. Immature females were equally dominant to older and younger females but a significant number of young females were dominant to older females of lower maternal rank (Table 4.3).

Table 4.3. Number of dyads where an immature of each sex (with a maternal dominance rank of high, middle or low) was dominant to another same-sex immature who was older or the same age, or younger, but whose maternal rank was higher or lower than its own.

	Dominant to			
Age of recipient	Same or older		Younger	
Maternal rank of recipient	Higher	Lower	Higher	Lower
Males' maternal rank				
High	0	12†	2	20†
Middle	0	6	10	20
Low	1	0	11	1†
Total	1	18†	23	41
Females' maternal rank				
High	0	6*	0	4
Middle	0	4	0	2
Low	1	0	0	0
Total	1	10*	0	6

$*P < 0.05$.
$†P < 0.01$.

The rank order of the immatures in the cross-sex, immature hierarchies again appeared to be influenced by age, maternal rank and sex. In two of the study groups, the relative rank orders of the immatures in the cross-sex hierarchies were the same as those in their same-sex hierarchies but immature males were not consistently dominant to immature females. For four of the ten immatures in the third group, their rank order with members of their own sex was different from that when they were ranked with all immatures. One older juvenile female was dominant to several older juvenile males of lower maternal rank, while the younger female who was dominant in the female hierarchy was subordinate to these older males. Neither age, sex nor maternal dominance could be used alone to predict the rank orders of the immatures in the cross-sex hierarchy but these variables presumably interacted to produce the overall rank orders (Lee 1981).

Reversals in outcome over different resources

In each group only a small proportion of dyadic reversals in the outcome of an interaction among immatures were against the expected direction. Each year of the study and each group was considered separately to examine developmental changes in the stability of the hierarchies. Thus, the total number of possible dyads where one immature could initiate an interaction against another immature was calculated for each group in each year of the study. In the first year, there was a total of 250 dyads in Group C and for 21 (8.4%) there were reversals from the predicted direction of interactions. There were 40 dyads in Group B and 13 (32.5%) had reversals. There were 124 in Group A and 6 (4.8%) had reversals. In the second year, there was a total of 248 possible dyads in Group C and 31 of these (12.5%) had reversals. In Group B there were 52 dyads, in 9 (17.3%) of which reversals occurred. In Group A, there were 84 dyads and 9 (10.7%) had reversals. Group B had the smallest number of immatures, all but one of whom were infants in the first year of the study. During that year, they had a higher proportion of reversals than the other two groups, which were larger and included more older immatures. The proportion of reversals declined in Group B in the second year, again suggesting that hierarchies were more stable with increasing experience. The highest proportion of reversals in all three groups occurred over grooming partners when compared with the total number of grooming supplants in each group (Table 4.4). The proportions of reversals which took place over grooming were significantly higher than those in other contexts for Group A and Group C.

For supplants over grooming partners where the success of the subordinate immature was high, the identity of the contested partner appeared to be the contributory factor in affecting the outcome of the interaction. Of the 221

Table 4.4. Number of reversals against the hierarchy between immatures for each context of competitive interaction.

| | Group A | | Group B | | Group C | |
	%	Total no.	%	Total no.	%	Total no.
Feed	4.4	135	4.7	106	7.5	290
Avoid	6.4	312	15.7	127	7.2	683
Groom	56.3	32	20.0	30	26.2	107
Place	30.8	13	18.2	11	10.9	29

supplants over grooming partners involving immatures, subordinates were able to supplant dominants primarily from their own mothers or siblings (Fig. 4.2). Dominant immatures often supplanted subordinates from the subordinate's mother but less often from their own mothers. This may have resulted from low frequencies of grooming between the dominant females and subordinate immatures and thus few opportunities for this type of interaction would occur. The dominant immatures may have initiated these interactions against both the subordinate immatures *and* their mothers.

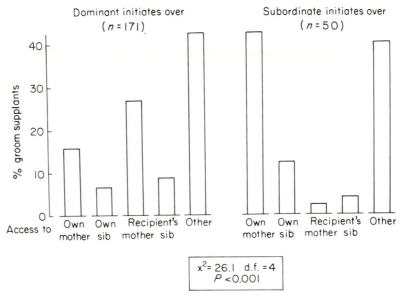

Fig. 4.2. Percentage of grooming supplants that were initiated by either a dominant to a subordinate or by a subordinate to a dominant, when the contested grooming partner was the initiator's mother or sibling, the recipient's mother or sibling or an unrelated group member.

When supplants were over food resources, the presence of a third individual appeared to affect which immature would gain access to the resource. Subordinate immatures were able to take food from a dominant in only 4% of 177 interactions that occurred when no other animal was within 2 m of the contestants. However, of the 243 supplants over food where other animals were close by, the subordinate 'won' in 9.1%. The presence of a third individual thus doubled the occurrence of supplants that went against the predicted direction. In Fig. 4.3, the relatedness of the neighbours has been considered and two factors appear to be important in the success of the subordinates. The first of these is that mothers and/or siblings were present in over 50% of the supplants initiated by

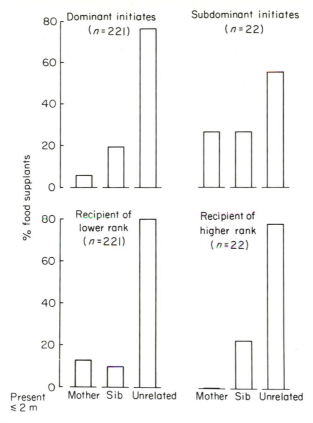

Fig. 4.3. Percentage of food supplants that took place when the mother, sibling or an unrelated animal was within 2 m. These have been separated into those present when the dominant initiated the supplant, when the subordinate initiated the supplant and when the dominant or subordinate received the supplant.

subordinates, while they were present for only 24% of supplants initiated by dominants. Second, when a subordinate immature was supplanted from food by a dominant, its mother was often nearby but this was never the case for the dominant immatures who were supplanted by subordinates. Thus, while reversals against the hierarchy were unlikely to occur over food, when they did, the subordinate appeared to assess its neighbours for potential alliance partners and for potential helpers of its dominant contestant as well.

Summary and conclusions

In general, the success of an immature in competitive and aggressive interactions was predicted by the dominance rank of that immature and the frequency of interactions between individuals did not appear to influence this relationship. Both competitive and aggressive ranks of the immatures were predicted by maternal rank in most cases and this was persistent through time. Dominance among the immatures was also relatively stable across a range of contexts but could change as a result of the nature of the resource, its quality, the age, sex and relative rank of the participants and the presence of a potential supporting individual.

The outcome of a dominance interaction for these immature vervets thus appeared to be the result of different qualities of the individuals, their age, sex and maternal dominance, and of factors which varied between the groups, such as the numbers of animals of each age and sex. The initial factor affecting the outcome was that of maternal rank, which predicted the rank of both immature males and females with their same-aged peers. Another factor was that of a difference in age, which affected relative rank among the males. A third factor was the context of the interaction—the type of resource involved and the presence of other individuals, particularly family members. In particular, the presence of the mother and her close proximity during a competitive interaction determined whether a subordinate would be able to take resources from a dominant: this effect was marked for interactions over grooming partners. The highest proportion of dyadic reversals in the hierarchies took place during supplants from grooming partners. This suggested that for grooming, immatures were willing to escalate against someone who was usually dominant, while this was less true for competition over food. Access to a grooming partner was not determined solely by relative rank but was affected by other factors, such as kinship, the prior grooming relationship between individuals or the willingness of the other to groom.

Chapter 5
Influence of Individual Characteristics upon Relationships

5.1 Individual Characteristics and Relationships

ROBERT A. HINDE

Some of the differences between species in social relationships and structure can be understood in terms of characteristics of the individuals of the species concerned. Mason (1973), for example, compared two species of South American monkeys: one, *Callicebus*, which is normally found in small family groups each consisting of an adult male, an adult female and their offspring, and another, *Saimiri*, which lives in groups containing adults of both sexes but with a tendency for individuals of the same sex to stay together within the group. In laboratory tests, Mason found that *Saimiri* females were strongly attracted to one another and also in some degree to males. The males were moderately attracted to females and also came to spend much time with particular other males if given sufficient exposure to them. By contrast, both sexes of *Callicebus* were wary of strangers. However, an individual male and female could form a close attachment to each other which endured in a group situation. Furthermore, if a strange male was introduced to a male–female pair, the male *Callicebus* showed increasing attraction to his mate and increasing antagonism to the intruder, the closer the latter approached. the male *Saimiri* showed no such 'jealousy' responses (Cubicciotti & Mason 1975, 1978; Mason 1974, 1975; see also Phillips & Mason 1976).

45

More usually, however, primatologists must deduce differences in individual propensities from the relationships they observe. For instance in some macaques (e.g. rhesus) male–infant relationships are rare and occur only occasionally (Redican & Mitchell 1973) while in others they are much more usual (e.g. Barbary macaque: Deag 1980; stump-tail: Estrada & Sandeval 1977): it must be presumed that there are interspecies differences in paternal or filial propensities. Sometimes relationships which appear superficially similar turn out on examination to have quite different bases. In many species that form mixed-sex troops the males either actively maintain proximity to the females or coerce the females to stay close to them. In the mountain gorilla, however, the stability of the group depends on the females' propensities to stay close to the male (Harcourt 1979a,b; section 13.3).

Within species, field observations provide much evidence that differences between relationships are due to differences between dyads in the natures of the participating individuals. The most obvious cases concern age and sex differences. That behavioural propensities change with age or differ between the sexes is obvious enough to every observer. Relationships of a particular type may differ according to the age of the participants. For instance in Japanese macaques, maternal behaviour differs between primiparous, multiparous and 'old' mothers (Hiraiwa 1981; see also Grewal 1980b; Hooley & Simpson 1981). However, ascribing differences in relationships to differences between the behavioural propensities of age/sex classes always requires caution. For example, the properties of a relationship depend upon the natures of both of the participating individuals and it is not always obvious who is primarily responsible. The rhesus mother–infant relationship, for example, changes as the infant develops: the infant spends more time off the mother and becomes gradually more independent of her. At first sight, these changes appear to be due to changes in the infant: they are correlated with physical growth, increasing locomotor skills and interest in the environment. Yet analysis shows that the rate at which the infant becomes independent is determined most immediately by changes in the mother (Hinde 1979a; Negayama 1981; section 6.1). Similarly, Simpson (section 5.3) shows that the rhesus mother–infant relationship differs according to the sex of the infant but detailed analysis is necessary to determine whether this is due to differences between male and female infants or to differences between their mothers' behaviour to them.

In addition to differences between species and between age/sex categories within a species, relationships of a similar type may differ because of the individual idiosyncracies of the individuals concerned. Harcourt (1979c) found marked differences among the relationships between adult males within gorilla groups. The evidence strongly suggested that the differences were related to the earlier experiences of the individuals concerned (Harcourt & Stewart 1981).

Since the differences also affected their mating strategies, they were far from trivial.

Finally, to anticipate an issue discussed in more detail in Chapter 8, the properties of relationships are influenced by the social network in which they are embedded. Differences in the play interactions and relationships between juvenile male–male and female–female dyads could be due to differences between males and females but it could also be due to differences between the behaviour of adults to juveniles of the two sexes, kin relationships with other individuals, the demographic structure of the group, etc. (Cheney 1978a; Chapters 6 and 8).

Some examples of differences between relationships of the same type within a species are given in the following contributions. In section 5.2 the individuals concerned differ in age (or parity), while section 5.3 provides some examples of the ramifying effects of an infant's sex on both mother and infant rhesus monkeys. In section 5.4, the participants in the relationship differ not in an intrinsic characteristic but in their dominance status within the group: their relative dominance is related to how they behave with each other.

5.2 Age-related Variation in the Interactions of Adult Females with Adult Males in Yellow Baboons

KATHLYN L. R. RASMUSSEN

Introduction

Interindividual variation in the interactions of adult females with adult males has been noted in a number of studies of baboons and macaques. In captive colonies, this variation has typically been found to be most strongly associated with female agonistic rank: high-ranking females are generally closest to the group male and exhibit the highest rates of social and sexual interactions with him (Goldfoot 1971; Rowell 1969). In larger, free-ranging groups, the 'monopolization' of males by high-ranking females has been observed primarily in species such as the gelada which form one-male harem units (Dunbar 1980b; section 13.2) or in multimale troops with a small number of adult males (Seyfarth 1975). Although dominant females in large multimale troops may show somewhat higher frequencies of grooming interactions and proximity to adult males than subordinate females, the majority of studies have found no relationship between female agonistic rank and the frequency of sexual invitations and copulation (Hanby *et al* 1971; Kaufmann 1965, Rasmussen 1980). Instead, variation in the sociosexual behaviour of females in such large groups appears to be more commonly associated with female age than with agonistic rank. Kaufmann (1965), for example, found an overall trend for increasing oestrus behaviour (following, grooming and copulation) with increasing age in free-ranging rhesus females, with a peak interest in males occurring when females were seven years old. Similarly, Hanby *et al.* (1971) found that Japanese macaque females five-and-a-half to six-and-a-half years old were more active on all behavioural measures during the breeding season than were younger or older females; however, there was no corresponding difference in the number of ejaculations by the male partners of those females.

Rasmussen (1980) studied the interactions of 13 adult males and 20 cycling adult females in a troop of yellow baboons in Mikumi National Park, Tanzania. Adult females were grouped by parity (and thus roughly by age) into the categories: nulliparous (seven females) and parous (13 females). In the following, the interactions of the two groups of females with adult males are compared in non-sexual contexts and during sexual consortships*. As affiliative and sexual behaviour are necessarily dependent upon the behaviour of both interactants, the behaviour of males towards the two groups of females will also be considered. Mann–Whitney U-tests (two-tailed) have been used to test the significance of differences between the two female groups.

Male–female interactions while not in consort

Outside of consortships, cycling adult females in the Mikumi troop approached adult males on average about twice as often as the males approached them. Nulliparous females were particularly active in seeking male attention: they showed a strong tendency both to approach and to present to males more frequently than did parous females ($P<0.10$) (Fig. 5.1a,c). There was no difference between the two groups of females in their frequency of grooming adult males (Fig. 5.1b).

Adult males, on the other hand, generally appeared to find parous females more attractive than the younger nulliparous females, since they showed a marked tendency to approach parous females more often than they approached nulliparous females ($P<0.10$) (Fig. 5.2a). There was no significant difference between the males' rate of grooming nulliparous versus parous females (Fig. 5.2b). Males inspected the perineal regions of nulliparous females significantly more than they inspected those of parous females ($P<0.05$) (Fig. 5.2c); however, since over three-quarters of these inspections occurred following approaches by females, the frequency of male inspections of nulliparous females was probably a reflection of their rate of approaching males, which was higher than that of parous females (see above). There was no significant difference in the rate at which males supplanted younger versus older females outside of consortships.

*Temporary pair bonds or consortships between a male and an oestrus female are characterized by unusually close proximity and exclusive social interaction between the pair. Consorting pairs are usually peripheral to the main body of the troop.

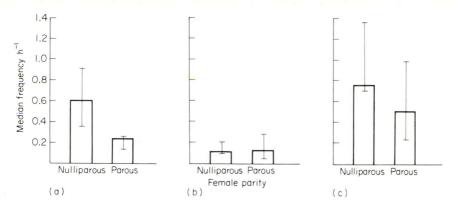

Fig. 5.1. Comparison of the frequency of behaviours directed by nulliparous and parous females towards adult males while not in consort. The medians and interquartile ranges are shown. (a) The female approaches the male; (b) the female grooms the male; (c) the female presents to the male.

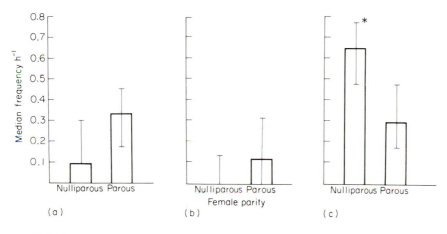

Fig. 5.2. Comparison of the frequency of behaviours directed by adult males towards nulliparous and parous females while not in consort. The medians and interquartile ranges are shown. The asterisked column denotes significant differences between the two groups of females. (a) The male approaches the female; (b) the male grooms the female; (c) the male inspects the female.

Male–female interactions during consortships

The greater activity of younger females in initiating interactions with adult males in non-sexual contexts was also apparent in their behaviour during sexual consortships. Nulliparous females showed a strong tendency to both follow and groom their consort partners more frequently than parous females ($P<0.10$) (Fig. 5.3a,b) and they presented to male consorts significantly more often than did the older parous females ($P<0.05$) (Fig. 5.3c). Whereas parous females presented to their consort partners significantly less than they presented to adult males outside of consortships ($P<0.01$, Wilcoxon test), nulliparous females presented to consort partners even more frequently than they did to adult males when they were not in consort (Figs 5.1c, 5.3c).

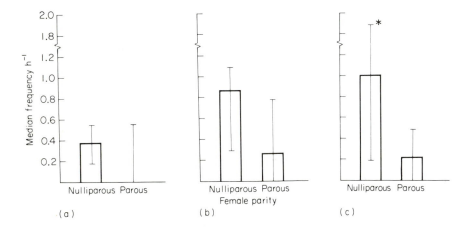

Fig. 5.3. Comparison of the frequency of behaviours directed by nulliparous and parous females towards male partners during sexual consortships. The medians and interquartile ranges are shown. The asterisked column denotes significant differences between the two groups of females. (a) The female follows the male consort; (b) the female grooms the male consort; (c) the female presents to the male consort.

Although younger females were more overtly attentive to their consort partners, those males generally did not respond any more intensely to them than they did to older females. Nulliparous females were not groomed, inspected or mounted by their male consorts any more often than were parous females. However, relative to their behaviour toward parous females, males did show a somewhat higher frequency of following nulliparous female consorts ($P<0.10$) (Fig. 5.4a) and a very significantly higher rate of running after them ($P<0.002$) (Fig. 5.4b). This latter result is probably a consequence of the greater physical activity of the young females, rather than a reflection of a lack of receptivity on the female's part: nulliparous females would often dash off suddenly in pursuit of another troop member and their male consort would be forced to follow quickly to keep from losing the female in heavy vegetation. (Bona fide running-avoidance of male consorts by their female partners was very rare; such avoidance was usually precipitated by an approach by the male and was nearly always characterized by visual orientation of the female towards the male during the ensuing 'chase'.)

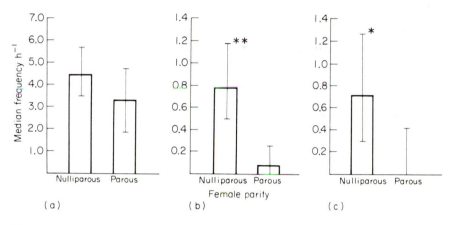

Fig. 5.4. Comparison of the frequency of behaviours directed by adult males towards nulliparous and parous female partners during sexual consortships. The medians and interquartile ranges are shown. The asterisked columns denote significant differences between the two groups of females. (a) The male follows the female consort; (b) the male runs after the female consort; (c) the male supplants the female consort.

One further difference between the consortships of younger versus older females was apparent in the frequency with which the two groups of females were supplanted by male partners. Males supplanted nulliparous female consorts significantly more often than they did parous female consorts ($P < 0.05$) (Fig. 5.4c). The frequency of females being supplanted by their consorts was not a simple function of proximity: there was no relation between the percentage of time females spent within 5 m of their consort and the number of times they were supplanted by him. Instead, the observer had the strong impression that the younger females often 'misinterpreted' approaches by their consort and reacted by moving away, whereas this was rarely the case for older females.

Summary and conclusions

Age-related variation in the interactions of Mikumi baboon females with adult males was apparent both in non-sexual contexts and during sexual consortships. Young, nulliparous females were consistently more active than older, parous females in maintaining proximity and presenting to adult males; this tendency was especially pronounced during consortships. However, the greater proceptiveness of nulliparous females did not appear to elicit higher rates of social or sexual responses by their male consort partners.

The interpretation of the behaviour of one participant in a dyadic interaction necessarily requires some knowledge of the interactions of that dyad in other contexts. Thus it would be unwarranted to conclude that the higher frequencies of proceptive behaviours* shown by young Mikumi females relative to older females were indicative of a greater degree of sexual receptivity on their part. Herbert (1969) noted that high rates of sexual soliciting by rhesus females sometimes occurred in response to male disinterest. As Mikumi males did appear to prefer parous females as consort partners (Rasmussen 1980), it is possible that young females found it necessary to stimulate male sexual behaviour more than did older females. Alternatively, nulliparous females might have been more 'nervous' than parous females because they were less accustomed to the continued proximity and following of male consorts: Kummer *et al.* (1974) reported that presenting was more typical of females in newly formed pairs of hamadryas baboons and in females who had been forced into a bond by the male. The impression that young females in the Mikumi troop misinterpreted approaches by their male consort and thus were supplanted more often does suggest that nulliparous females may have been more nervous while in consort than were older females.

A sexual consort bond is qualitatively distinct from most male–female relationships observed in non-sexual contexts, and the ability of both partners to maintain spatial proximity and coordinate their sexual interactions while foraging and moving with their social group is undoubtedly enhanced by experience. In the study troop, the relative inexperience of young females appeared to be reflected in a somewhat lesser degree of coordination with male consort partners. Although nulliparous females were clearly attracted and receptive to their partners, the high frequency with which their partners had to run after them to avoid losing them implies that their attention towards the males often may have lapsed. It is also not unlikely that the experience of attending to and adjusting their activity to an infant may have enabled parous females to coordinate their

---

*Behaviours including active solicitation of male sexual attention.

movements with that of a consort partner more easily than was possible for the younger females.

5.3 Effect of the Sex of an Infant on the Mother–Infant Relationship and the Mother's Subsequent Reproduction

MICHAEL J. A. SIMPSON

The sex of a rhesus infant can affect its early history of interaction with its mother and social companions in such a way as to influence its further behavioural development. Mothers in the Madingley colony rated by observers as more Confident, for example, had yearling daughters who tended to be Confident, while their yearling sons were not necessarily so. This difference in the effect of maternal Confidence could be explained in terms of the effect of the infant's sex on early patterns of interaction between it and its mother (Stevenson-Hinde & Simpson 1981; section 7.2): with more Confident mothers, mother–daughter but not mother–son dyads share patterns of interaction which give the infants more security as they begin to explore their surroundings.

That the sex of an infant could affect its mother's subsequent reproductive history is suggested by the tendency of mothers in the Madingley colony to be slower to breed again after giving birth to daughters than after giving birth to sons (Simpson *et al.* 1981; Table 5.1). The effects of the infants' sexes on their mothers therefore have further consequences for the infants themselves: sons are more likely than daughters to be presented by their mothers with a new sibling in their second year, and presence or absence of a sibling in an infant's second year could make further differences to the infant's subsequent development (section 8.5). Some of the ways in which an infant's sex has effects on its own and its mother's destiny are considered below.

Different patterns of mother–infant interaction could stem from differences in appearance and behaviour inherent in the infants. Male rhesus monkey infants have large scrotums which are conspicuous from birth. This sex difference is all the more striking for being absent in the infants of some other primate species, such as vervet monkeys. The mothers and social companions of male rhesus infants often handle the infants' genitals and thereby have ample opportunity to notice an infant's sex. Behavioural sex differences (Mitchell 1979) include the males' greater activity and greater readiness to initiate bouts of chasing play (Harlow 1965). Prenatal hormonal conditions have been shown to affect individual rhesus monkey's psychosexual orientations (Goy & Goldfoot 1973).

Mothers do in fact treat male and female infants differently. In the Madingley social groups 18 of the 23 male infants' mothers were seen to play with their infants at some time during their first year, compared with eight of the 21 mothers

Table 5.1. Contrasts between male and female rhesus monkey infants according to their mothers' social ranks.

Infant sex	High-ranking mothers		Other mothers	
	Male	Female	Male	Female
No. births§	15	38	54	32†
Days from birth to mother's next conception●	198*	517	232	289‡
Aggression received by** mother from other adults	0.3*	3	1.5	3‡
Combined score of aggression received by mother from other adults and submissive behaviour given by mother to others	3	8	9	10
'Passive preventions' by mother of nipple access	14*	2	5	7
Total rejection by mother	32	9	21	22
Restrains by mother	2	7	5	16.5‡

*$P < 0.05$, Mann-Whitney U-test for differences between males and females of high-ranking mothers.

†$P < 0.01$ for the difference in sex ratios according to maternal rank.

‡$P < 0.05$ for sex differences for the whole sample of mother–infant dyads studied after 1972.

§Birth sex-ratio data come from the colony's whole history of infants conceived in social groups, including still births.

●Infants born after 1972, when the main programme of separation experiments had ceased and the colony seemed less disturbed, provide the remainder of the data in this table (see Anderson & Simpson 1979).

**The behavioural scores are median rates 6 h^{-1} observation in the infants' eighth week.

of females. Mothers caged singly with their infants also play more often with sons than with daughters (Mitchell 1968). Mothers who have been reared without surrogate or real monkey mothers are more likely to subject their male than their female infants to violent abuse at and soon after the birth (Ruppenthal *et al.* 1976).

Group companions other than the infants' mothers may also respond differently according to the sex of the infant. In the 'nuclear family apparatus' which gives infants but not their parents access to a common play area, male infants are more likely to initiate play, to be the recipients of play overtures and to respond with play if the initiators are males (Ruppenthal *et al.* 1974). The sex differences presented so far could be explained in terms of how group companions respond to males as opposed to females and in terms of the consequences of the male infants' greater early activity (in the second week) (Simpson *et al.* 1981) and initiative in social play. Other sex differences can be related to differences in the responses of other adults to the infants' mothers. In rhesus (Table 5.1) as well as bonnet macaques (Silk *et al.* 1981b), other adults tend to be more aggressive to mothers with infant daughters than to those with infant sons and in pig-tail macaques even to females pregnant with a female fetus than to those pregnant with a male (Sackett 1982). Not surprisingly, the Madingley mothers restrain their eight-week-old infant daughters more often than their sons (Table 5.1).

Apart from variables obviously reflecting maternal protectiveness, few simple sex differences in aspects of the mother–infant relationship have been found in the Madingley colony. However, the *rates* at which mothers restrain and reject male/female infants may be differentially affected by the rates at which mothers receive aggression from other adults. Thus, mothers who receive more aggression restrict their daughters more ($r_s = 0.51$ compared with 0.16 for sons) and reject their sons less often ($r_s = -0.63$ compared with -0.30 for daughters). Thus, it seems that mothers who receive high levels of aggression and have eight-week-old daughters are usually unwilling to let them move off, while mothers who have sons and receive high levels of aggression are not necessarily unwilling to let them go but are more ready to let them regain contact without rejecting them. In high-ranking mothers this last contrast can be seen in Table 5.1: high-ranking mothers of sons reject them relatively frequently and more often than the high-ranking mothers with daughters. The Madingley high-ranking mother–son dyads also receive low levels of aggression, as do the remainder of the mother–son pairs (Table 5.1). It is thus possible that they can better afford to promote their infants' early independence.

This may have relevance to another important issue. Most rhesus monkey sons eventually leave their maternal families (e.g. Lindburg 1971; cf. Missakien 1972), whereas daughters remain in the groups. Little is known of the origins of this sex difference (but see section 9.4). At 52 weeks some important differences between mother–son and mother–daughter dyads are already apparent: sons spend less time in contact with their mothers than do daughters (Stevenson-Hinde & Simpson 1981) and in the Madingley colony mothers of sons are more likely to be pregnant again (Simpson *et al.* 1981). One factor that could explain both these sex differences is the lower rates of aggression received by the mothers of sons from other adults. This results in mothers of sons being more free to permit and promote (Hinde 1974; Simpson *et al.* 1981) the independence of their offspring without harmful consequences, than is the case with mothers of daughters. Functionally, this permits mothers to give more extended attention to their daughters and thus protect them in the face of the higher levels of aggression that they provoke.

Very high rates of rejection, however, could be counter-productive, making the infants more rather than less likely to try and stay close to the mother and cling

(Rosenblum & Harlow 1963). It has been suggested (Simpson *et al.* 1981) that a mother tests and promotes her infant's capacity for independent activity by playing with it and rejecting it relatively gently, often beginning as early as the second week and not usually persisting in rejecting an obviously protesting infant in its first two or three months. The following findings are consistent with this suggestion:

1   High-ranking mothers with sons reject their infants more often and are quicker to conceive than those with daughters.

2   Throughout the infants' first 20 weeks, mothers who are 'quick' (in the sense of going on to bear an infant in the next breeding season) reject their infants more than those who are to bear an infant later (Simpson *et al.* 1981).

3   Infants whose mothers are to be 'quick' are more active throughout their first 12 weeks and two-week-old males are more active than two-week-females.

4   Mothers are more likely to play with sons than with daughters.

In one common mother–infant game (Hinde & Simpson 1975), the mother leaves the infant, then looks back, often lip-smacking and presenting as she does so, as if to encourage the infant to follow, and returns to the infant if it fails to follow. It is difficult to resist the impression that the mothers are trying out their infants' capacities for independent movement and their abilities to follow their mothers as they move about separately. Moreover the repetitively 'rejecting' sequence of events in which a mother breaks contact with her offspring is often a way of encouraging the infant to follow on its own feet.

Greater independence consequent upon maternal rejection of an infant would be associated with less frequent suckling and this also could hasten the onset of oestrus cycling in the mothers, for suckling influences maternal prolactin levels. In human mothers high frequencies of suckling can delay the resumption of menstrual cycling and affect prolactin levels (Delvoye *et al.* 1978). In the Madingley colony, infant suckling, as opposed to simply holding onto the nipple with the mouth, was impossible to record reliably but in mother–daughter dyads in which the mothers went on to give birth again in the next year, less time was spent by the daughters on the nipple (Simpson *et al.* 1981).

Further complexities arise from the fact that high-status mothers were more likely to have daughters than sons (Table 5.1; Simpson & Simpson 1982). This is unlikely to be an effect of the infant's sex on maternal status, for status seldom changes from conception, during the resulting pregnancy or during the infant's first year; furthermore, the effect of sex ratio on status is also present in primiparous mothers. This effect of maternal status on birth sex ratio ensures that lower-ranking mothers who have sons rather than daughters are therefore less likely to suffer the high rates of aggression from other adults associated with having daughters. High-ranking mothers, who are more likely to have daughters than sons, are presumably better able to afford the aggression associated with daughters, especially if, as a result of having daughters, they are likely to delay their next pregnancy and thus not be encumbered by a neonate during their daughter's first two years. While the mechanism involved results in subordinate mothers being less likely to become involved in aggression with other adults because they are less likely to have daughters, it seems to work independently of the levels of aggression received by mothers. Thus, there was not a tendency for mothers, who both received high rates of aggression and who often directed submissive behaviour to others, to have sons rather than daughters. It makes functional sense that mothers with sons should *not* respond to the consequent low levels of involvement in aggression by going on to have daughters, for subordinate mothers who currently had sons would then have daughters and consequently experience unsupportably high levels of aggression.

Another likely consequence of being a dominant mother is having successive pairs of daughters. That this combination can be especially demanding follows not only from the aggression attracted by daughters but also because an older daughter can be particularly troublesome to both mother and her youngest female infant (Spencer-Booth 1968; section 8.5). High-ranking mothers would be best able to withstand the extra burden of such intrafamily strife.

In summary, these data show that an infant's sex can be a potent force on its own and its mother's destinies. It may affect not only how the mother treats the infant but also how other social companions treat the mother–infant dyad. Since some of the effects can be indirect, operating through the behaviour of other individuals in the group, they interact with the mothers' social status, which in itself also affects the sex of the next infant. Thus, the sex of the infant has effects which ramify through its own and subsequent generations and are closely interwoven with the effects of status.

5.4 Rank Influences Rhesus Male Peer Relationships
JOHN COLVIN

Considerable evidence has now accumulated to indicate the importance of dominance rank as a characteristic of relationships with a pervasive influence over their structure. Rank is predictive not only of the direction of agonistic interactions but also, in many cases, of the nature and direction of affiliative interactions, particularly grooming (see also section 10.1): Seyfarth (1976, 1977, 1980) has shown that in several species females of the highest rank receive the most grooming, although in other species this appears not to be the case (Silk 1982). Similarly, among male chimpanzees the more dominant individuals tend to be involved in more grooming bouts than their inferiors (Simpson 1973). Among immature baboons, individuals whose mothers are of high rank are preferred as play partners to those of low rank (Cheney 1978a). The ways in which individual differences in rank are associated with differences between relationships: and how this involves relations between rank and interindividual attractions and avoidance are discussed in detail below.

In many studies, the effects of rank are easily confounded with those of age and relatedness. In the study to be described, which concerns relationships between immature male rhesus monkeys on Cayo Santiago (see also section 3.2), age effects were removed by restricting attention to relationships between males within single age-cohorts. The confounding effects of relatedness remained, however, and will be discussed further in section 10.3. Since in the present study the most closely related males were also those closest in rank, it will be assumed here that the effects of close relatedness and of close rank reinforce one another, even though there are some exceptions to this rule (section 10.3).

On Cayo Santiago, male age-cohorts vary in size between one and ten males, with a mean size of 5.4 males. The results reported here are taken from a study of two cohorts of three-year-old males in Group I (Colvin 1982), one cohort of three-year-old males in Group L and one cohort of two-year-old males in Group M (Tissier *et al.* in prep.). Each of the Group I cohorts was studied in the spring (birth season) and the summer (breeding season). One cohort (the 1974 cohort) contained five males in the spring but had lost one (through emigration) by the summer. The other cohort (the 1975 cohort) contained nine males in the spring, reduced to six by the summer. The Group L and Group M cohorts (the 1978 and 1979 cohorts) were studied in the summer only and contained six and seven males respectively.

Males within each cohort ranked in a stable, linear hierarchy*, with the ranks of the males following those of their mothers (Loy & Loy 1974; Sade 1967). Almost all males associated most with those of adjacent rank (Fig. 5.5). Overall, in some cohorts it was the males of highest rank who spent the most time in association with their peers (the 1975 cohort [spring], the 1978 cohort and the 1979 cohort), whereas in others, it was the males of intermediate rank (the 1974 cohort and the 1975 cohort [summer]) (see also Fig. 10.2, section 10.3). Thus, within cohorts of males, rank correlated with both the identity of the partner with whom each male spent most time and also with the total time spent with peers.

Seyfarth (1980) found that among adult female vervet monkeys the preferred partners for (non-grooming) proximity, who were ranked adjacently, were also the preferred partners for grooming and for alliance formation. A similar pattern emerged in the present study, extending also to playful and feeding activities (Fig. 5.6). Thus, closely ranking dyads not only spent more time in passive association

---

*Note however, that there was a reversal of ranks within the dyad 630–631 of the 1975 cohort between the spring and summer.

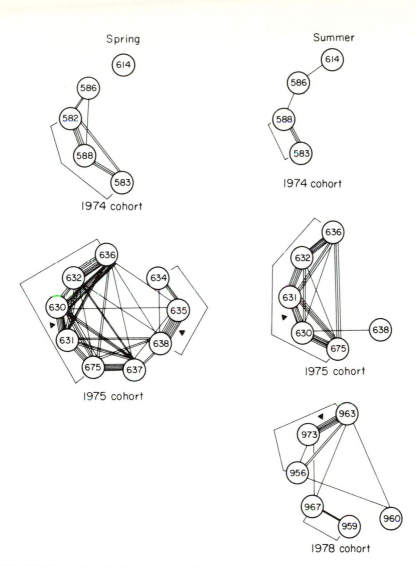

Fig. 5.5. Networks of male peer associations among four cohorts in two seasons. Males are shown in decreasing order of rank running in a counter-clockwise direction. Related males are bracketed and closely related dyads ($r > 1/8$) are indicated by a solid triangle. Most association is between males of adjacent or close rank.

— 1.0–1.9 min h^{-1};
= 2.0–2.9 min h^{-1};
≡ 3.0–3.9 min h^{-1}; etc

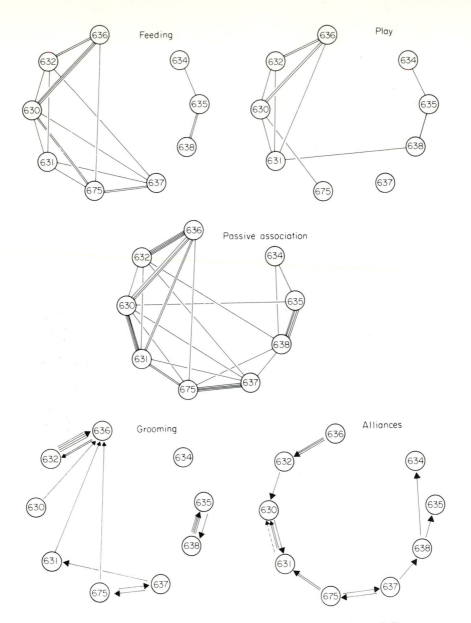

Fig. 5.6. Distributions of feeding, play, passive associations, grooming and alliances among males of the 1975 cohort (spring). The majority of each of these types of associations/interactions occurred between partners of adjacent or close rank. Males are shown in decreasing order of rank in a counter-clockwise direction. The top three parts of the figure show the time spent feeding in proximity, playing and in passive association (including resting, walking and grooming). (— 1.0–1.9 min h⁻¹; = 2.0–2.9 min h⁻¹; ≡ 3.0–3.9 min h⁻¹; etc.) The bottom two parts of the figure show the number of grooming bouts and alliances. All alliances were coalitions except for the single case shown by the broken line (aid) (one line per interaction).

than other dyads but also spent more time feeding and playing together. These dyads were also more likely to groom together and to form alliances. These latter results have already been encountered in a different form in section 3.2. There it was shown that grooming and alliance formation are common within 'strong' peer dyads but rare within 'weak' peer dyads. The two sets of findings are consistent with each other, since the mean rank difference among strong peer dyads is one, while between weak peer dyads it is two (Colvin & Tissier in prep.). The finding that males spent the most time in play with those of adjacent rank is in harmony with the findings of Caine and Mitchell (1979), who reported that rhesus infants play most with those of adjacent rank, although no such selectivity could be found among infant bonnet macaques.

Clearly then, proximity in rank is associated with a similar pattern of preferences across a number of activities, including feeding, play, grooming and alliance formation. However, while immature males interact preferentially in a variety of ways with those closest in rank, the initiative to do so is not necessarily balanced evenly between the higher- and lower-ranking partners. Dominant partners were responsible for a significantly greater percentage of approaches and a significantly smaller percentage of leavings than their subordinate partners. Thus, it is the dominant partners who are primarily responsible for maintaining proximity. Interestingly, this appears not to be the case in the more intimate sibling relationship, where both elder brother (dominant partner) and younger brother (subordinate partner) contribute approximately equal proportions of approaches and leavings (Fig. 5.7).

The imbalance of approaches and leavings in peer relationships could reflect either an imbalance of attraction between dominant and subordinate partners or greater fearfulness on the part of one partner. There is some support for the idea of an imbalance of attraction from data on grooming. Of 23 grooming dyads in the two Group I cohorts, in 17 the subordinate groomed the dominant partner more frequently than the dominant groomed the subordinate, in three the number of bouts exchanged was equal and in only three dyads did a subordinate receive grooming more often from a dominant partner than give it. Such results indicate a greater attraction for higher- than for lower-ranking partners.

However, this runs counter to the imbalance of attraction expected from the approach–leave data. The patterning of the approach–leave data taken over the entire peer network in fact suggests that males are somewhat fearful of approaching those who rank above them. Thus most approaches are by domin-ant partners while most leavings are due to subordinate partners. The idea that immature males are to some extent fearful of their higher-ranking peers re-ceives further support from data on agnostic interactions. Here the difference between sibling and peer dyads in their approach–leave indices is reflected in a difference in the rate at which threats were received. Whereas males were

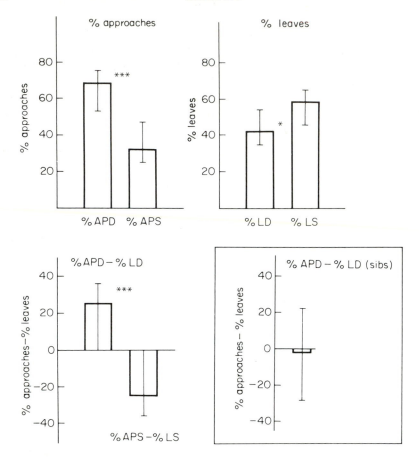

Fig. 5.7. Relative roles of dominant and subordinate members of peer dyads in maintaining proximity. Dominant members were responsible for a greater percentage of approaches (% APD) than were subordinate members (% APS), whereas subordinate members were responsible for a greater percentage of leaves (% LS) than were dominant members (% LD). As a result, dominant partners were primarily responsible for the maintenance of proximity (% APD−% LD) ($n = 45$ dyads, spring data only [***$P < 0.001$, *$P < 0.05$, Mann-Whitney test]). For comparison, the proximity indices of sibling dyads ($n = 9$) are shown in the boxed figure. In these dyads, responsibility for the maintenance of proximity is far more evenly balanced between dominant and subordinate partners. The medians and interquartile range are shown in all the figures.

threatened by their older brothers at a median rate of 0 threats h^{-1} in proximity, they were threatened by their higher-ranking peers at a median rate of 0.74 threats h^{-1} in proximity (spring data only).

Not only were males inhibited (through fear) from approaching single higher-ranking peers on their own, they were also inhibited from approaching two more highly-ranking peers as a dyad. However, not all inhibitions appear to be based on fear. Thus, the degree of inhibition involved in approaching two lower-ranking peers was related not to fear but to the relative preferences of the 'target' male for the approaching male ('interferer') and for his partner (the interferer's 'rival') and to the degree of familiarity between interferer and rival (see also Bachmann & Kummer 1980; Kummer *et al.* 1978). As a result, high-ranking males showed, overall, higher rates of interference with other peer dyads than did low-ranking males.

In the peer relationships of immature male rhesus monkeys, further evidence for the pervasive correlates of rank are therefore found. On the one hand, the distribution of partner preferences over a range of activities is a function of attraction between individuals close in rank. This in turn has consequences for social structure. On the other hand, the relative rank of partners influences the degree of reciprocity/complementarity in a relationship because, through its basis in agonistic interactions, rank is associated with fearfulness on the part of the subordinate partner towards the dominant partner. It must be borne in mind, however, that just as rank is not the only factor associated with attraction between partners, neither is it the only one associated with inhibition, as has been seen.

Although only the relations between rank and peer relationships are illustrated here, similar effects are in fact found in many of the immature male's other relationships. Immature males, for example, appear to be attracted to mature females who rank above their mothers (HRFs) rather than to those who rank below their mothers (LRFs). When HRFs are sexually receptive, immature males spend considerable time observing at close range these females in consortships with adult males. By contrast, immature males tend to groom and occasionally copulate with LRFs. Similar findings have been reported for immature male bonnet macaques (Glick 1980; Silk *et al.* 1981) and baboons (Cheney 1978b). The most likely explanation for this is that while males might be attracted to groom and even to copulate with HRFs, they are inhibited from doing so due to the females' high rank. In support of this interpretation, it was found that HRFs are more aggressive to immature males than are LRFs (see also Silk *et al.* 1981a; sections 10.2 and 10.3).

Rank thus affects relationships in important ways. To close, however, to what extent is rank a characteristic not only of relationships but also of individuals, and particularly of groups of individuals? By definition, rank is used here as an intervening variable used to explain consistencies in the directionality of complementary interactions within a relationship (Hinde 1978; Hinde & Datta 1981). Since most individuals participate as the more highly ranking partner in some dyadic relationships and as the lower-ranking partner in others, rank cannot be used as a conceptual shorthand for a set of characteristics of the individual. However, there is stronger evidence to suggest that rank can act as a 'group' characteristic, i.e. as consistently characteristic of the behaviour of a group of individuals, because 'lineage rank' provides a category which, combining the group characteristics of rank and kinship, allows for a consistent distinction between kin groups which also differ from one another in rank. This suggests the possibility that under certain conditions lineage members might be categorized by others in terms of their lineage rank and treated accordingly.

Chapter 6
Development and Dynamics of Relationships

6.1 Development and Dynamics
ROBERT A. HINDE

As stressed in Chapter 1, the problem of the dynamics of interindividual relationships, of how interactions affect interactions and are affected by influences from outside the dyad, merges with the problem of the development of relationships. The participants in any relationship are likely to change with time through the processes of growth and decay, through their mutual influences on each other or as a consequence of external influences. Their relationship is thus prone to change also; the cumulative effects of interactions upon interactions may contribute to that change or may have a steadying influence, maintaining the relationship within limits despite perturbations, as with the gestures of reconciliation and consolation which often follow fights between chimpanzees (de Waal & Van Roosmalen 1979). The stability which many interindividual relationships in primates in fact have (e.g. Dunbar 1979b) must thus be seen as dynamic in origin and any distinction between the study of the dynamics of relationships and that of their development is inevitably artificial.

The relationship whose development has been most studied is that between mother and infant. It is therefore important to stress that other relationships may differ from the mother–infant relationship, both in their mode of inception and in their dynamics. Where the mother—infant relationship develops through the physiological processes of pregnancy and parturition, other relationships may start through gradual acquaintance, while in yet other cases interindividual attraction may lead to specific acts which disrupt the current status quo and initiate new relationships (see below).

While the mother–infant relationship develops relatively smoothly, at least to weaning, other relationships may be totally disrupted by a single dramatic event, such as when a fight leads to a reversal of dominance relations. In some cases there is a sequence of interactions whose invariance suggests that it is necessary for establishing a relationship. Thus initially strange gelada baboons on meeting go through a fixed sequence of interactions—fighting, presenting, mounting and grooming. Although any one stage may be omitted, the sequence is remarkably consistent. The speed of progression through the sequence is an index of the 'Compatibility' between the two animals, which varies with their absolute and relative status. The initial fighting may serve to establish the relative dominance and is omitted where there is a clear difference, as in encounters between males and females (Kummer 1975; see also Welker *et al.* 1980).

It is tempting to suppose that the development of a relationship could be analysed simply in terms of a series of successive actions by each partner on the other. There are, however, a number of reasons why this is too simple a view. For one thing, when a relationship is studied, each act observed is a product of both partners. A mother suckles and an infant is suckled; a subordinate approaches and a dominant threatens; and the behaviour of two females in interaction today may depend on who bossed whom the previous day. Thus, each measure taken is a measure of the relationship: from their actions the behavioural propensities of one partner cannot be assessed independently of those of the other (see section 4.1).

In addition, the partners in every relationship are changing gradually and are influenced by interactions outside as well as within the relationship. The effect of each interaction depends upon those of previous ones and perhaps involves expectations of interactions in the future. Furthermore, changes in relationships may be much less simple than appears at first sight. Thus, over the period in which the rhesus infant is becoming more independent of its mother, the relationship is also becoming more intimate, in the sense that the behaviour of each individual is becoming more finely attuned to that of the other. Teazing apart the roles of the participants in a dynamically changing relationship is thus no easy matter.

Indeed, the bases of changes in a relationship are often quite different from what appears at first sight. During development, for example, rhesus infants spend less and less time on their mothers. It might seem obvious that the change is due to the development of the infant: it is growing, acquiring locomotor skills, becoming more and more interested in its physical and social environment. Yet the data show that the increase in time spent off the mother is associated with an increase in the frequency with which the mother rejects the infant's attempts to gain contact: if changes in the infant were primarily responsible for the increase in time off, we would expect the reverse to be the case. It is changes in the mother's behaviour that determine the rate at which the infant achieves independence.

Work on the rhesus mother–infant relationship also emphasizes another point about studies of the development of relationships: the questions asked must be stated with great precision if progress is to be made. Thus, one question which can be asked is, Who is responsible for a particular character of the relationship at any one age? As an example, we might be concerned with the maintenance of proximity between mother and infant. From data on the frequency with which each approaches and leaves the other the difference between the percentage of approaches for which the infant is responsible and the percentage of leavings for which the infant is responsible can be calculated:

$$\left(\frac{Ap_I}{Ap_I + Ap_M} \times 100 \right) - \left(\frac{L_I}{L_I + L_M} \times 100 \right)$$

This index which, after a certain point, is reasonably independent of the total number of approaches and leavings, gives a direct indication of responsibility for proximity. If it is positive, the infant is contributing a higher proportion of the approaches than leavings and is thus primarily responsible for proximity. If it is negative, the reverse is the case. In fact, the data show that the mother is primarily responsible for the maintenance of proximity in the early weeks and the infant is later.

Alternatively, it might be asked whether it is changes in one partner or the other that are responsible for changes in proximity with age. The argument is identical with that given earlier. As the infant gets older, it spends less time near its mother. If this were associated with a decreasing relative role of the infant in the maintenance of proximity as indicated by the above index, it would be concluded that the decrease in proximity was primarily due to a change in the infant. In fact, it is associated with an increase in the index. Thus, the decrease in proximity is due primarily to a change in the mother's behaviour. Over an age-range when the mother is primarily responsible for *maintaining* proximity with the infant from moment to moment, it is *changes* in the mother which are primarily responsible for a decrease in proximity with age.

Similarly, it might be asked, At any one age, is it differences between mothers or differences between infants that are responsible for differences in mother–infant proximity between dyads? Again a similar argument can be used: if those dyads which remain in closest proximity are those in which the index of the infant's relative role in the maintenance of proximity is highest, then differences between infants are likely to be primarily responsible for the interdyad differences in proximity, and vice versa. In fact, early on it is differences between mothers and later on differences between infants that are primarily responsible for interdyad differences. Thus, the bases of individual differences at one age are not necessarily the same as the bases of developmental changes over age: the several questions may lead to different answers (Hinde 1974, 1979a). Furthermore, the answers may differ according to the measures used: one partner may be primarily responsible for close proximity, the other for maintaining general proximity (see also Hoff 1982).

In attempting to understand the mechanisms by which interactions affect subsequent ones, we may expect to draw heavily on the laws of learning, including the principle of reinforcement, as studied in laboratory situations. Yet it is clear that laboratory studies have often grossly underestimated the intellectual capacities of non-human primates. Abilities to cope with such conventional laboratory problems as cross-modal matching have been demonstrated only relatively recently in macaques and often require large numbers of trials. It is therefore important not to neglect evidence of the sophistication which monkeys sometimes display in natural situations. The apparent knowledge of food and water sources shown by both monkeys (Sigg & Stolba 1981) and apes (e.g. Galdikas 1979) indicate considerable cognitive complexity. Tool use by chimpanzees has been widely documented in the field (Goodall 1968; McGrew *et al.* 1979): Boesch and Boesch (1981) have suggested that females are more highly motivated and skilled at learning the complex manipulations involved. There are many indications of even greater complexity in social behaviour (Humphrey 1976, 1980). For example whether or not a male hamadryas baboon attempts to appropriate a female consorting with another male depends on his assessment of the female's preference for her current owner (Bachman & Kummer 1980; see section 8.1), and at least some apes have the ability to lie in order to manipulate the behaviour of others (Menzel 1971). Some carefully documented examples of the complexity of chimpanzee social behaviour have recently been given by de Waal (1982).

In the light of such observations, it must be considered whether the rarity of triangular relationships in dominance hierarchies may not sometimes be due to similar processes: if A knows he can boss B, and sees B bossing C, he may then know that he could boss C. Such a possibility suggests that the principle of 'dissonance', applied successfully to certain aspects of human relationships (see section 13.6), may not be wholly irrelevant to non-human species.

The probability that monkeys, like people, may incur costs in a relationship if it is likely to bring future gains must also be allowed for. It has been suggested that some types of grooming and sociosexual presenting in primates are analogous to the giving of social approval: costs are incurred at one stage in a relationship if they are likely to bring later rewards (Hinde & Stevenson-Hinde 1976). Thus, there is at least a superficial resemblance between the way in which one worker may praise another more skilled than himself in the (conscious or subconscious) hope of later obtaining assistance with his own work (Homans 1951, 1961) and the manner in which monkeys groom selectively those individuals with whom a relationship is most likely to bring later benefits (e.g. Seyfarth 1976; section 10.2). Cheney (1978b) found that immature female baboons, who remain throughout their lives in the troop in which they were born, direct much of their grooming to the more dominant females in their troop, while males tend to groom the more subordinate adult females. She suggests that it is adaptive for females to form relationships with the more dominant females, who may subsequently aid them in aggressive encounters, and for males to form relationships with those females with which they are most likely to be able to practice mating without interference from the adult male (see also Mori 1975; Packer 1977; Seyfarth 1977).

There may be no *need* to regard human exchange and these examples from non-human primates as more than analogies and nothing is being said here about mechanism. In the human case, social approval may be given deliberately as part of a Machiavellian plot to exploit the relationship later, or more or less unconsciously as one of the ways in which individuals have come (by nuture or nature) to foster relationships with others. In non-human primates, it is usual (and perhaps safer) to assume that individuals are adapted to foster relationships that will subsequently be of value to them than that they have foreknowledge of the possible advantageous consequences of their actions. It could be that the mere proximity involved in grooming, or the consequent familiarity, furthers the development of the relationship and thus increases the probability of future rewards (see Chapter 12). However, we must not be deceived by scientific necessity to be economical with our hypotheses: evidence is accumulating that at least some non-human primates may engage in more complex manipulations of their social environment than they have hitherto been given credit for (e.g. Goodall 1975, 1978; de Waal 1982).

Since relationships are likely to change with time it might be surmised that, in the absence of group norms specifying what relationships of particular sorts should be like, the relationships of non-human primates might show even less stability than those of humans. Nevertheless, some are remarkably persistent and there are regularities in the ways in which others change with time. Some of these can be described in terms of negative or positive feedback of the effects of interactions upon interactions. If a relationship is temporarily diverted from its normal state or course of development, it may exhibit a capacity to recover—an example is given in section 6.2. Other cases can be described as involving positive feedback: if A grooming B leads to greater proximity and familiarity between A and B, it may lead to A grooming B more; while if a mother and infant start to spend more time apart, this may permit the infant to learn to be even more independent. Of course, positive and negative feedback are being used here only as descriptive labels but they do often capture the ways in which one individual is adapted to respond to changes in another. Often processes describable as positive and negative feedback proceed simultaneously.

Not surprisingly, it is mother–infant relationships whose development has been studied most. For that reason, little space has been given to them in this chapter and attention has been confined to some processes which can be described in terms of feedback (section 6.2) and the consequences of being orphaned (sections 6.3 and 6.4).

The role of play in developing relationships between juveniles is discussed in section 6.5 and section 6.6 describes the expanding network of relationships of young rhesus monkeys. Sections 6.7 and 6.8 discuss the dynamics of rank acquisition and maintenance. They place emphasis on the role of third parties on dyadic relationships, an issue which is the main focus of Chapters 8, 9 and 12 and thus give a reminder that relationships cannot be fully understood in isolation from the social group in which they are embedded. The remaining contributions focus on male–female relationships in and out of the mating season.

6.2 Feedback in the Mother–Infant Relationship
ROBERT A. HINDE

The concepts of positive and negative feedback can be applied in a descriptive way to a number of aspects of interindividual relationships. The case of the rhesus monkey mother–infant relationship will be considered in this section.

The normal mother–infant relationship depends on the stimulation provided by each partner to the other. The infant's ability to participate in the relationship depends on the mother's cooperation: with a mother who, by virtue of her own rearing history, does not behave maternally, the infant is likely to die (Harlow & Harlow 1965; Hopf 1981). Conversely, clinging by an infant monkey can elicit and maintain maternal responses in non-lactating females. Harlow and Harlow (1965) described how two infant monkeys, 78 and 38 days old, were placed with adult females whose own infants had been removed at birth some months previously. The babies were adopted—in one case immediately and in the other after some delay—and both mothers eventually lactated and 'apparently provided biochemically normal milk'. The Harlows believed that infant–mother contacts maintained maternal responsiveness in the females. Thus, in so far as positive responding in each partner produced positive responding in the other, this could be described as positive feedback. Of course it does not proceed indefinitely: in the normal mother–infant dyad, excessive positive responsiveness by either partner can produce rejecting or aversive responses by the other.

Of more interest are processes that can be described as negative feedback. If the mother of a 20–30-week-old infant is removed, the infant shows a brief period of hyperactivity and much distress calling and then becomes depressed. If after a few days mother and infant are reunited, the infant is usually very clinging. It gives frequent distress calls and may follow its mother around continuously. If the mother responds by picking the infant up and cuddling it frequently, the relationship is likely soon to be restabilized. Over a period of a few weeks the infant's demands decrease and the relationship returns to its original course (Hinde & Spencer-Booth 1971b; Mineka & Suomi 1978). This could be described as negative feedback.

The detailed course of the readjustment is of considerable interest. Figure 6.1 shows changes in the time spent off the mother before and after a six-day separation period for four infants in the Madingley colony. On the day of reunion, each infant spent less time off its mother than before separation. Two of them were very clinging, the other two much less so. In each case, however, the return of this measure towards its initial trajectory involved a temporary recovery, followed by a period of regression to a trough, and then (in three cases) more permanent recovery. These changes reflected oscillations in the nature of the mother–infant relationship. The mother initially acceded to the infants' demands, then seemed to become 'fed up' with them, but finally relented.

More detailed examination threw further light on these changes. In general, if the time the infant spent off its mother were to increase from one day to the next and this was accompanied by an increase in the frequency with which the mother

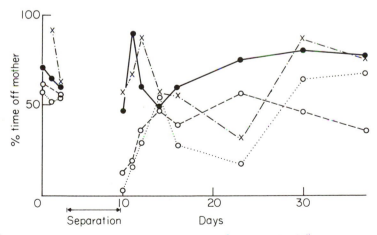

Fig. 6.1. Percentage of half minutes in which four rhesus monkey infants were recorded off their mothers before and after a six-day separation period. Each line refers to one individual. (After Spencer-Booth & Hinde 1967.)

rejected the infant's attempts to gain contact, responsibility for the change in 'time off' could be ascribed primarily to the mother. If, however, an increase in time off were accompanied by a decrease in the frequency of rejections, responsibility should be primarily ascribed to the infant. Similar arguments apply to decreases in time at a distance (cf. section 6.1). The decrease in time off between the preseparation watches and the day of reunion was accompanied by an increase in the frequency of rejections, indicating that the change in time off was due primarily to a change in the infants. The day-to-day changes in time off before separation and after the trough, however, were due primarily to changes in the mother: changes in time off and frequencies of rejections tended to be in the same direction. However, between the day of reunion and the trough there was no clear relation between these two measures: apparently changes in both the mothers' and infants' behaviour were responsible for the changes in time off.

Furthermore, before separation the frequency of rejections and the measure of the infants' responsibility for the maintenance of proximity (% Ap – % L, see p. 67) tended to be positively correlated with each other; those infants who were frequently rejected by their mothers were the ones who played a large role (or had to play a large role) in the maintenance of proximity with their mothers. Similarly, before separation the day-to-day fluctuations in these two measures tended to be in the same direction. After reunion, however, this was no longer the case: perhaps the (still depressed) infant then responded to an increase in the mother's rejections by approaching her less (Hinde 1969).

One of the four infants for whom data are presented in Fig. 6.1 failed to show recovery on this measure. The negative feedback thus failed to operate adequately. Further experiments showed that the speed or degree of recovery could be affected by the way in which the separation was carried out. If the separation were accomplished by removing the infant from the home environment and isolating it in a strange place, it would continue to protest with frequent distress calls for several days and 'despair' and depression would set in only gradually. However, if the mother were removed, leaving the infant in the physical environment and with the social companions to which it was accustomed, the phase of acute protest was more brief and 'despair' set in more rapidly. On reunion, such depressed infants were less effective in gaining their mothers attention than infants still protesting vigorously. The same manipulations affected also the mothers' responsiveness. If she was left in the home pen, she would usually be ready to go a long way to meeting her infant's demands when it was restored to her, but if the separation were accomplished by removing the mother from the home pen, then on reunion she also had to re-establish her relationships with her other social companions and was less tolerant of the demands of her infant. Thus, the most acute effects were found when the severity of the infant's separation-induced depression reduced its overt demands and the mother's responsiveness was low. The lack of responsiveness of the mother exacerbated the depression of the infant (Hinde & McGinnis 1977). In any case, recovery was closely linked to the nature of the mother–infant relationship: the infants showing most distress after reunion were those whose mothers (before and after separation) rejected them often and played a small role in maintaining proximity to them when they were off them (Hinde & Spencer-Booth 1971b).

These data incidentally demonstrate the dynamic nature of the mother–infant relationship and the manner in which its nature is affected by both of the participants. Follow-up work demonstrated also the dialectic between the nature of the relationship and those of the participants: in the group of monkeys in which the mother–infant relationship was most affected by the separation experience (i.e. those in which separation was achieved by removal of the mother), effects of the six-day separation could be detected six months and even two years later (Spencer-Booth & Hinde 1971b).

6.3 Effects of the Loss of the Mother on Social Development
PHYLLIS C. LEE

The dependency of infant monkeys on their mothers is highlighted by the fact that the most frequent consequence of the loss of a mother is the death of the infant. While the nurturing and protective function of mothers gradually declines after weaning, the mother is still the focus for many of the varied social interactions in which juveniles participate (Cheney 1978b; Pusey 1978).

The effects of the loss of a mother on the subsequent social behaviour of young primates has generally been explored experimentally through brief periods of artificial separation imposed on the pair (e.g. Hinde & Davies 1972; Mineka & Suomi 1978). Some infants have responded to these temporary separations with persisting difficulties in forming stable relationships, while for others the effect appears to be merely a short-term disruption of the mother–infant bond which in time is re-established (Hinde *et al.* 1978; Hinde & Spencer-Booth 1971b). In the wild, it is generally difficult to assess the effects of the permanent loss of the mother on the dynamic processes of social development, since the death of the mother with the survival of the young is a relatively rare event.

Lately, several authors have described the adoption of orphans by other members of the social group. Experimental loss of the mother in langurs resulted in infants taking the initiative in finding an adult female to act as a substitute caretaker during the mother's absence (Dolhinow 1980). In free-ranging baboons and Japanese macaques, adult and adolescent males associated with and cared for orphans although many, especially those under one year of age, did not survive (Hamilton *et al.* 1982; Hasegawa & Hiraiwa 1980). For older juveniles, adoption seldom occurs if the immature is capable of surviving on its own. In any case, since the mother is generally such an important social partner, her death may have both immediate and long-term effects on the relationships formed by the juvenile. Although all animals lose their mothers at some time, the consequences for social development may depend on when this occurs and the sex of the individual concerned.

Among primates where competitive dominance is determined by the support and assistance of the mother (reviewed in Cheney 1977; Walters 1980; see e.g. sections 6.7, 6.8 and 12.6–12.8), the loss of the mother may entail a subsequent loss of rank. The juvenile who lacks a mother also no longer has a consistent partner for certain types of interactions, such as grooming. Does it find new partners who can substitute for its relationship with the mother and are the qualities of these relationships similar to those between mothers and offspring? Relationships between siblings may be affected as well. Thus, an understanding of the changes in the behaviour and quality of relationships of immature primates after the death of the mother facilitates understanding both of the dynamics of the mother–offspring relationship and the development of the relationships between immatures and other members of their group.

Five orphans and three orphaned sibships (defined as two or more siblings who had lost a mother), making a total of 13 yearlings and juveniles, were observed during a 22-month study of the social development of vervet monkeys in Amboseli National Park. The orphans were found in two of the three study groups (Groups A and C) and all but three lost their mothers during the course of the study. They ranged in age from 13 months to five years. Four other young infants (of less than one year) lost their mothers during the course of the study but all died shortly after their mothers. Nine of the orphans were observed for at least three months before and after their mothers' deaths. One was observed for only two weeks, when he was injured by baboons and died. The data were taken from focal samples on the immatures. The behaviour in the periods before and after the mother's death will be compared for ten orphans and the general partner preferences for all 13 assessed.

The number of orphans was small and, in the majority of analyses, no statistically significant differences between age/sex classes were found. However, differences in the behaviour of individuals, apparent to the observer, led to this attempt to assess quantitatively the responses to the mother's death.

Changes in activities and near neighbours

Seven of the orphaned juveniles lost their mothers within one of the six-month seasonal periods of dry or wet weather and for these there were enough data to examine activity patterns within the single season in order to minimize the overriding effects of seasonal changes (Lee 1981; section 10.8). The percentage of time that they spent in feeding and foraging activities before and after the death were thus compared within the single season only. There were no consistent changes in the percentage of daily time spent feeding, suggesting that any trauma resulting from the death which would be likely to disrupt basic maintenance activities was either short-lived or non-existent. Of ten orphans, half showed an increase in the percentage of time spent feeding after the death and half showed a decrease. The magnitude of the change was small (median % change: +1%; range: −12 to +18%). For juveniles who were self-sufficient in feeding, the consequences of the loss of the mother on maintenance activities were thus not noticeable, but those on social interactions were more conspicuous.

Of the total of 13 orphans, there was at most a slight difference from other juveniles in the amount of time that was spent close (within 2 m) to other individuals in the social group. Eight spent less time close to others than was expected from the seasonal median for all immatures and the difference from the expected was small. Neither age nor sex appeared to influence which of the orphans spent less time close to others and which spent more. However, of the four offspring of high-ranking mothers, only one increased the amount of time spent near others. Both the daughters of the highest-ranking mother in Group C decreased the time spent close to others and they may have suffered from increased harassments after their mother's death (see below). The offspring of middle- and low-ranking mothers showed no consistent pattern.

Offspring of all ages tended to have stronger associations with their mothers than with any other individual (Lee 1981). Thus, with the death of the mother, the major associative partner was lost. Half of the orphans ($n = 10$) spent more time with another, unrelated adult female and half of the orphans less time, after their mothers' death. It thus appeared that at most only some of the orphans found a substitute, specific associative partner among the adult females. However, relative to the general median for immatures, eight spent more time with adult females after the mother's death, suggesting some general tendency to increase the degree of association with adult females.

Among the immatures who made up the long-term orphaned sibships, there was a tendency to cluster together within the group. Siblings were considered as dyads, A and B, B with A, A with C, etc. and these dyads were assessed for their preference for each other over and above that for peers; this was called the index of sibling preference (Lee 1981). Each sibling dyad was scored for the time spent together as a percentage of the mean time spent with all unrelated peers. Sibling dyads were then assessed for the difference in age between the pair. Six of the eight sibling dyads in the three orphaned sibships were above the median for non-orphaned siblings of the same difference in age. Thus, the long-term orphan siblings tended to have higher than average associations compared with those of other siblings.

Frequencies of social interactions

The hourly rate of grooming with other group members decreased after the mother's death for six of ten orphans. Of the four who showed an increase, three were females. All four of the female orphans with siblings increased the rate at which they groomed with their siblings and four of five females increased the rate at which they groomed with peers. Only two of the five male orphans increased the rate at which they groomed with peers and one of four had increased rates with siblings (Fig. 6.2). For both sexes, the greatest decline in grooming rates was with adult males (five of six decreased such grooming, four never groomed with males) and with unrelated adult females (seven of ten decreased such grooming). Some orphans appeared to have fewer opportunities to groom with different age/sex classes.

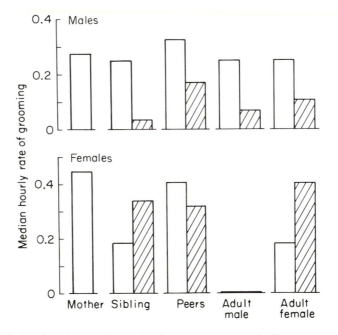

Fig. 6.2. Median hourly rate of grooming between orphans and other group members before (open columns) and after (hatched columns) the mother's death. Males (*n* = 5 before, 7 after) and females (*n* = 5 before, 6 after) are shown separately.

All eight long-term orphaned dyads groomed more with their siblings than with peers and their index of sibling preference was greater than that of the non-orphaned sibling dyads, where ten of 42 groomed more with peers than with siblings. Five of the dyads were above the sibling median for their difference in age. Thus, the patterns of orphan grooming do not appear to reflect a shift to find an adult female grooming partner to replace the mother but an increased strength of the interactions between siblings, at least for females.

The mother assists her offspring during aggression and is one of the major partners in tripartite interactions (Lee 1981). Upon her death, the major source of aid given to immatures during aggression is no longer available. Some orphans may suffer from increased aggression and competition as a result of their increased vulnerability. However, for six orphans the rates of aggression, both initiated and received, were either unchanged or lower after the mother's death. Of the four who had higher rates, three were females.

Only two orphans had increased rates of competitive interactions with other group members after their mother's death. These were the two daughters of the highest-ranking female in Group C. All other orphans were involved in fewer competitive interactions after the death. It has been shown that the presence of the mother is particularly important in enabling a subordinate immature to take resources from a dominant (section 4.4). It appears that with the mother no longer present, orphans engaged in fewer of the types of interactions where the support of the mother could have been important for success. For the two high-ranking sisters, one of whom was approaching sexual maturity, the increased rates were due either to increased levels of competition directed towards them from others who recognized the loss of the mother or to more frequent interactions initiated by them against others. The younger daughter initiated ten of 12 interactions after her mother's death, while in the previous months she had initiated six of 11. She may have initiated competitive inter-actions somewhat more frequently in order to maintain her position relative to the adult females. The older daughter initiated five of her ten interactions after the death, while previously she had initiated 11 of 15. Thus, the daughter who was approaching sexual maturity was being challenged more frequently after her mother's death. The change in the aggressive behaviour of these two females was very noticeable in the first few days after the mother's death. The older female was constantly under attack from other females and, indeed, even from her younger sister. It subjectively appeared as though the typical alliance structure of the group had broken down and chaos among the females was the result. After several weeks, aggression between the females became less common and some stability was regained.

Little aggression or competition was observed between siblings during the study and all the long-term, orphaned sibling dyads competed less with their siblings than did those siblings with mothers alive. Only one of the orphans competed more with its sibling than with its unrelated peers, compared with 18 of the 26 non-orphaned sibling dyads.

Taking the measures of proximity to others, grooming and the combined rates of aggression and competition (called harassment) and dividing the orphans into those who increased or decreased on these measures, a clustering was apparent (Table 6.1). Only one orphan declined on all three measures and he was the yearling who was killed by baboons shortly after his mother's death. As the youngest animal in the sample, he appeared to suffer the most social disruption from the death and, indeed, this may have increased his vulnerability to predators. Only one orphan increased on two social measures, while all the others declined on at least two.

Table 6.1. Clusters of orphans on three different social measures which increased in frequency or decreased in frequency after their mother's death. The sex of each is indicated with M (male) or F (female) together with the age in years.

Harassments							
Increase Grooming				Decrease Grooming			
Increase Proximity		Decrease Proximity		Increase Proximity		Decrease Proximity	
Increase	Decrease	Increase	Decrease	Increase	Decrease	Increase	Decrease
F4			F4	F3	M4	M1	
				M2	M5		
				F4	M3		
					F2		

Thus, the orphans did appear to suffer from the loss of their mothers during immaturity, in terms of the rates at which they engaged in social interactions with other group members (see also Hinde & Davies 1972). Their relationships with their siblings appeared to be intensified and orphaned sibships appeared to be somewhat more tightly clustered spatially and more interactive than the non-orphan sibships. With the loss of the mother as the focal point of the family, the siblings continued to maintain themselves as a unit within the social group as a whole. The most serious consequence of the loss of the mother was death for one weaned juvenile. For the female approaching sexual maturity, the loss of the mother resulted in increased challenges to her dominance from others and she responded by initiating somewhat more aggression (see also Walters 1980). After the study ended, several other young males were orphaned and these males subsequently transferred out of their natal group into a new group along with their brothers (sections 9.4 and 11.3). This also suggests that both mothers and siblings together can be viewed as the focus of immature social development and that with the loss of the mother relationships between siblings intensify. Some alternative social partners were found, particularly for interactions such as grooming, and these often tended to be peers. Males and females showed similar responses to their mother's death but the dominance rank of the mother appeared to be a variable that affected the subsequent patterns of interactions for females more than it did for males.

6.4 Effects of being Orphaned: a Detailed Case Study of an Infant Rhesus

CAROL M. BERMAN

Unweaned orphans rarely survive to maturity in the wild and thus rarely contribute to the next generation's gene pool. Nevertheless, their study in free-ranging but provisioned and predator-free colonies like Cayo Santiago (where they occasionally survive to adulthood) can give important insights into the nature of early social relationships with mothers and with other group members (Berman 1982c). A detailed description of the social development of one such infant from birth to orphaning and adoption during its third month and then to thirty weeks of age suggests: (1) that typical mother–infant patterns of interaction between an orphan and a nulliparous foster-mother may not be established immediately but can develop gradually through adjustments on the parts of both partners; and (2) that interest and care for orphans by males may be immediate, skilful and competitive but less likely to lead to a long-term foster-parent–infant relationship than interest by close female relatives.

The orphan under study was the son of a middle-ranking twelve-year-old multiparous female who disappeared a few days after suffering severe wounds. The orphan, a few days short of eleven weeks of age, showed symptoms of depression similar to those described for captive infants artificially separated from their mothers: a day-long period of agitation and apparent searching behaviour, followed by several weeks of decreased activity, hunched posture and frequent whoo-calling (e.g. Hinde 1977b; Kaufman & Rosenblum 1967; Mineka & Suomi 1978).

Male care and adoption

During the first seven to ten days following the mother's disappearance, the orphan received care primarily from three males: (1) the alpha male who has been excluded as a possible father (Duggleby pers. comm.; Sade pers. comm.); (2) a middle-ranking adult male who associated closely with the mother and who has not been excluded as a possible father; and (3) a four-year-old brother. At first, the males competed for access to the orphan and each displayed typical maternal behaviours such as ventro-ventral carrying and 'gaming' (Hinde & Simpson 1975). However, they appeared gradually to lose interest in him over the ten-day period; time spent near and in contact with the males (estimated with two-minute point time samples) decreased dramatically. The infant probably contributed to this by resisting contact with them. After about a week, the orphan's three-year-old nulliparous sister became his principal caretaker and remained so, at least until the orphan was thirty weeks of age (when focal-animal sampling ended).

Foster-mother–infant interaction

Before her disappearance, the orphan's relationship with the mother could have been described as 'typical'; when several measures of mother–infant interaction for the pair were compared with those for 20 other mother–infant pairs from the same social group, the values for the mother and future orphan lay close to the median values for the other pairs (see Fig. 6.3). However, for about two months after the mother's disappearance, values for the same measures of interaction between the orphan and the foster-mother were at the extremes of the range or outside the range of values for mothers and infants of the same age.

The foster-mother and orphan spent more time in ventro-ventral contact than did mothers and infants of the same age. Measures of contact initiation suggest that this was primarily because the foster-mother sought contact more with the orphan than did mothers with infants; she initiated a larger proportion of

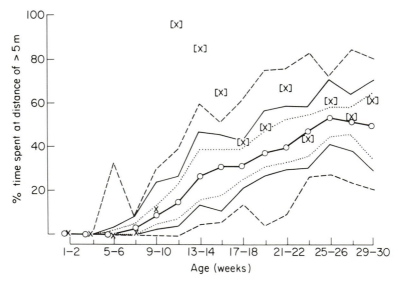

Fig. 6.3. Percentage of time spent at a distance of more than 5 m from the mother. X = scores for the orphan-to-be-with its mother: [X] = scores for the orphan with his foster mother; open circles = the median values for 20 mother–infant pairs of the control sample. Dotted, solid and broken lines = the interquartile (50%), 80% and full (100%) ranges, respectively, for the control sample.

nipple contacts than did mothers and she rejected a smaller proportion of the orphan's attempts to get on the nipple. Indeed, field notes indicate that at first the orphan resisted her attempts to make contact, as he had done previously with the male caretakers. It is suggested that the foster-mother's persistent seeking of contact led to his gradual acceptance of her and to a mutual attachment.

On the other hand, for about two months after the mother's disappearance, the foster-mother and orphan spent much more time at a distance (some of it out of each other's sight) than did any mother–infant pair (Fig. 6.3). Again, this difference was due primarily to differences between the foster-mother and the mothers of infants of the same age; the orphan took relatively more responsibility for maintaining proximity with her (% AP_I–% L_I, see p. 67) than did other infants of the same age. Had differences in time spent at a distance been due primarily to differences between the orphan and other infants, one would have expected the orphan to have taken relatively less responsibility for maintaining proximity than did other infants. These results are consistent with field notes which indicate that, after bouts of contact, the foster-mother would abruptly leave the orphan and often go out of its sight for 20 minutes to over an hour. The orphan would become distressed but frequently would not follow. At this stage, he showed symptoms of depression primarily when the foster-mother was out of sight.

Apparent recovery

At the end of two months, the values for all three measures of contact for the pair had gradually become similar to those for mothers with infants of the same age; time on the ventrum had decreased, the proportion of 'maternally' initiated contacts had decreased and rejections had increased. This pattern of change is consistent with the suggestion that changes in the relationship were due primarily to decreases in the foster-mother's tendencies to seek contact with the orphan. As the orphan became gradually more attached to her, she adjusted by seeking contact less. Time spent more than 5 m away also gradually decreased to 'typical' levels, though the infant's relative responsibility for maintaining proximity (% AP_I – % L_I) remained high compared with mother–infant pairs. Thus, changes in time spent at a distance were probably due to the orphan's increased abilities to keep up with the foster-mother.

6.5 Play as a Means for Developing Relationships
PHYLLIS C. LEE

The friendly relationships of immature primates are characterized by a variety of interactions, initially those of spending time together, grooming, lack of competition or aggression and, most frequently, play (reviewed in Lee 1981). Although the costs and benefits resulting from play in both the short- and the long-term can be debated (e.g. Baldwin & Baldwin 1973, 1977; Fagen 1981), the diversity of play patterns and the frequency with which these patterns are observed suggest that an understanding of immature play could be valuable for interpreting the complexity and diversity of the social relationships of immatures.

The causal links between play and social development have been discussed by many authors (e.g. Chalmers 1980; Dolhinow & Bishop 1970; Loizos 1967; Symons 1974) but the majority of the associations have been so vague as to be of little predictive value (discussed in Beckoff & Beyers 1981). Most researchers now agree that play cannot be seen as a category of behaviour with a single function (e.g. Bateson 1981a; Beckoff & Byers 1981; Fagen 1981) and Bateson (1981a) raised the question of whether different systems of play are causally distinct during development.

In this paper, data are presented on the social play of free-ranging vervets which suggest that its frequency and timing during ontogeny are not rigidly patterned but depend on opportunities for play, themselves limited by ecological conditions and group composition. In addition, the link between the acquisition of adult skills (e.g. sex roles and dominance) may be less straightforward than has been proposed. The nature and timing of social play may have more to do with the conditions under which play is seen than with long-term consequences of the behaviour. Nevertheless, play can still be viewed as contributing to social development, and understanding the unpredictable character and quality of play may enhance understanding of the ways individuals learn to express and maintain social relationships.

Factors influencing play

In most and perhaps all species, play between immatures generally appears in relaxed contexts, when environmental and social stresses are minimal (reviewed in Fagen 1981). During times of bad weather or intergroup tension, play is noticeably reduced in frequency (Bernstein 1972, 1976b; Stewart 1981; White 1977). When the availability of food and water are reduced, play again declines (Baldwin & Baldwin 1973, 1976; Berger 1979a, 1980; Hall 1963; Loy 1970; Oliver & Lee 1978; Richard 1974). The sensitivity of play to ecological con-

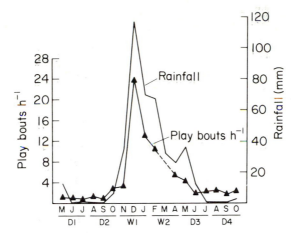

Fig. 6.4. Monthly rainfall plotted with the mean number of play bouts per hour across all three groups. The broken line connects February and April when no focal data were available.

ditions was demonstrated during a study of the development of immature vervet monkeys in Amboseli National Park. During annual, six-month-long dry periods when the quality and quantity of food was low, vervets of all ages and both sexes played very infrequently. As the rains began and there was a related increase in the abundance and quality of food, rates of play increased (Fig. 6.4).

This sensitivity to ecological and social conditions suggests that frequent play is not an essential component of the ability to form and maintain relationships on the part of the young, for periodic play deprivation does not necessarily produce a socially incompetent adult. It also suggests that play need not occur within a fixed time or in a rigid sequence during development. However, when play does occur, the social contexts in which it appears and the nature of the play interactions may both enhance physical skills and the complexity of the relationships between immatures.

Frequency of play and types of play partners

In vervets, as in many species, younger animals generally play more frequently than older ones and males often play more than females (Bramblett 1978; Fedigan 1972; Raleigh *et al.* 1979). The frequency of play has been related to maternal rank, in that animals of high rank may be more playful and have a wider range of partners (e.g. Cheney 1978a; Tartabini & Dienske 1979), and to the amount of time that immatures spend with peers, their potential play partners (Cheney 1978a; Owens 1975a).

Table 6.2. Median and range of the number of play bouts per hour during the six-month wet season for males and females of each age-class.

	Median	Range	No. of subjects
Infants			
(4–12 months)			
Males	15.1	6.0–22.5	4
Females	Insufficient data		
Young juveniles			
(12–36 months)			
Males	10.8	2.8–36.9	11
Females	5.7	3.4–10.5	7
Old juveniles			
(36+ months)			
Males	8.9	0.1–13.6	7
Females	3.3	2.4– 6.7	4

Among the immature vervets in this study, the effect of age on rates of play was marked for females ($r_s = -0.608$, $n = 16$, $P < 0.02$) but less so for males ($r_s = -0.296$, $n = 34$). Young males appeared to play at the highest hourly rates and older females at the lowest rates (Table 6.2) and across all immatures, males tended to play somewhat more frequently than females ($P < 0.06$). However, when a small sample of each sex was matched for age, there were no significant differences between the sexes (the median for two-year-old males was 10.1 bouts h^{-1} [$n = 6$]; the median for two-year-old females was 7.1 h^{-1} [$n = 4$, $P = 0.26$]). The sex difference appeared to be due not to differences in play rates throughout immaturity but to the earlier decline in rates for females as they aged. Maternal rank did not correlate with how frequently an immature would play but males who spent the most time with peers were also those who played at the highest hourly rates ($r_s = 0.570$, $n = 34$, $P < 0.01$).

In captive vervets, as well as other species of primates, play tends to be most frequent between animals of the same sex and age (reviewed in Cheney 1978a). However, this was not apparent for most of the immatures in the three study groups and there were few marked preferences for same-sex peers (Table 6.3). Old juvenile males (those between three and five years old) played with each other more than with other age/sex classes but this was the case only in the one group where many different partners from the entire range of ages and sex were available. In the smaller groups, one playful age/sex class could dominate the partner preferences of the entire group. Indeed, the most frequent play partner for immatures in the three groups changed through time (Lee 1981). Thus, consistency in the choice of play partners did not appear to be a necessary aspect

Table 6.3. Rank order preferences for each age/sex class as a play partner, based on the mean percentage of bouts between different partners in each group. 1 = highest; — = none available.

		Old juvenile		Young juvenile		Infant		(No. of
		Male	Female	Male	Female	Male	Female	subjects)
Old juvenile								
Male	Group A	2	5	1	6	3	4	(4)
	Group C	1	5	2	3	4	6	(5)
Female	Group A	4	3	1	2	5.5	5.5	(2)
	Group C	2	5	1	4	3	6	(2)
Young juvenile								
Male	Group A	3	4	1	2	5.5	5.5	(3)
	Group B	—	—	1	3	4	2	(5)
	Group C	3	4	1	5	2	—	(3)
Female	Group A	3	2	1	—	—	—	(1)
	Group B	—	—	3	2	1	—	(2)
	Group C	4	6	1	5	3	2	(4)
Infant								
Male	Group A	3	5	1	—	4	2	(2)
	Group B	—	—	2	4	3	1	(5)
	Group C	5	4	3	2	1	6	(6)
Female		Insufficient data						

of immature friendly relationships. There was a marked tendency in Groups A and C, at any time, for those individuals who spent the highest proportion of time together and, less often, those who groomed each other most frequently, to play the most with each other (Table 6.4).

While environmental factors, age and, to some degree, sex affect how frequently immatures will play, the choice of partners appears to be constrained by the demographic composition of the group, i.e. whether or not certain age/sex classes are present and available as play partners (see also Cheney 1978a). In addition, the choice of partners for play depends on how frequently a partner is nearby and thus available and perhaps on its level of playfulness or its willingness to play.

Quality of play

Since the quantity of play appears to vary with a variety of external factors, play may not be related directly to the expression of friendly relationships between

Table 6.4. Percentage of immatures whose top play partner was their top grooming partner and their most frequent immature associate. The mean expected was based on the number of available immatures (Group A: $n = 10$; B: $n = 8$; C: $n = 17$).

	Mean percentage
Grooms other	
Group A	20.0
Group B	12.5
Group C	29.4
Proximity	
Group A	40.0
Group B	12.5
Group C	52.9
Mean expected	
Group A	11.1
Group B	14.3
Group C	6.3

immatures. Nevertheless, aspects of the quality and the timing of play during ontogeny may have important consequences for the physical and social development of immatures.

In many species of primates and in other animals, there is sexual differentiation in the types of play observed. Aggressive play tends to be more common among males (Berger 1980; Cheney 1978a; Owens 1975b) and this sex difference has been causally related both to physical skills such as fighting ability and to an increased ability to assess dominance positions (Symons 1978). Thus, when males play, they may be practising those physical skills which will be most important in later contests for access to mates, as well as assessing their own abilities of the moment against a variety of opponents of differing levels of skill, sophistication and dominance. Female play often consists of approach–withdrawal rather than rough-and-tumble play (Cheney 1978a; White 1977) and has been associated with the acquisition of caretaking skills through gentle play with infants (Bramblett 1978; Fedigan 1972; Lancaster 1972).

Among the vervets in this study, both males and females engaged in the same types of play (Fig. 6.5). The most frequent type of play for both sexes was wrestling, where two immatures would tussle with each other using hands, feet and mouths. The mean length of a play bout was similar for all immatures (7 s). Play patterns became more complicated with age and consisted of more vigorous

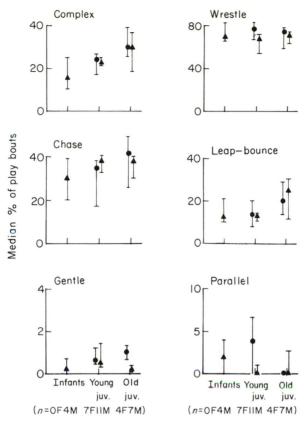

Fig. 6.5. Median percentage of play bouts, which included play of different motor patterns, for infants (between four and 12 months of age), young juveniles (between 12 and 36 months) and old juveniles (over 36 months). The interquartile ranges are shown with the solid lines. Solid triangle = males; solid circle = females.

rough-and-tumble play. A single short bout of play was likely to contain several different motor patterns joined together in a single fluid bout which increased in complexity with age.

While females did tend to concentrate some of their play on infant partners (Table 6.3), it was by no means exclusively restricted to infants nor did the quality of their play differ from that of the males. There was, however, a tendency for males to have slightly higher proportions of bouts of chasing when they were playing with other males rather than with females ($P = 0.06$). There was only one type of play where there was a statistically significant difference between males and females, even when the sex of the play partner was taken into consideration. When the proportion of those bouts of play that consisted of 'leap-bounces' between males was compared with the proportions between females, females in their single-sex play tended to have fewer bouts ($P < 0.05$). Leap-bouncing play was similar to the motor patterns used by males when they display to conspecifics during territorial encounters; adult females only rarely display (Cheney 1981; Lee 1981). However, when females played with males, the proportion of their bouts that consisted of leap-bounce play was no different from that of males with other males. It was only in same-sex play that this one difference emerged.

The lack of observed differences between the sexes in the complexity, quality and timing of play could result from the coarse-grained nature of these data on a sophisticated and complicated interaction. Alternatively, since males were more numerous in these groups and may have been somewhat more playful, when a female played she was likely to have a male as a partner. When a female played with a male, she may have been constrained to play in a male-fashion or risk having the bout terminated.

In conclusion, since play among these vervets was restricted to a brief period of the year when the quality of the diet was high, their play cannot be seen as a continuously necessary part of the establishment of social relationships. Neither can the components of play be seen as appearing in a fixed time period during development. None of the skills used in play are unique; they can be practised in a variety of different interactions (see also Loizos 1967). Play does, however, offer an interaction where there are few social risks. Play can become aggressive but unlike play, aggression is always risky. A playing animal expends some additional energy (e.g. Martin 1982) but it does not directly lose food to a competitor.

However, timing of play may be relevant to the ability of a developing animal to maintain persisting relationships. As the animals mature, new skills relevant to particular social and physical competences may be developed and certain skills which were previously important have been acquired and become less frequent in the repertoire. Thus, one element in the importance of play may be related to the appearance of different types of play during development.

The lack of a marked sex difference in play was surprising in light of previous reports. Since the physical and social needs of each sex can be met through a range of interactions not limited to play, and the meeting of these needs may change as a function of the age and sex of the partner (e.g. Lorenz 1970), sexual differentiation in play may not be the most efficient way of meeting these needs.

As Bateson (1981b) has pointed out, there may be many ways for a developing animal to reach the end-point of a socially competent adult. Play may be only one of these pathways but it is one that provides a changing and unpredictable arena for the relaxed practice of motor skills and the imparting and acquiring of information about a wide range of other individuals. Thus, while an immature may not need to play to form a relationship with others in its social group, those individuals who do play have a social environment that changes to meet physical and social needs as they mature. Certain aspects of behavioural flexibility may be more easily learned in play, others may be more efficiently learned in the course of other interactions where unpredictability is not as important.

6.6 Differentiation of Relationships among Rhesus Monkey Infants

CAROL M. BERMAN

Infant rhesus monkeys begin to interact with a wide range of group members other than the mother within a few days after birth. Group members are attracted to new infants and their mothers, whom they approach and attempt to touch or groom. The amount of interaction and the kinds of interaction the infant experiences is thought to be governed primarily by the infant's attractiveness and by the mother's abilities and propensities to control its interactions (e.g. Caine & Mitchell 1979; Spencer-Booth 1968). The infant's social relationships within the group are assumed to develop out of this early interaction and to reflect broadly those of the mother. However, the precise ways in which infants develop their own social networks and relationships within large, lineage-based groups are not fully understood. Data from Cayo Santiago (Berman 1982a) suggest that the process resembles one of differentiation. Through their early control of infants, mothers pass on their patterns of distributing interaction among group members (i.e. social networks); however, as infants become independent of their mothers, each of their relationships within the network becomes increasingly independent from those of the mother.

Social networks and differentiation

From the beginning, infants distributed positive interactions among the group in similar ways to their mothers and in this sense their social networks mirrored those of their mothers. Thus, like their mothers, infants spent more time near (within 5 m) close kin than distant kin or unrelated individuals. This was shown by comparing the percentage of two-minute point time samples each infant spent near each age/sex class of companion, as a function of the companion's degree of relatedness to the infant through maternal lines (Fig. 6.6). For each age/sex class of companion, median scores for the time spent near infants correlated positively with the degree of relatedness. The results of the same analyses were similar for measures of more intimate interaction: frequencies of approaches and departures at 60 cm and frequencies of friendly contact initiations (touching, grooming, embracing and playing).

Like their mothers, infants associated more with female companions than with male companions. Infants not only spent more time near female than male

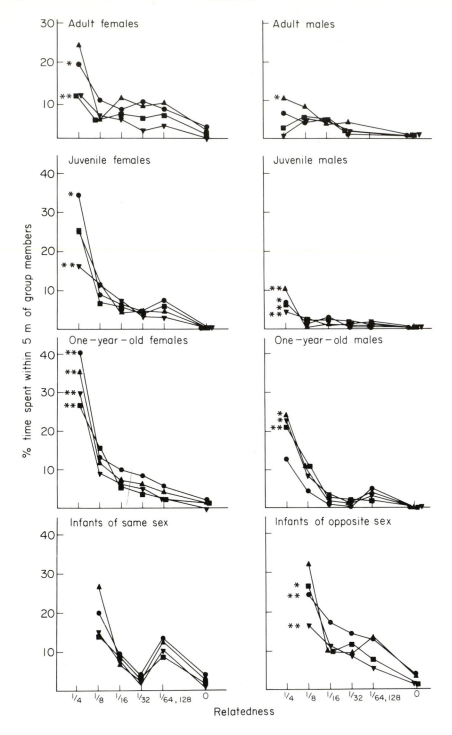

Relatedness

companions of the same age and maternal kinship category, they also approached and departed from them more frequently and engaged in friendly contact more frequently. A third pattern which was shared by mothers and infants was the tendency to associate more with younger immatures than with older immatures: infants spent more time near other infants than near yearling companions of the same sex and maternal kinship relationship and they spent more time near yearling companions than near juvenile (two- and three-year-old) companions of the same sex and maternal kinship relationship. Again, results were similar for frequencies of approaches and departures at 60 cm and for frequencies of friendly contact initiations.

Finally, mothers and infants both followed patterns of interaction characteristic of their lineages, patterns which appeared to be based on the dominance status of the lineage. Infants in the top-ranking lineage spent more time near their own kin (lineage members) than did infants in either other lineage. However, infants in this lineage did not necessarily interact more frequently with their kin in more intimate ways than did infants in other lineages (see section 8.3).

These four patterns of distributing interaction among companions were shared by infants and mothers from birth to at least 30 weeks of age. Of course, social networks which mirror those of the mother are not surprising in new infants because new infants are almost totally under their mother's control. They follow these patterns simply by virtue of their dependence on their mothers. However, the persistence of these patterns in infants as old as 30 weeks suggests that mothers' early influence has long-term consequences on the development of social networks. By 30 weeks infants no longer have the attractive physical characteristics of new infants (Sade 1971) and they are considerably more independent from their mothers (e.g. Berman 1980a; Hinde & Spencer-Booth 1967). They spend about 50% of their time more than 5 m from the mother, some of it out of her sight. Moreover, it is infants rather than mothers who take primary responsibility for maintaining contact and proximity. Nevertheless, infants continue to distribute their affiliative interactions largely according to maternal patterns. In this sense, mothers pass on their networks to their infants.

Fig. 6.6. Time infants spent within 5 m of group members as a function of the degree of relatedness (r) through the maternal lines. The median scores are shown for the time infants spent near companions in each age–sex–kinship category. The sample sizes within the age–sex–kinship categories range from 3 to 20 infants with a mean of 12. Separate curves are shown for eight age/sex classes or companion during four four-week age periods (solid circle = 3–6 weeks; solid triangle = 11–14 weeks; solid square = 19–22 weeks; solid inverted triangle = 27–30 weeks). * = $P < 0.05$ and ** = $P < 0.01$, Spearman rank-order correlation coefficients, for the time spent within 5 m with the degree of r within each age/sex class.

It is necessary to distinguish clearly between the development of the social network and the development of each of its component relationships. Although the infant's network remains more or less stable as it gains independence from the mother, the nature and qualities of each of its relationships change dramatically. At first the nature and quality of the infants' relationships are largely reflections of those of their mothers. Indeed, interaction between mothers and companions is primarily responsible for sustaining their relationships. However, by 30 weeks, infants spend much more time interacting with companions than previously and, in many cases, take primary responsibility for maintaining contact and proximity with them (Berman 1978a). Mothers no longer attempt to prevent this interaction, although they and other close female relatives aid the infants in agonistic encounters (Berman 1980b; Cheney 1977). In this sense, infants' relationships can be described as differentiating from those of their mothers; while maternally derived networks remain largely intact, infants gradually develop relationships which they sustain through their own interactions and which may have qualities different from those between their mothers and companions.

Identifying differentiated relationships

One way to describe the extent to which infants have formed differentiated affiliative relationships with companions is to determine the extent to which companions who initiate interaction are seeking to interact specifically with the infant rather than its mother (Berman 1982b). The distribution of approaches (or other measures of positive interaction) to infants by companions can be considered, for example, as a function of the distance between the mother and the infant when they occurred. This distribution can then be compared with an expected distribution based on the proportions of time the mother spent within each distance category. If a companion approached more frequently than expected when the infant was near the mother (e.g. within her arm's reach), the companion may have been seeking proximity primarily with the mother rather than the infant. To this extent, subsequent interactions with the infant could be described as coincidental. However, if the companion approached the infant more frequently than expected when the infant was at a distance from the mother, one can be more certain that the companion was seeking proximity specifically with the infant.

When an analysis of this sort is made for young infants (1–14 weeks) who are still physically attractive to others and are highly dependent on their mothers, it is found that all age/sex classes of companions tend to seek proximity with infants more than expected when they are at a moderate distance (between 60 cm and 5 m) from the mother and less than expected when the infants are near their mothers. Thus, the companions are attracted specifically to infants and not simply to their mothers. However, this does not necessarily mean that infants of this age have formed differentiated relationships with their companions. Although infantile attractiveness which is due to the special physical characteristics of new infants cannot be distinguished definitively from that which develops out of social interaction, it is likely that physical characteristics play the larger role during the first several weeks of life. Since these distinctive physical characteristics typically disappear by about three months of age, an analysis of the extent to which infantile relationships are differentiated must focus on older infants.

If the older infant's attractiveness is only a temporary phenomenon which does not lead to differentiated relationships, one should find evidence that companions no longer necessarily seek proximity with infants when they approach. However, if social relationships develop out of early interaction and acquire a degree of independence from those of the mother, companions should continue to approach infants more frequently than expected when infants are at a distance from the mother. Indeed, this tendency should become stronger over time.

The results of this sort of analysis for older infants (17–30 weeks) on Cayo Santiago suggest that the rate at which infants form differentiated relationships differs greatly according to the age and sex of the companion. In general, the degree to which relationships with females have differentiated by 30 weeks is negatively related to the companion's age. Infants showed early and increasingly marked tendencies over time to seek out one another specifically. In contrast, they showed no evidence that they had formed differentiated relationships with adult females by 30 weeks. Hence, it is possible that early attractiveness and early regulated interaction with others does not lead inevitably to well-differentiated relationships. Juvenile and yearling female companions showed patterns which were intermediate; although they showed increasing tendencies to seek infants during the first 14 weeks of life, these tendencies became weaker and in a few cases reversed between 17 and 30 weeks.

6.7 Relative Power and the Acquisition of Rank
SAROJ B. DATTA

Introduction

Monkeys often have clearly defined dominance relations with others in their social group. The dominant of a pair is regularly able to boss the subordinate, displacing it from food, drink and other objects of competition or common attraction (Clark & Dillon 1973; Dittus 1977; Packer 1979b; Seyfarth 1977,

1980; Stammbach 1978). This asymmetry in the relationship can be profoundly disadvantageous for the subordinate: not only is its access to resources restricted but it may experience physiological stress (Sassenrath 1970) and may produce, raise or even conceive fewer offspring (Drickamer 1974a; Dunbar & Dunbar 1977; Keverne *et al.* 1982; Rowell 1970, 1972b). Moreover, among those species in which adult females maintain stable dominance hierarchies and daughters inherit maternal rank relative to other group females, the disadvantages of subordinance may be perpetuated over several generations (Kawai 1958; Kawamura 1965; Koyama 1967, 1970; Sade 1967, 1969).

Clearly, individuals would 'prefer' to be dominant and if dominance reversal is profitable (i.e. if the potential benefits exceed the potential costs) subordinates may be expected to attempt it. Conversely, however, dominants may be expected to 'prefer' to remain dominant and to try and maintain status. Changes in the direction of dominance do occur (see below) although dominance relations can also be highly stable (see section 6.8).

This section is specifically concerned with the conditions (both social and non-social) which enable some young rhesus monkeys to reverse dominance with older and larger opponents. However, a general condition is also proposed which may need to be satisfied before the direction of dominance–subordinance can alter in most (if not all) dominance dyads in this and other primate species.

Study site and animals

The data in this section come from a 16-month study of the social behaviour of rhesus monkeys in the dominant (031) matriline of Group J on Cayo Santiago.

In this study, the subjects were the related adult females and their pre-adult offspring in the matriline. They could be split into 11 'mother-families', each consisting of an adult female and her pre-adult offspring. Nine adult and nine pre-adult females were focal animals but ad lib data on these and other members of the matriline were freely collected.

Long-term studies have shown that on Cayo Santiago offspring inherit maternal rank relative to other females and their offspring (Missakian 1972; Sade 1967, 1972b). (This 'rule' is more clearly applicable to daughters than to sons, since the latter generally leave the group of their birth at adolescence. Those males who do stay into adulthood are quite likely to dominate females dominant to their mothers, since adult rhesus males typically dominate most adult females.)

Definitions

These are as follows:

Focal dyad: one for which at least some data on interactions were collected focally, one or both members being focal animals.

Dominance reversal: recorded if a clear period of consistency (in the direction of dominance–subordinance) in one direction was followed by a clear period of consistency in the opposite direction. A period of inconsistency could separate the two.

High-born and low-born: A is high-born (HB) relative to B if A's family (mother) ranks above that of B. B is of course low-born (LB) relative to A.

Interference: said to occur when a third animal or animals interrupt a dispute between two individuals, normally by threatening one of the original antagonists.

Target: the individual to whom the interferer directs its aggression.

Beneficiary: the individual the interferer, by threatening the other, sides with or supports.

Effective support: intervention in which the interferer and/or its supporters cause submission in the target and its supporters (if any).

Direction of dominance–subordinance at the end of the study

When the study ended, the direction of dominance was known in 532 dyads between members of different mother-families (from now on referred to as families) in the 031 matriline. Both focal and ad lib data were used to determine the direction of dominance, the advantage of this being that many dyads without focal data could be included in the sample. Only dyads with three or more agonistic interactions were considered. One hundred and seventy (32% of 532) dyads did not conform to expectations based on maternal rank: in these, the HB individual was subordinate to the LB. In all of these exceptions the former was an immature, younger and smaller than the latter.

Table 6.5 shows that the pattern of observed dominance between members of different families was best explained by an interaction between maternal rank and the relative age/size of opponents. When HBs were older than, or the same age as, their LB opponents they almost invariably dominated them. However, when younger than LBs, HBs dominated a greater proportion of those opponents close to them in age than older.

Changes in relationships during the study

Two kinds of (related) changes were seen to occur in dominance relationships during the study. First, some subordinates began to 'challenge' dominants by initiating, joining in or returning aggression to the latter. Such behaviour appeared and intensified in 76 out of 532 dyads. In all of these the subordinate was higher-born and younger than the dominant. In no dyads where the subordinate was *lower-born* and younger than the dominant did challenging occur.

Table 6.5. Median percentages of low-born opponents dominated by high-born individuals in different age-class combinations. The figures below the age-classes are the numbers of males (M) and females (F) involved in agonistic interactions. The figures in brackets are the number of dyads in which agonistic interaction was seen.

High-born age-class (years)	Low-born age-class (years)						
	0–1 / 5F / 6M	1–2 / 7F / 7M	2–3 / 7F / 5M	3–4 / 4F / 3M	4–5 / 6F	>5 / 14F	Mean %
0–1 / 2F / 5M	100% (12)	87.5% (11)	41.6% (10)	0% (4)	0% (6)	0% (14)	38.2
1–2 / 5F / 4M	100 (12)	100 (46)	50 (33)	0 (17)	0 (19)	0 (47)	41.6
2–3 / 5F / 3M	100 (11)	100 (33)	100 (27)	66.7 (18)	75 (17)	28.7 (40)	78.4
3–4 / 2F / 3M	100 (15)	100 (24)	100 (17)	100 (11)	100 (14)	66.7 (37)	94.4
4–5 / 2F	100 (2)	100 (9)	100 (7)	100 (4)	100 (5)	100 (12)	100 $r_s = 1$
Mean %	100	97.5	78.1	53.3	55	39.1	$P < 0.05$, two-tailed

$r_s = 0.943$
$P < 0.05$,
two-tailed

Second, in 32 dyads, subordinates managed to reverse dominance with their opponents during the study. (In 30 of these [which came from the group of 76 above] a period of challenge preceded reversal.) In 30 out of the 32 the subordinate was HB and younger than the dominant. In the two exceptions two one- to two-year-old males outranked two one- to two-year-old females. Males, immature and otherwise, are more likely to break the maternal inheritance pattern (Cheney 1977; Hausfater 1975; Missakian 1972).

Clearly, the *direction* of change is strongly influenced by the relative family ranks of the individuals concerned. However, the *timing* of change is affected by the relative age/size of opponents. Not all HB subordinates challenged LB dominant opponents and those that did so did not invariably challenge them all. Neither was there an absolute age at which HBs initiated challenge. It was rather the case that HB subordinates were likely to challenge those older LB dominants closest to them in age/size (Table 6.6). This pattern was also reflected in

Table 6.6. Mean percentages of high-born subordinates' dyads with low-born dominants in which the former challenged the latter. Means are used because the number of dyads per high born was generally less than three. Blank cells are those with 100% dominance in Table 6.5. The figures in brackets are the number of dyads in which low borns were dominant at the beginning of the study.

High-born subordinates (years)	Low-born dominants (years)						
	0–1	1–2	2–3	3–4	4–5	>5	Mean %
0–1		100% (7)	100% (8)	50% (4)	33.3% (6)	0% (14)	56.7
1–2			79 (19)	64.3 (14)	60 (15)	49 (47)	63.1
2–3				100 (6)	100 (4)	85 (40)	95
3–4						100 (16)	100
Mean %		100	89.5	71.4	64.4	58.3*	

*Spearman $r_s = 1$, $P < 0.05$, two-tailed.

Table 6.7. Mean percentages of high-born subordinates' dyads with low-born dominants in which high-born subordinates managed to reverse dominance by the end of the study. Blank cells are those with 100% dominance. The figures in brackets are the number of dyads in which low borns were dominant at the beginning of the study.

High-born subordinates (years)	Low-born dominants (years)						
	0–1	1–2	2–3	3–4	4–5	>5	Mean %
0–1		42.8% (7)	12.5% (8)	0% (4)	0% (6)	0% (14)	11.1
1–2			42.1 (19)	21.4 (14)	13.8 (15)	0 (47)	19.2
2–3				50 (6)	50 (4)	17.5 (40)	39.2
3–4						43.7 (16)	43.7
Mean %		42.8	27.3	23.8	21.1	15.3	

$r_s = 1$
$P < 0.05$,
two-tailed

the success with which dominance reversal was achieved against LB dominant opponents (Table 6.7).

Next, the nature of the influence(s) of age/size and maternal rank will be considered.

Nature of the influence of age/size

It has been suggested for some primate females (e.g. gelada baboons: Dunbar 1980b) that 'intrinsic aggressiveness' may be a function of age, peaking in young adulthood and declining thereafter, and that it may be a major determinant of rank within and between families. There are also suggestions, implicit or otherwise, that subordinates reverse rank with their dominant opponents at some 'critical' age such as adolescence (e.g. Walters 1980) when they may be most 'aggressive'.

These suggestions are complicated by the consideration that individuals may assess (in the sense of Parker 1974) their potential opponents. Individuals may be expected to be readily aggressive only to those against whom their chances of winning make fighting more profitable than retreat (Parker 1974). If relative age/size were a major determinant of fighting ability then individuals at their physical peak would be expected to initiate aggression with a higher proportion of their opponents than smaller individuals. This might well give the false impression that the former were 'intrinsically more aggressive', false because observed patterns of aggression were in fact due to assessment by individuals of their potential opponents.

In the present study, no difference in 'aggressiveness' was found between age-classes, using as a measure of aggressiveness the median and range of rates at which individuals threatened their LB subordinate age-mates (Datta 1981). Indeed observation showed that even very young individuals (0–1 years old) could be highly persistent and aggressive (chasing, grappling, nipping) in their encounters with older LB dominants close in age/size. The same individuals, however, yielded without resistance to LB dominants much older and bigger than themselves.

Nature of the influence of maternal rank

Immature monkeys behave as if they have a clear idea of whom, on the basis of maternal rank, they 'ought' to outrank in the social group. Apparently acting on this awareness, those immatures whose original rank is lower than their potential rank strive against LB dominants and eventually outrank them (as shown above).

Two main questions are considered here. First, How may individuals acquire 'awareness' of their potential rank? Second, How do younger, smaller, HB subordinates manage to reverse dominance with LB dominants in spite of resistance (see below) by the latter?

In discussing the first question, only the experiences of HB subordinates *before* any challenges are made to LB dominants will be considered. These suggest that HB subordinates could have used several sources of information about their 'potential' rank relative to others in their matriline. These are as follows:

1   Interactions of the mother with others in the matriline. The ranks of adult females on Cayo Santiago remain remarkably consistent and they dominate all female members of families below them in rank. Immatures probably have ample opportunity (being closely associated with their mothers) to observe the agonistic interactions of their mothers and the identity of those who win or lose against their mothers.

2   The aggressive behaviour of LB and HB dominants towards subordinates. In threatening immatures, LB dominants were significantly more likely to take into account the proximity of the immatures' mothers. Ten out of 12 immatures threatened by LB dominants were threatened at a higher rate (median % threats h^{-1}) when the mother was more than 2 m away than when she was less than 2 m away ($P = 0.038$, Sign test, two-tailed). HB dominants did not show this bias.

3   The direction and nature of support in disputes. HB subordinates were significantly more likely to be supported in dyadic disputes with LB dominants than were the latter. One-hundred-and-fifty-three interventions (focal plus ad lib sampling) were observed. In 63 out of 67 pairs with focal data and nine out of ten pairs with ad lib data only, HBs received greater support than LBs (who typically received none). ($P = 0.002$, Sign test, two-tailed [focal dyads]; $P = 0.022$, two-tailed [ad lib dyads].)

Depending on the consequences for HB subordinates, LB dominants and interferers, interference can be said to have certain functions or potential future effects. Could the kind of support HB subordinates received have encouraged them to rebel? Did LB dominants receive support which could have helped maintain their dominance, or not?

The consequence of interference in a given triad (a unique combination of three individuals—interferer, HB subordinate and LB dominant) was taken to be that sequela which exceeded in frequency any other sequelae. The proportion of triads in which interference had a given consequence was then determined. Of principal interest here are the proportions of support to HB subordinates and LB dominants which were effective, i.e. which terminated in submission by the target and its supporters (if any). This kind of support is expected to be potent in its effect on the stability of dominance between the two original contestants: subordinates receiving it may be encouraged to rebel, while dominants receiving it may be strengthened in their dominance.

In 91 out of 98 interference triads in which HB subordinates were supported, interference was ultimately effective in causing submission by the target and its supporters. However, in only two out of 11 triads in which LB dominants were supported did the target and its supporters submit. Supporters of both HB subordinates and LB dominants were likely (in over 80% of triads) to be dominant to their targets and to elicit submission from them. However, HB subordinate targets were: (1) more likely to solicit aid against interferers, possibly because they 'knew' they had powerful opponents who could defeat the interferer (see below); and (2) more likely to be aided against the interferer than were LB dominant targets. Moreover, all aids to HB subordinate targets were effective in causing the interferer to submit, while none to LB dominant targets were. This is because the supporters of HB subordinate targets were dominant to interferers (the supporters of LB dominants), while the supporters of LB dominant targets were subordinate to interferers (the supporters of HB subordinates).

To conclude, therefore, the supporters of HB subordinates, by the combination of primary and secondary interference, were largely effective in their interference and succeeded in reducing the effectiveness of support to LB dominants.

The second question of dyadic agonistic interactions once HB subordinates have begun to challenge LB dominants shall now be considered. Challenges by HB subordinates against LB dominants were seen in 76 out of 532 dyads with LB dominants during the study. In 45 dyads, challenges appeared in the latter two-thirds of the study, while in the rest they were present from the beginning. Interference before challenges began was seen in 30 out of the 45 dyads in which challenging developed during the study; in all, overall effective support was for the HB subordinate. While this is not 'proof' of a causal link between effective support and the initiation of challenging, it is certainly what would be expected if there were such a link.

When they initiated challenge HB subordinates were always younger and distinctly smaller than the LB dominants they threatened. Thus, they were probably in a position of considerable potential vulnerability. Dyadic and triadic (see below) interactions between HB subordinates and LB dominants during the period between first challenge and dominance reversal (if it occurred) were examined, to find out how HB subordinates fared during challenge and how they managed to reverse dominance with LB dominants.

HB subordinates gave the impression of actively seeking confrontation with LB dominants. Moreover, they actively solicited potential supporters (such as the mother) while doing so. Such solicitations were sometimes successful. A manipulative element also seemed to be present in the choice of occasions to threaten dominants. In 11 focal dyads, HB subordinates and LB dominants spent some time within 2 m of each other both within 2 m and more than 2 m away from the mother or other known supporter of the HB subordinate. In ten dyads the rate of threats initiated against LB dominants was greater near the mother or other protector than farther away ($P = 0.012$, Sign test, two-tailed).

LB dominants often did not accept threats from HB subordinates passively. In 53 out of 76 focal dyads where HBs challenged LB dominants, the latter returned some threats. Retaliation was often mild, possibly because of the frequent proximity (see above) during agonistic interactions of known supporters of the HB subordinate. However, it could on occasion be fierce.

Finally, in agonistic interactions involving interference by a third party, HB subordinates were, as before, significantly more likely to get overall support: they did so in all 45 focal dyads in which interference was seen during challenges. In fact, LB dominants received support in only two dyads. Once again (in 42 out of 45 triads) and for similar reasons, the majority of supports to HB subordinates were 'effective', as was not the case for the few supports to LB dominants.

Discussion

In the present study, relative maternal rank was an excellent predictor of the *direction* of changes in agonistic interaction between members of different families. Only those individuals dominated by members of lower-ranked families (relative family rank being defined as relative maternal rank) were likely to initiate challenges or to reverse rank. As a result, immatures from higher-ranked families came to dominate more and more of their opponents in lower-ranked families. However, while maternal rank predicted the *direction* of changes, it did not predict their timing, which was markedly affected by the relative age/size of opponents. Thus, although HB subordinates were always younger and smaller than their LB, dominant opponents at challenge and reversal, they did not begin challenge or reverse rank until the discrepancy in age/size between themselves and their opponents was reduced. Any attempt to explain how immatures (on Cayo Santiago at least) acquire rank must, therefore, describe the nature of the effects and interaction of maternal rank and age/size.

One way to visualize the process of rank acquisition is in terms of an 'assessment of relative power' between two antagonists. This view is justified if there is evidence to suggest that the dominant does not willingly give up status but does, on the contrary, resist outranking. On theoretical grounds alone (see Chapais & Schulman 1980; Datta 1981), rhesus females are expected to be unwilling to cede rank; indeed 'voluntary' submission probably does not occur except under certain rare circumstances determined by the structure of relatedness within societies (Datta 1981). The question has been neglected in observational studies of rank acquisition but the present study shows that LB dominants do indeed resist attempts at outranking, sometimes vigorously.

In these circumstances, i.e. an opponent unwilling to submit and prepared to resist outranking, the initiation of challenge and reversal would be expected to depend on the relative fighting capacities of challenger and challenged. Subordinates should refrain from challenge until their fighting capacity is sufficient to minimize the risk in tackling a larger opponent and dominants should avoid submitting until their fighting capacity is exceeded by that of the subordinate and resistance becomes more costly than submission. The fact that relative age/size is taken into account in agonistic interactions and that immatures therefore probably assess (see Parker 1974) their opponents before challenging them, supports this idea. However, assessment based on individual (intrinsic) fighting capacity cannot be the only factor determining challenge and reversal for two reasons. These are as follows:

1   HB subordinates are normally younger and distinctly smaller than LB dominants during challenge and, indeed, at reversal. Thus, they would be expected to be potentially more vulnerable during dyadic disputes and to be wary of dyadic challenge.

2   In pairs between LB subordinates and HB dominants, challenge to, and reversal with, dominants is almost absent, whether LB subordinates are smaller than, similar in age/size to or older and bigger than HB dominants.

As described elsewhere (Datta 1981), an individual's total fighting capacity (or power) can be regarded as some combination of its intrinsic and extrinsic fighting capacities, the latter being due to effective support from third individuals. Data which are consistent with some such combination and with the idea that individuals take relative power into account are as follows:

1   HB subordinates receive more, and more effective, aid against LB dominants than do LB dominants against them. Thus, the power of LB dominants is practically equivalent to their intrinsic fighting ability, while that of HB subordinates is augmented by supports and may well equal or exceed that of the larger LB dominants. This may explain why HB subordinates smaller than LB dominants are willing to challenge and are able to reverse dominance with the latter.

2   HB subordinates receive more, and more effective, aid against LB dominants than do LB subordinates against HB dominants. Hence, the power of an HB subordinate of given age/size is likely to exceed that of an LB subordinate of similar age/size. This may explain why the former are much more likely than the latter to challenge dominants of a given age/size.

3   Subordinates challenge only those dominants against whom they receive, or know they are likely to receive, effective aid. Only HB subordinates received regular and effective support against (LB) dominants, both before and during challenge. (Interventions *before* challenge may be crucial in informing an individual whom it is likely to receive support against.) Only these subordinates regularly challenged and reversed rank with dominants.

4   There is evidence to suggest that HB subordinates seek support against those they are challenging (an indication that they may need outside help to overcome dominants). While threatening, HB subordinates actively solicited support and maintained proximity to known supporters. Furthermore, their threats were highly vocal and may have attracted the attention of supporters to disputes. Finally, support appeared crucial when LB dominants resisted, particularly when they did so vigorously.

In Group J, therefore, support appeared to play an important role in the acquisition of rank by immatures. The influence of maternal rank appeared largely explicable in terms of it: individuals whose mothers outranked others were significantly more likely to receive effective support than the latter. Moreover, support was given throughout immaturity, both before and during challenge: its influence was arguably pervasive.

6.8 Relative Power and the Maintenance of Dominance
S A R O J   B .   D A T T A

Introduction

Among rhesus monkeys* on Cayo Santiago, reversals in dyadic dominance–subordinance are typically restricted to those dyads in which the subordinate belongs to a higher-ranked family (and so is higher born) than the dominant (see section 6.7). Through a combination of intrinsic power (related to age/size) and effective support from allies, high-born (HB) subordinates apparently outstrip their low-born (LB) dominant opponents in 'power' and so challenge the latter successfully.

Does the 'relative power' hypothesis also apply to dyads in which the direction of dominance is maintained? On Cayo Santiago, these are dyads in which the HB individual is also the dominant. If older than (or the same age as) its LB opponent, the HB is likely to have been dominant from the start of agonistic interaction; if younger, it probably became dominant after reversing a previous subordination (section 6.7). Obviously, the maintenance of dominance is not simply a matter of the bigger and stronger individual dominating the other.

Attempts to account for these cases, which appear contrary to common sense, have included the idea that defeated individuals develop 'psychological inferiority' or 'learn to lose' (cf. Ginsburg & Allee 1942) or that subordinates are inherently more nervous and fearful animals (e.g. Rowell 1974). It is notable that these suggestions invoke factors *intrinsic* to the individuals concerned. Indeed,

*Neither adult males nor close relatives (mothers, offspring or siblings) are considered here.

the possible role of extrinsic or extradyadic influences in the maintenance of dominance has usually either been ignored (e.g. Deag 1977; Popp & DeVore 1979; Rowell 1964, 1966) or given only passing consideration (e.g. Bramblett 1970; Seyfarth 1980; de Waal 1977). However, data from Cayo Santiago and elsewhere suggest that such influences may often be critically important.

Data on interventions (by third individuals) in disputes between HB dominants and LB subordinates were available for 176 dyads, involving 42 different individuals. All were members of the 031 matriline of Group J on Cayo Santiago and consisted of adult females and their pre-adult offspring of both sexes.

The data were obtained by a combination of focal and ad lib sampling. In a given dyad, the overall direction of support (i.e. for HB dominant or LB subordinate) was determined by which individual received most supports.

Dyadic interactions

Previous studies have somewhat simplistically assumed that how dominants and subordinates behave towards one another gives direct information about the way in which dominance–subordinance is maintained. However, this approach often confounds *how* an individual behaves with *why* it behaves, or continues to behave, as it does, e.g. by suggesting that dominants are dominant because they are more aggressive individuals and subordinates subordinate because they are intrinsically more fearful individuals. Thus, Rowell (1966) observing that subordinate baboons readily submit to dominants (who had often made no obvious threat), concluded that the status quo was maintained because subordinates had such a strong tendency to submit, not because they were forcibly kept in place. However, as Deag (1974, 1977) pointed out, submissive behaviour might be a learned response to aggression (both past and present) from dominants.

Cause and consequence are of course often difficult, if not impossible, to separate in social behaviour. When dominant A threatens subordinate B because the latter has approached A's infant, is she simply exercising the power she already has over B to prevent B approaching the infant, is she reinforcing her own dominance, or both? Yet another difficulty lies in having too restricted a view: in a complex social network, such as that of non-human primates, attempts to understand the dynamics of dyadic relationships without reference to extradyadic influences seem doomed to distortion. Such influences appear to be important in the maintenance of dominance–subordinance (see below; Datta 1981). Hence, only a brief consideration of dyadic agonistic behaviour and proposed dyadic mechanisms of maintenance will be made in this section.

In Group J, the direction of submission in dyads between HB dominants and LB subordinates remained very consistent during the study (Datta 1981). HB dominants submitted (temporarily) in only two out of the 362 dyads for which data were available, while subordinates normally submitted readily, often without obvious aggression from dominants.

LB subordinates appeared, therefore, to 'accept' subordinance. An examination of the direction and nature of threats shifted this impression only a little. As in previous studies (e.g. Deag 1974; Rowell 1966), the direction of threats was less consistent than that of submissions because subordinates sometimes (in 79 out of 320 dyads) threatened dominants. However, unlike dominant threats, most subordinate threats (108 out of 127 observed during focal sampling) were scream-threats, which appear motivated by fear as well as aggression (e.g. Datta 1981; de Waal *et al.* 1976). About half (60 out of 127) were made in defence of close relatives being threatened by dominants and hence could not be considered direct challenges to dominants. The remainder, made in retaliation to direct aggression from dominants, could, however, be considered so. They were made by: (1) females recently outranked by younger HB females (37 threats), although such females in time 'accepted' their new status (Datta 1981); (2) adult females in sexual consort with, and sometimes actively supported by, adult males (20 threats) although the same females were never seen to challenge HB dominants outside the mating season; and (3) non-adult females who had a consistent supporter (usually the mother although in one case an adult male) dominant to the HB individual they threatened (10 threats). Like the evidence of

Table 6.8. Relation between rank in adult female hierarchy and rates of absolute aggression given to and received from all adult female non-close relatives. Non-close relatives are all relatives except the mother, offspring and sibs. Adult females are ranged from top to bottom in order of rank. The absolute rate of aggression is the number of threats h^{-1} of focal observation.

Focal female	Threats given h^{-1} (G)	Threats received h^{-1} (R)	$\dfrac{G \times 200}{G+R}$
320	0.392	0.016	96.04
420	0.297	0.014	95.5
7H	0.292	0.021	93.2
9H	0.246	0.111	68.9
XJ	0.104**+	0.046**−	69.1**+
417	0.098	0.145	40.3
375	0.122	0.120	50.5
Z1	0.078	0.159	32.9
416	0.037	0.229	13.4

**Spearman $r_s = 0.95$; $P < 0.02$, two tailed
−Negative correlation between rank and measure of aggression.
+Positive correlation between rank and measure of aggression.

resistance to outranking in section 6.7, (1) suggests that LB subordinates were reluctant to submit, while (2) and (3) suggest that powerful support made subordinates more likely to rebel.

In agreement with some previous studies (Bernstein 1970; Bernstein & Sharpe 1966; Biernoff *et al.* 1964) but not others (e.g. Seyfarth 1976) there was a significant positive correlation between the rank of an individual in its age group (for simplicity only adult females are considered here) and the rate at which it threatened group members (Table 6.8). This result, however, need not imply that individuals were high ranked and able to maintain high rank *because* they were intrinsically more 'aggressive': the more individuals a female can boss the higher the overall rate at which she threatens can be expected to be. The crucial point is that if high rank depends to a significant extent on non-dyadic factors such as alliances (see below), then rates of dyadic aggression may largely reflect status rather than the means by which such status was achieved or is maintained. In fact, there was no evidence for a positive correlation between rank and the rate at which adult females threatened LB subordinates of adjacent rank (Datta 1981).

Direction and consequences of interference

Data on the direction of interference showed that during disputes between the two, HB dominants were not more likely to be aided than LB subordinates. Out of 176 dyads, HB dominants received more supports than LB subordinates in 78, the reverse was true in 84 and 14 were neutral ($P = 0.76$, Binomial test, two-tailed).

This seems contrary to prediction: if support is important in maintaining dominance, HB dominants might be expected to receive significantly more. However, it turns out that the kinds of support received by the two types of antagonist are not equivalent, in that they have different consequences (Table 6.9). HB dominants are significantly more likely to receive support which elicits submission from their (LB subordinate) opponents. This kind of 'effective' support was more often given to HB dominants than LB subordinates in 159 out of 170 dyads in which the two antagonists received different amounts of effective support ($P < 0.001$, Binomial test, two-tailed). (LB subordinates typically received *no* effective support.)

Another major difference lay in the success with which the original dispute was ended by interference. Since those who supported the HB dominant in effect supported the aggressor, it is not surprising that the original dispute often continued (Table 6.9). However, in supporting LB subordinates, interferers supported victims and generally managed to divert or halt the original aggression, often because they themselves became the objects of aggression by the HB dominant or its supporters (Table 6.9; see also Kaplan 1977). Although they

Table 6.9. Consequences of support to high-born dominants and low-born subordinates. An interference triad is a unique combination of interferer (I) dominant and subordinate.

High-born dominant supported			Low-born subordinate supported	
% triads	No. triads	Interferer (I)	No. triads	% triads
86.3	88	1. I dominant to target	61	50.8
13.7	14	2. I subordinate to target	59	49.2
	14	A. I subordinate to target	59	
	0	a. Target submits to I	0	
14.3	2	b. Target ignores threat by I	12	20.3
	0	c. Target retaliates against I	47	79.7
—	—	d. Retaliation successful	45/47	95.7
7.1	1	e. I supported against target	2	3.4
100	1/1	f. Target yields to supporter of I	0/2	0
42.8	6	g. Original dispute halted	45	76.3
	88	B. I dominant to target	61	
87.5	77	a. Target submits to I	27	44.3
11.4	10	b. Target ignores threat by I	16	26.2
1.1	1	c. Target retaliates against I	3	4.9
0	0	d. Target solicits aid against I	15	24.6
15.9	14	e. Target supported against I	39	64
0	0/14	f. I yields to supporter of target	38/39	97.4
47.7	42	g. Original dispute halted	58	95.1
		C. Effectiveness of aid		
87.2	89/102	Target submits to I or supporters of I	11/120	9.2

therefore interfered at considerable risk to themselves, their intervention may have had the useful effect of preventing possible injury to the (usually related) LB subordinates they supported.

Function interference

The suggestion (above) that support to LB subordinates may have protected them from potential injury receives added force from the finding that the younger and smaller the LB subordinate relative to its HB dominant opponent (and hence presumably the more vulnerable), the more likely it is to receive support (Table 6.10).

What about support to HB dominants, however? Dominants do not appear to

Table 6.10. Effect of the relative age of opponents on the support given to low-born subordinates: percentage of dyads with interference in which low-born subordinates were supported at all. The figures in brackets are the number of dyads with intervention. Dyads in different age-class combinations were not independent so no statistical test was used. The line connects same age-class combinations.

High-born dominant age-class (years)	Low-born subordinate age-class (years)					
	0–1	1–2	2–3	3–4	4–5	>5
0–1	50 (6)	62.5 (8)	0 (1)			
1–2	100 (5)	52.9 (17)	60 (5)	0 (1)		
2–3	50 (4)	88.3 (12)	60 (10)	50 (4)	9 (2)	
3–4	100 (6)	77.7 (9)	62.5 (8)	0 (1)	0 (2)	20 (5)
4–5	100 (1)	100 (4)	100 (2)	67 (3)	0 (1)	33.3 (3)
>5	100 (9)	93.3 (15)	62.5 (8)	20 (5)	40 (5)	33.3 (9)

Summary

Age of subordinate relative to dominant	No. dyads with intervention	No. and % dyads in which subordinate is supported at all
Younger	96	75 (78.1%)
Same age	44	21 (47.7%)
Older	31	12 (38.7%)

'need' support, which seems like 'token' aid to an individual already winning a dispute. The nature of the support (it is 'effective') is certainly compatible with the idea that it reinforces the prevailing direction of dominance. However, it could be argued that interferers intervene mainly for purposes of their own: (1) to gain closer association with dominants, perhaps leading to improved access to resources (e.g. Cheney 1977; Seyfarth 1977, 1980); (2) if subordinate to the target, to gain greater safety in challenging it; and (3) if dominant to the target, to gain safety in reinforcing personal dyadic dominance over an individual clearly being beaten. If this were so, the pattern of support would not be expected to conform to any pattern of 'need' dominants might have for support. If a need exists it might be expected to vary according to the relative vulnerability of the dominant in a dispute. This follows from the idea that in dyadic disputes the relative intrinsic vulnerability (or relative intrinsic fighting ability) of an indiviual depends on some factor such as age/size. Dominants would thus be expected to be most vulnerable against opponents larger/older than themselves and least vulnerable against those smaller/younger than themselves.

Table 6.11 shows that the pattern of support to dominants fits these expectations: they were most likely to receive support when younger and smaller than subordinates, less likely to receive it when the same age/size as subordinates and least likely to receive it when older and larger than subordinates. Although support to dominants may well have been motivated by a variety of factors (Datta 1981), therefore, the postulated 'need' seems to have had a major effect on the pattern of support to dominants in Group J.

Table 6.11. Effect of the relative age of opponents on the support given to high-born dominants: percentage of dyads with interference in which high-born dominants were supported at all. The figures in brackets are the number of dyads with intervention. Dyads in different age-class combinations were not independent and so no statistical test was used. The line connects same age-class combinations.

High-born dominant age-class (years)	Low-born subordinate age-class (years)					
	0–1	1–2	2–3	3–4	4–5	>5
0–1	50 (6)	50 (8)	100 (1)			
1–2	60 (5)	64.7 (17)	80 (5)	100 (1)		
2–3	50 (4)	25 (12)	60 (10)	100 (4)	100 (2)	
3–4	16.7 (6)	44.4 (9)	75 (8)	100 (1)	100 (2)	100 (5)
4–5	0 (1)	0 (4)	50 (2)	33.3 (3)	100 (1)	67 (3)
>5	11.1 (9)	40 (15)	75 (8)	80 (5)	60 (5)	100 (9)

Summary

Age of dominant relative to subordinate	No. dyads with intervention	No. and % dyads in which dominant is supported at all
Younger	31	25 (80.6%)
Same age	44	31 (70.4%)
Older	96	41 (42.7%)

Discussion

Section 6.7 showed that dominance reversal between members of different families could be most adequately explained on the basis of a real asymmetry in relative power between antagonists. In considering how dominance might be maintained, two main approaches seem possible, one acknowledging the importance of relative power, the other not. The second will be considered first. Perhaps the backbone of this approach is the idea that one defeat makes another more likely because the loser psychologically 'accepts' its inferiority (cf. Ginsberg & Allee 1942; Rowell 1974). It assumes that subordinates do not necessarily continue to assess (in the sense of Parker 1974) or respond to the 'real' power of their dominant opponents and apparently explains cases where subordinates appear intrinsically capable of defeating dominants yet continue subordinate to them. There are at least two problems with this approach. First, it seems non-adaptive to submit when you could win and at least some selection against such behaviour would be expected (see also Datta 1981). Second, there is evidence to suggest that subordinates do assess relative power and behave accordingly. Thus, in group J, subordinates were readier to challenge those dominants against whom their power was increased by effective aid. More striking, there are many instances in the literature of subordinates in well-established groups reversing rank with dominants when the latter lost power due to injury or physical infirmity resulting from age or illness (Missakian 1972, 1976), loss of allies (Bernstein 1969; Marsden 1968; Smuts 1982; Varley & Symmes 1966) or when subordinates gained powerful allies (Berstein 1969; Chance *et al.* 1977; Hasegawa & Hiraiwa 1980; Koyama 1970; Marsden 1968).

The other approach acknowledges the importance of relative power and the possibility that it may be continually assessed but may or may not ignore the contribution of extrinsic power due to alliances. Where alliances are ignored it may be argued that subordinates are (or become, as a consequence of defeat and its attendant deprivations and debilities) *intrinsically* less powerful than dominants. It is conceivable that dominants are intrinsically more powerful in pairs where they are larger than subordinates but this cannot apply to pairs where they are similar in age/size or where subordinates are larger. On Cayo Santiago there was not any obvious concentration of debilities among low-ranked individuals. On the other hand there were numerous cases of individuals affected by injury, sometimes extreme, retaining their status relative to others (Datta 1981).

The other alternative approach is that dominants have access to a source of extrinsic power which not only buffers them against diminution of intrinsic power but may in many, if not most, cases be essential to the maintenance of dominance.

There is widespread evidence that third individuals intervene in disputes between individuals who have stable dyadic dominance relations. Such interventions have been described in vervets (Lee pers. comm.; Seyfarth 1980; Struhsaker 1967b), savannah baboons (Cheney 1977; Walters 1980), crab-eating macaques (de Waal 1977, 1978), rhesus (Kaplan 1977, 1978), pigtail macaques (Massey 1977) and others. That no serious attempt has been made to see if the pattern of such supports is compatible with the maintenance of dominance—subordinance may be largely due to the fact that dominants, in their dyadic behaviour, appear powerful enough not to 'need' support to sustain their positions. Furthermore, interventions are comparatively uncommon, a fact which has sometimes led to the assumption that they are relatively uninfluential (e.g. Lee & Oliver 1979). As argued elsewhere (Datta 1981), however, there is no logical justification for such a view.

Some authors (e.g. Cheney 1977; Seyfarth 1980) have seen supports to dominants as primarily 'selfish' acts which lead to closer relationships with dominants and hence privileged access to resources, or which enhance the dominance position of the interferer, rather than making an important contribution to the maintenance of the dominant's position.

The first two of these effects may well exist (Datta 1981) but in Group J at least, the pattern of supports to dominants suggested that the third was of major importance. Dominants received effective and powerful support against subordinates while the converse was not the case. Moreover, support to dominants seemed commensurate with a 'need' for it: those dominants who (on the basis of relative age/size) were expected to be most vulnerable to subordinates were also most likely to receive support. If interferers intervened primarily for their own ends this would not be expected—indeed the opposite would be.

Finally, two important points about the fundamental necessity for, and mode of action of, alliances must be made. First, it may now be conceded that dominants smaller than or the same age/size as subordinates need effective alliances to match or exceed the power of the latter and hence maintain dominance over them. However, is this also the case for dominants larger/older than subordinates? The answer to this question (which falls into the trap of a dyadic perspective) is likely to be yes: dominants need to consider not only their individual subordinate opponent but the potential—or actual—allies of such opponents (Datta 1981). If subordinates do have allies, dominants may only be able to maintain dominance by having even more or more powerful allies themselves. Even a dominant larger/older than a subordinate may, therefore, 'need' alliances to maintain dominance.

Here it may be countered that evidence from the present study itself shows that the allies of subordinates are 'ineffective'—they do not cause submission in their targets or the supporters of their targets but are, in fact, likely to be defeated by the latter. Why then should they be taken seriously? Related to this 'problem' is an apparent circularity of effect: effective support comes from interferers dominant/HB to those they intervene against; dominance/family rank is maintained by effective support.

Both the question and the circularity spring, once again, from a dyadic perspective on dominance–subordination. The confusion disappears if it is realized that individuals belonging to subgroups within social groups may, by consistently supporting fellow members against members of other subgroups, maintain a collective superiority over rival subgroups. Each member of a subgroup, therefore, has the potential power of that subgroup.

That the allies of LB subordinates are themselves generally subordinate to opponents of the latter is simply due to the prevailing power relations between subgroups. If the relative power of these subgroups should change in favour of the subordinate subgroup, e.g. by loss of members from the dominant subgroup, then widespread rebellion by subordinates would be expected. There is evidence that such rebellions have occurred (e.g. Burton 1980; Gouzoules 1980; Koyama 1970).

The *composition* of these subgroups is another issue, considered elsewhere (Datta 1981, in prep.).

6.9 Dynamics of 'Special Relationships' between Adult Male and Female Olive Baboons
BARBARA B. SMUTS

Introduction

Most studies of male–female interactions in non-human primates have focused on sexual behaviour. Several recent studies, however, have shown that in savannah baboons and macaques, adult males and females may form long-term friendly bonds that persist in the absence of any immediate sexual relationship (Altmann 1980; Chapais 1981, section 10.4; Ransom & Ransom 1971; Ransom & Rowell 1972; Rasmussen 1980, section 6.10; Seyfarth 1978a,b; Strum 1975). Some of the findings of a detailed study (Smuts 1982) of male–female affiliative relationships in a large troop of savannah olive baboons are described in this section.

In August 1977 the author began an 18-month study of Eburru Cliffs, at Gilgil, Kenya. At that point there were 115 troop members, including the 34 adult females, 14 adult males and four subadult males who were the subjects of this study (the subadult males weighed about 80% of adult weight). The results described here are based on 1000 hours of focal samples on all adult females while they were pregnant or lactating and *not* undergoing sexual cycles, supplemented by *ad libitum* observations. All references to females below refer to non-cycling adult females and all references to males refer to adult and subadult males only.

Do 'special' relationships exist?

Observations suggested that each adult female baboon had particularly strong bonds with a small subset of the adult/subadult males in the troop. To test this hypothesis the distribution of two measures that have been used repeatedly as indicators of affiliation between individual non-human primates were examined: frequency of grooming and proximity (e.g. Hinde 1977a; Seyfarth 1978 a,b).

Grooming bouts between females and males were recorded on an ad lib basis. For each of the 34 females, the proportion of all independent grooming bouts with males that involved each of the 18 possible male partners was calculated. (The direction of grooming was ignored in this analysis: most bouts involved females grooming males.) Every female but two (who were very rarely seen grooming with males) showed a statistically significant preference for one male grooming partner (the favourite or F male), and half of the females also showed a significant preference for a second male partner (the S male) ($P < 0.01$ in all cases). Under the null hypothesis, it was assumed that for each grooming episode or 'trial' the probability that a female would groom with male 'm' was equal to the proportion of all grooming episodes between males and lactating or pregnant females that involved male 'm'. Using the binomial expansion, the probability that female 'f' would groom with male 'm' x times or more over n trials, where x = the observed frequency with which female 'f' groomed with male 'm' and $n =$ the total number of grooming episodes of female 'f' with all males was then determined. Different females tended to have different male grooming partners. All females combined performed 65% of all grooming with males with their individual F males and 86% of all grooming with their F and S males combined. In other words, the vast majority of a female's grooming with males was restricted to one or two out of 18 possible partners.

At five-minute intervals during 30-minute focal-female samples the identities of all males found within one of four distance categories from the female were recorded: 0–1 m, 1–2 m, 2–5 m and 5–15 m. A composite proximity score for each male–female dyad was then calculated by combining the *weighted* proportions of time the male spent within each distance category (the shorter the distance for the category, the more heavily the male score was weighted in order to avoid swamping of data from short-distance categories by the inevitably much larger values from the longer-distance categories) (see Smuts 1982, p. 54 for details). The results for four typical females are shown in Fig. 6.7. For nearly all females most of the 18 males had extremely low proximity scores but one, two or occasionally three had much higher scores. The males with higher scores were usually the same males the female groomed with most often: for 93% of the females the F grooming partner ranked either one or two in proximity. Since proximity scores were based on data collected mainly when females were foraging (approximately 80% of their time) and grooming with males accounted for less than 2% of a female's time, the high proximity scores were not simply the result of a high frequency of grooming between members of the dyad.

The striking similarities between the grooming scores and proximity scores

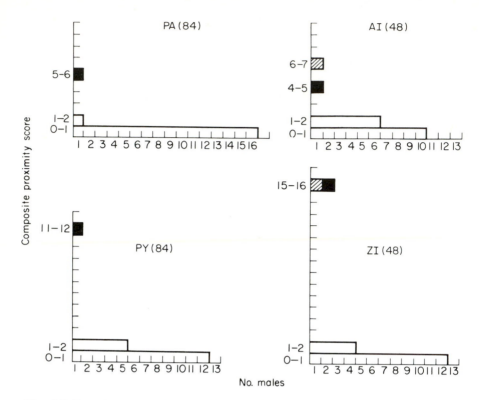

Fig. 6.7. Distribution of male composite proximity scores for four females. Each graph shows for a particular female (initials shown) the number of males whose composite proximity score fell within a given interval. In each, the first and second most favourite, male grooming partners of the female are indicated with a solid bar and a cross-hatched bar respectively (females PA and PY did not have a second male grooming partner). The number of one-half-hour focal female samples from which the composite proximity scores were derived is indicated in parentheses next to the females' initials.

allows for most females, a particular male or males who scored either first or second on *both* measures to be distinguished. These males are called 'friends'. Other males, who were judged subjectively to share a strong bond with a particular female, were not identified by these criteria. These males, however, always scored at least third on one measure and often on both, and were grouped into a second category—'associates'. In none of the measures described below were the results for friends and associates significantly different and so they were lumped together under the term 'special (S) males'. In what follows, 'special dyads' refers to any female–male pair involving a special male (12% of all possible female–male pairs), and 'non-special (NS) dyads' refers to all other pairs.

Dynamics of special relationships: a comparison of special and non-special dyads

Responsibility for maintaining proximity between males and females was determined using Hinde's index: $\%Ap_f - \%L_f$, where Ap_f = approaches by the female and L_f = leaves by the female (see p. 67). Since the frequency of approaches and leaves involving NS males was small, data for all NS males were combined for each female.

The index was calculated first for movements into and out of the longer-distance categories 2, 3 and 4 (1–15 m). For 85% of the S dyads, the index was positive, indicating that the female was primarily responsible for maintaining proximity ($P < 0.001$, Sign test). The reverse was true for females and NS males: for 79% of the females the males were more responsible for maintaining proximity ($P < 0.01$, Sign test). These results suggest that females preferred to remain within 1–15 m of males with whom they had S relationships but preferred to be at distances greater than 15 m from all other males.

Second, in close proximity (0–1 m), in the majority of S dyads (64%) and in all NS comparisons, the *male* was primarily responsible for maintaining proximity. However, for 80% of the 15 females for whom adequate data were available, the female was relatively more responsible for maintaining proximity to the S male than to the NS male ($P < 0.05$, Sign test). Thus, in terms of their movements females showed a consistent preference for S males relative to NS males.

All females spent more time in close proximity to their S males than to NS males. This was partly because they approached and were approached by S males more often but it also reflected the fact that once an approach had occurred, members of S dyads remained in proximity for more than a minute significantly more often (22%) than did members of NS dyads (12%). The longer duration of close proximity reflected two further differences between S and NS dyads: (1) females were more than twice as likely to avoid (move away immediately) the approach of an NS male than that of an S male (50% and 22% of the time, respectively); and (2) when the animal who left was the same as the animal who approached, he or she moved away within five seconds of the approach significantly more often in NS dyads than in S dyads.

These results suggest that baboons may approach S and NS individuals for different reasons. While it is not possible to determine the motivation of an approaching animal, a closer look at the types of interactions that occurred following a close approach sheds further light on the nature of the relationships.

Five types of interactions occurred at a significantly higher frequency when an S male was involved: feeding together, travelling together, grooming, presenting for grooming and interactions between the male and the female's infant. Three types of interactions occurred at a significantly higher frequency when NS males were involved: the male supplanting the female, the female presenting her perineum to the male and the female exhibiting submissive gestures to the male. Males also inspected the perineums and mounted NS females more often; it is significant that these females were not in oestrus. These results indicate that members of S dyads were more likely to engage in either routine activities or relaxed friendly interactions, while members of NS dyads were more likely to engage in agonistic interactions or tense appeasement interactions.

NS and S males also differed in the roles they played when the female or her offspring were attacked by another troop member, in their relationships with the female's infant and in their sexual relationships with the female. These differences, which suggest possible benefits to the male and female of S relationships, are discussed later in this volume (section 12.3) and elsewhere (Smuts 1982, 1983).

6.10 Influence of Affiliative Preferences upon the Behaviour of Male and Female Baboons during Sexual Consortships

KATHLYN L. R. RASMUSSEN

Introduction

In recent years, a number of field studies have investigated factors influencing the mating patterns of savannah baboons. Many of these studies have been concerned with selective pressures which have shaped the development of the baboon mating system and have, therefore, examined the importance of such variables as dominance rank and the general preferences of each sex. For instance a prominent finding has been that high-ranking males are reproductively more successful than subordinate males (Hausfater 1975; Packer 1979a). Females prefer to mate with dominant males (especially if the males have recently transferred from another troop) and show higher frequencies of grooming and presenting to such males during sexual consortships (Packer 1979b; Rasmussen 1980). Male baboons compete for parous females who are highly proceptive (i.e. females who are active in seeking male sexual attention) and they behave more 'possessively' towards those females while in consort (Rasmussen 1980). Thus, the general characteristics and sexual preferences of male and female baboons appear to influence male reproductive success, the level of male–male competition for female consort partners and the frequency of some types of social behaviour observed in consorting pairs.

On another level, the analysis of dyadic relationships may provide further insights into the quality and patterning of male–female interactions in the context of mating. Rasmussen (1980) studied the individual preferences of 13 adult males and 20 adult females in a troop of yellow baboons in Mikumi National Park, Tanzania. The frequency with which subjects approached each member of the opposite sex was used as a measure of affiliative preference in a non-sexual context; the level of affiliative preference between male–female dyads was then related to the behaviour of those dyads during sexual consortships. In the following, the spatial relationships and frequency of socio-sexual behaviours of male and female consort partners have been analysed according to whether each member of the pair expressed a high preference, an intermediate preference or no preference for his or her partner when they were not in consort. The five females that each male approached most frequently were defined as being his high-preference females, females he approached less frequently were his low-preference females, and females he did not approach were his no-preference females. Similarly, the three males that a female approached the most were defined as high-preference males and the remaining males were either low- or no-preference males.

Spatial relationships between consort partners

Consort partners in Mikumi typically fed, travelled and rested in close proximity to each other. As this spatial coordination was primarily effected by the male following his female partner, the average distance apart of consort partners was almost certainly a function of male rather than female behaviour. Figure 6.8a shows the relationship between the affiliative preference of males for females and

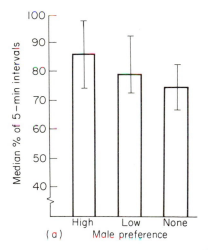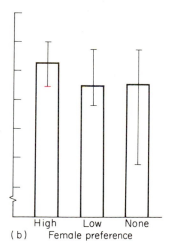

Fig. 6.8. Percentage of five-minute interval samples where consort partners were less than 5 m apart, according to the level of preference shown by the male (a) and the female (b) for their partners when they were not in consort. The medians and interquartile ranges are shown.

their distance from those females during consortships. Males maintained significantly closer proximity to highly preferred females than to non-preferred females ($P < 0.05$, Wilcoxon test). (Note that in this and the following, tests between different preference categories were not always possible because not all subjects had consortships in each category or because ties reduced the number of matched pairs to less than five.) Since females were also closer to their highly preferred male consorts than to less-preferred partners (Fig. 6.8b), it is likely that pairs who shared a high mutual preference maintained closer proximity than other pairs.

Social interactions between consort partners

Males herded female consorts for whom they showed a strong preference significantly more than they herded female consorts for whom they did not ($P < 0.02$, Wilcoxon test). Male herding did not appear to reflect any lack of cooperativeness on the part of the female, since females who strongly preferred their male consorts were herded more frequently than females who showed no such preference. Indeed, 57% of all herding occurred in pairs with a high mutual preference, even though such pairs comprised only 19% of all consort dyads.

The frequency with which males groomed female consorts was not significantly related to their preference for those females while not in consort, although highly preferred females were groomed somewhat more than other females (Fig.

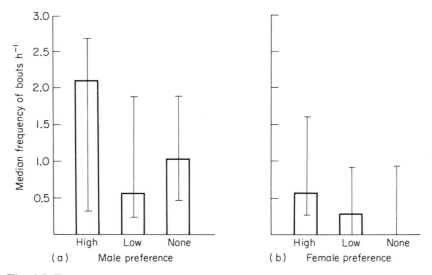

Fig. 6.9. Frequency with which males groomed their female consorts (a) and females groomed their male consorts (b), according to the level of preference shown for those partners outside of consortships. The medians and interquartile ranges are shown.

6.9a). There was also a tendency for males to groom female consorts who showed a strong preference for them more than they groomed females who preferred them less or not at all. Females groomed male consorts they highly preferred significantly more than they groomed males for whom they showed no preference ($P < 0.02$, Wilcoxon test) (Fig. 6.9b). Similarly, males who showed a high degree of preference for their female consorts were groomed by those females significantly more than were males who showed only a weak preference for those females ($P < 0.05$, Wilcoxon test). High levels of grooming by consort partners thus appeared to be a reflection of a high degree of mutual preference. There was no indication that males groomed females more who might have been expected to be less cooperative (i.e. females who showed no preference for them).

Females presented significantly more to highly preferred male consorts than they did to males for whom they showed a low degree of preference ($P < 0.05$, Wilcoxon test). There was no apparent relationship between the frequency of female presenting and the level of male preference for the females.

Sexual behaviour in consort pairs

The frequency with which males voluntarily inspected the perineum of their female consorts (i.e. inspections which were unsolicited by female presentations) was not related to the degree of either male or female affiliative preference.

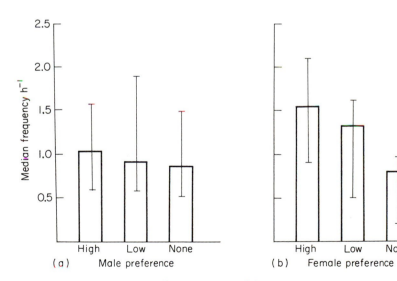

Fig. 6.10. Frequency with which males mounted their female consorts, according to the level of preference shown by the male (a) and by the female (b) for their partners when they were not in consort. The medians and interquartile ranges are shown.

Males did not mount highly preferred, female consort partners more frequently than they mounted less preferred partners (Fig. 6.10a). However, the effect of female preference for males upon mounting behaviour was quite clear: males mounted females who strongly preferred them significantly more than they mounted females who showed only a weak preference for them ($P < 0.01$, Wilcoxon test) (Fig. 6.10b) and there was a marked tendency for them to mount females with a weak preference for them more than females with no preference for them (Fig. 6.10b). The level of female receptivity thus appeared to be closely related to the level of female preference for individual males.

Summary and conclusions

The influence of individual affiliative preferences in Mikumi baboons was apparent in a number of behaviours observed during their sexual consortships. Males who showed a strong preference for their female consorts stayed significantly closer to those females, herded them significantly more frequently and tended to groom them more often than females they preferred less. Reciprocally, females in consort with highly preferred male partners groomed and presented significantly more to those males and stimulated a significantly higher rate of mounting by those males than by less-preferred males.

The relative roles of Mikumi males and females in maintaining the consort bond was more clearly outlined in the analysis of dyadic preferences than in that of general characteristics or preferences. Although competition between males for preferred females may to some extent have determined which dyads were able to form consortships (Rasmussen 1980), it was evident that female preferences influenced the behaviour of consort pairs more than has been assumed: females who strongly preferred their male consorts actively facilitated both social and sexual interactions with their partners. Indeed, preferences of females for specific males appeared to be more important in determining levels of female receptivity and hence rates of male mounting than were general characteristics of the female, e.g. her overall level of proceptivity or attractiveness to males, or of the male, e.g. his agonistic rank or attractiveness to females.

The quality of certain interactions was likewise apparent from the dyadic analysis of consort behaviours. Male herding, for example, was probably not the 'coercive' behaviour it appeared, since Mikumi males herded females who strongly preferred them (and who were therefore likely to have been the most cooperative) far more frequently than they herded females who did not prefer them. Similarly, there was no indication that males used grooming to induce their female partners to cooperate. The data on spatial proximity, social and sexual behaviour all suggest that the greatest degree of coordination observed between consort partners was attained by dyads who expressed a strong mutual preference in a non-sexual context.

Chapter 7
Effects of Interactions and Relationships upon the Individual

7.1 Social Experience and Individual Differences

ROBERT A. HINDE

As discussed in Chapter 1, not only do the characteristics of an individual affect the interactions and relationships in which he is involved but he or she is in turn changed by them. Most of the experimental work on this topic has involved the rearing of animals in situations artificially structured so that access to social companions of various types was controlled. Harlow and his collaborators reared rhesus monkey infants in bare wire cages, with inanimate surrogate mothers, with their natural mothers, with their mothers and controlled access to peers, with peers only and in more or less natural groups. Not surprisingly, the conditions of rearing have profound effects on the behavioural characteristics of the developing animals (Suomi *et al.* 1973; Suomi & Harlow 1978). For example, rhesus monkeys reared without mothers or peers have markedly aberrant social behaviour and can be mated only with difficulty. If they become pregnant, they are likely to be abusive to their own infants. Rhesus infants reared with male and female parents and with controlled access to peers develop more sophisticated patterns of social behaviour than those shown by animals reared under more socially impoverished conditions and they maintain high levels of interactive play for longer than feral-reared monkeys (Ruppenthal *et al.* 1974).

Close examination of the behaviour of monkeys brought up under different conditions can also reveal how a complex social environment permits the

121

development of sophisticated social skills, such as the ability to take part in triadic interactions (Anderson & Mason 1974; Capitanio 1982).

Of course, such data do not mean that the effects of an early impoverished environment are totally irreversible; even monkeys reared in social isolation can be 'treated' by confining them with younger, socially reared peers (Cummins & Suomi 1976; Suomi & Harlow 1978) and abusive, motherless mothers improve with social experience (Ruppenthal *et al.* 1976). The depression shown by young rhesus monkeys temporarily separated from their mothers usually lifts after reunion, especially if the mother–infant relationship is a warm one (see sections 6.2 and 6.4), though in some cases the effects persist (Hinde *et al.* 1978; Spencer-Booth & Hinde 1971).

Evidence about the more subtle effects of naturally occurring differences in relationships on the development of individual characteristics inevitably comes mostly from observational data, so the conclusions that can be drawn are regrettably limited. One study going somewhat beyond this is described in section 7.2.

As yet little is known about the mechanisms by which social relationships affect the behavioural characteristics of individuals. One route involves the endocrine system (e.g. Bernstein *et al.* 1974; Gordon *et al.* 1979; Rose *et al.* 1975; Sassenrath 1970) or, in the longer term, the age of puberty (Epple 1981). Discussion of such mechanisms is beyond the scope of this book.

7.2 Individual Characteristics of Mothers and their Infants
JOAN STEVENSON-HINDE

The close mother–infant relationships of rhesus monkeys have been discussed in Chapters 5 and 6. It has also been shown that ratings of the monkeys can provide quantitative assessments of individual characteristics in terms of scores on the three dimensions: Confident to Fearful, Excitable to Slow and Sociable to Solitary (Chapter 4). These two sets of data provide a means for understanding how dyadic relationships affect and are affected by the characteristics of the participants—what Hinde (1979a) has called the 'dialectic' between personality and relationships.

Looking just at the personality side of this dialectic, mothers' characteristics were correlated with those of their infants, when the infants were just over a year old (58–85 weeks). However, the correlations between mothers' and daughters' scores were not the same as those between mothers' and sons' scores (Fig. 7.1). In particular, 11 daughters tended to be rank ordered in a similar way to their mothers, so that mothers with high Confident scores had daughters with high Confident scores ($r_s = 0.79$), Excitable mothers had Excitable daughters (0.66),

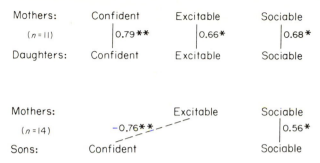

Fig. 7.1. Spearman correlation coefficients between mothers' and infants' Confident, Excitable and Sociable scores that reached $P < 0.05$, two-tailed. All remaining coefficients had an associated P value greater than 0.10. $* = 0.05$; $** = P < 0.01$.

and Sociable mothers had Sociable daughters (0.68). Yet for 14 mothers and sons there was essentially no correlation between either their Confident scores (0.07) or their Excitable scores (-0.29). However, Sociable mothers did have Sociable sons (0.56), and Excitable mothers produced sons with relatively low Confident scores (-0.76). In addition, four mothers who had both a son and a daughter produced correlations similar to those in Fig. 7.1, with one pattern for their daughters and another for their sons (Stevenson-Hinde & Simpson 1981).

Some understanding of how these correlations arose was gained by asking how mothers' characteristics correlated with what they did with their sons and daughters over their infants' first year. We shall consider nine measures of mother–infant interactions, recorded via a WRATS computer-compatible keyboard system (Chapter 2; White 1971) when the infant was 8, 16, and 52 weeks old. The high correlations between these interactions and mothers' Confident scores (i.e. $P < 0.10$, two-tailed) are presented in Table 7.1 (top half). First, there was no high correlation in common between sons and daughters at any particular age. Second, Confident mothers neither left (-0.62) nor rejected (-0.73) their 16-week old daughters. In terms of the mother–daughter relationship, these Confident mothers could be described as 'accepting' (Hinde 1979a; Hinde & Simpson 1975) at an age when weaning is normally in progress (Hinde 1974, pp. 194–8). With sons, on the other hand, Confident mothers rejected them at eight weeks in both absolute (0.63) and relative (0.70) terms, and no high correlations occurred between mothers' Confident scores and interactions at 16 weeks.

By 52 weeks, significant differences in interactions in the colony emerged, with mothers and sons interacting less than mothers and daughters. Although mothers left daughters *more* frequently than they did sons, those mothers who were more Confident left their daughters *less* (-0.62) than those who were Fearful, as they had done at the earlier age of 16 weeks. Daughters of Confident

Table 7.1. Mother–infant interactions that highly correlated with mothers' Confident and Excitable scores. M = mother; I = infant.

	Interaction	Age of infant	
	8 weeks	16 weeks	52 weeks
Confident M			
Sons	M reject*		−M reject*
	% M reject*		−time contact*
Daughters		−M leave*	−M leave⁺
		−M reject*	I leave
			− % Ap − % L*
Excitable M			
Sons	−time contact	M restrict*	I approach⁺
	M leave*	M approach*	% Ap − % L
	% Ap − % L*	M leave*	
Daughters	M restrict	M restrict*	M approach*
	M approach*	M approach*	M leave
		M leave*	
		I approach*	

No asterisk $P < 0.10$; *$P < 0.05$; †$P < 0.01$ (Spearman, two-tailed).

mothers left them more (0.57) and played less of a role in establishing proximity with mother (% Ap−%L:−0.69, see p. 67). Thus, the daughters themselves could be described as 'confident' at 52 weeks, which was indeed the case when they were rated a few weeks later.

Whereas mothers' Confident and Sociable scores were highly correlated ($r_s = 0.59$, $n = 25$), their Confident and Excitable scores were not (−0.25). Excitable mothers both approached (0.66; 0.70) and left (0.63; 0.71) their 16-week-old sons and daughters more than Slow (i.e. not Excitable) mothers did. In addition, they restricted sons (0.64) and daughters (0.63) (Table 7.1, bottom half). It is possible that this coming and going, coupled with restricting at the age of weaning, was in part responsible for Excitable mothers having older infants who were relatively not Confident (−0.76 for sons; −0.43 for daughters).

Since this correlation was particularly high for mothers and sons, the remaining correlations between Excitable scores and mother–son behaviour might throw some light on this. As early as eight weeks, Excitable mothers and their sons spent less time in contact (−0.49) than Slow mothers and sons. Excitable mothers left (0.65) their sons more and, in turn, their sons played a relatively high role in maintaining proximity with them (0.60). The latter continued at 52 weeks (0.50), when the absolute frequency of approaches to the mother was higher than for sons of Slow mothers (0.73).

Finally, Excitable mothers had daughters, but not sons, who themselves were Excitable (Fig. 7.1). Looking at the mother–daughter interactions (Table 7.1, bottom half), there was consistency across ages, in that one measure had a high correlation at all three ages and two at two successive ages. These involved Excitable mothers restricting, approaching and leaving their daughters. By 52 weeks, when more interactions were occurring between mothers and daughters than between mothers and sons, Excitable mothers approached (0.75) and left (0.54) daughters more than did Slow mothers. It is therefore possible that this extremely high level of coming and going by Excitable mothers contributed to the daughters' Excitable scores a few weeks later (for further details see Stevenson-Hinde & Simpson 1981).

Thus, the correlations between mothers' scores and early mother–infant interactions suggest how the resulting correlations between mothers' scores and infants' scores might have arisen. Although absolute levels of mother–infant interactions did not differ between the sexes at the early ages of eight and 16 weeks, correlations between those interactions and mothers' scores did differ. The implication is not that mothers in general treated their sons or daughters differently early on but that differential treatment arose in relation to the mothers' personality. By the end of the first year, when mothers did spend more time with and did more with daughters than with sons, daughters were like mothers, while the sons' characteristics were less directly linked to the mothers'.

Last, over the infants' first year of life, the mothers' Confident, Excitable and Social scores were highly consistent ($r_s = 0.81$, 0.84 and 0.74 respectively for 25 mothers) and stable (Wilcoxon matched-pairs tests were non-significant). It is therfore tempting to view the mother as primarily responsible for the characteristics of her one-year-old son or daughter. The correlations could thus be interpreted as showing how a mother's rather stable characteristics, acting through the medium of mother–infant interactions, affect the characteristics of her developing infant.

As infants got older and the mother's influence decreased, the correlations between mother and infant characteristics also decreased (Table 7.2). Surprisingly, the exception was the negative correlation between mothers' Excitable scores and *sons'* Confident scores, which remained high over the first three years. It is sons, not daughters, who leave their natal group and therefore must rely on their own physical condition and fighting ability (e.g. Sade 1972b). With the Madingley monkeys, sons' behaviour out of the colony was correlated both with their own characteristics (Stevenson-Hinde *et al.* 1980b) and with the mothers' (Stevenson-Hinde & Simpson 1981). Thus, while not incompatible with the view

Table 7.2. Correlation coefficients between Confident, Excitable and Sociable scores of mothers and their infants, taken in the November after an infant was one, two or three years old. M = male; F = female.

	Confident mother versus Confident infant		Excitable mother versus Excitable infant		Sociable mother versus Sociable infant		Excitable mother versus Confident infant	
	Sons	Daughters	Sons	Daughters	Sons	Daughters	Sons	Daughters
1 year (14M, 11F)	(0.07)	0.79†	(−0.29)	0.66*	0.56*	0.68*	−0.76†	(−0.43)
2 years (10M,10F)	(−0.07)	(0.50)	(0.01)	(0.22)	(−0.01)	(0.06)	(−0.50)	−0.59
3 years (10M, 8F)	(0.21)	(0.14)	(−0.20)	(0.55)	(0.11)	(0.14)	−0.61	(−0.21)

Brackets $P > 0.10$; no asterisk $P < 0.10$; *$P < 0.05$; †$P < 0.01$ (Spearman, two-tailed).

that sons must eventually rely on their own ability when they leave the group, these findings suggest that they may take with them characteristics ultimately derived from their mothers.

Chapter 8
Influence of the Social Situation
on Relationships

Dyads Embedded in a Group
ROBERT A. HINDE

In the preceding chapters, dyadic relationships have been discussed as though they existed *in vacuo*. However, as was seen in Chapters 1 and 6, the participants in every relationship are likely to have relationships with third parties and this may affect their relationship with each other. Such influences are in fact ubiquitous among non-human primates. The mother–infant relationship is affected by the presence of other individuals and the mother's relationship with those other individuals is affected by their interest in the infant (section 8.2.) Peer–peer play is affected by the possibility of maternal interference and by the availability of other peers (Cheney 1978b, section 6.5). Dominance status is affected by alliances (sections 6.7, 6.8 and 12.6–12.8); and male–female consort relationships are affected by the presence of competing males (sections 12.3 and 12.4). Hamadryas baboon dyads (see section 6.1) may regress to an earlier stage in the presence of conspecific individuals (Kummer 1979).

This has an important implication for studies of natural groups: while, as we have seen, differences between the social structures exhibited by different species in nature can sometimes be understood in terms of differences in the propensities of individuals of the species concerned (section 5.1), it may also happen that relationships are so much modified by the group situation that the affinities of individuals are obscured by the social situation. An important study of squirrel monkeys (*Saimiri*) by Vaitl (1977, 1978; Vaitl *et al.* 1978) exemplifies this.

Squirrel monkeys normally live in groups which in the non-breeding season consist of a cohesive, central core of females and young and a less cohesive, peripheral subgroup of males. In harmony with this, in tests of preference for individuals previously familiar to them, males preferred males and females preferred females. During the breeding season greater social integration occurred and the tendency to approach like-sexed individuals decreased. However, there were certain anomalies, e.g. males appeared to be attracted to familiar males almost as strongly as were females to familiar females, although in nature interactions between males are much less common than those between females. In the non-breeding season the attraction to a particular individual of the opposite sex was only slightly less than that to an individual of the same sex, although the natural troops are clearly subdivided. When tested with unfamiliar incentive animals, both males and females preferred females. Apparently the male preference for females is reversed as a consequence of living in the group. Furthermore, the influence of familiarity varied according to whether the individuals concerned had been living in pairs or in a group.

Some of these anomalies were resolved by Vaitl with data from experiments in which the effects of removing individuals from the troop were examined. When the females were removed from the troop, the males interacted much more frequently with each other and huddled together in a manner quite unlike their behaviour when females were present. Apparently the presence of females partially inhibits the males' tendencies to interact with each other. Vaitl explained this on the basis of observations that a male approaching a female–female pair might be attacked by one of them and the sight of a female screeching at a male might induce other females and also males to attack him. Males can inflict dangerous wounds on each other.

These removal experiments also revealed that the relationships between females could be classified into those that involved a high frequency of positive interactions and those involving a low. The former close female–female bonds apparently pre-empted the formation of close male–female bonds, which in the non-breeding season occurred only in groups containing odd numbers of females.

Similarly, in both natural and artificial groups, males avoid infants. Yet this is not because they are averse to them: if females are removed, males form relationships with specific infants, interacting positively and playing with them. These relationships, initiated by the infants, are normally prevented because female–male relationships inhibit frequent interactions between infants and males.

Thus, Vaitl's experiments demonstrate that in some circumstances the observed relationships between individuals may provide a much distorted picture of their basic affinities. This conclusion extends that of Chapter 4: the behaviour an animal shows may depend not only on its nature and that of the individual to whom that behaviour is directed but also on the social context of their interaction.

Section 8.2 establishes the fact that the social situation affects the mother–infant relationship and section 8.3 concerns the influence of the mother and related females on the expanding network of relationships of young rhesus monkeys. Section 8.4 shows how the arrival of an infant affects the mother's relationship with an older sibling and section 8.5 demonstrates how an older sibling affects the mother–infant relationship. The mother–infant relationship differs between primiparous and multiparous mothers—an effect due in part to the presence of an elder sibling in the latter case (section 8.6). Finally, unrelated individuals may affect the mother–infant relationship by attempting to handle the infant (section 8.7). These studies demonstrate the diversity of social influences on one type of relationship. Examples involving relationships of other types are presented in later chapters.

8.2 Differences in the Mother–Infant Relationship between Socially Living and Segregated Mother–infant Dyads

ROBERT A. HINDE

A straightforward demonstration that the social situation does affect the mother–infant relationship comes from a comparison of four mother–infant dyads, each living alone in one of the Madingley cages, with nine dyads each living in groups consisting of a male, two to four females and their infants. The infants of the segregated mothers spent more time off and at a distance from their mothers than was the case with the group-living dyads (Fig. 8.1). Since these differences were associated with higher frequencies of maternal rejections and more effort on the part of the infants in the maintenance of proximity, they must be ascribed to differences between the mothers of the two groups rather than to differences between the infants. The differences can be understood as a consequence of the attraction that infants have for other females: mothers in a group must protect their infants from the attentions of other females and are thus possessive, while mothers living alone can be more permissive. Observational evidence suggested that the female's possessiveness was affected by whether or not her relationship with the male was such that he would intervene in disputes on her behalf.

However, the differences observed could not be wholly accounted for in these terms: although the segregated infants were recorded out of arm's reach of their mothers in more half minutes than the group-living infants, they came and went more often and after the first ten weeks they spent fewer whole half minutes at a distance than did the group-living animals. This presumably reflects the fact that the group-living animals played with other infants and had more distractions to keep them at a distance from their mothers.

Thus, the presence of social companions can affect the mother—infant relationship because of both the mother's and the infant's relationships with other individuals.

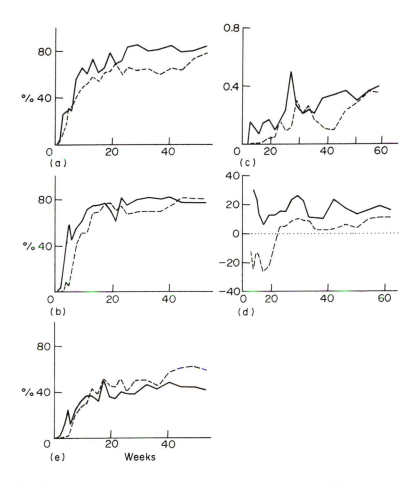

Fig. 8.1. Comparison of mother—infant interaction in segregated (solid line) and group-living (broken line) rhesus mother—infant pairs. (a) The time the infant was off the mother as a percentage of the time observed. (b) The time the infant was more than 0.5 m from the mother as a percentage of the time off her. (c) The frequency with which the mother rejected the infant as a proportion of the number of occasions on which it made contact or attempted to make contact with her. (d) The infant's role in the maintenance of proximity: the percentage of approaches due to the infant minus the percentage of leavings due to the infant. (e) The number of whole half minutes in which the infant was continually more than 0.5 m from the mother as a proportion of the number in which it was off her.

8.3 Matriline Differences and Infant Development
CAROL M. BERMAN

Matrilineages within groups of rhesus monkeys can differ in a number of inter-related ways, all of which are likely to have consequences for infants. For example matrilineages may differ in size and this is likely to affect the number and kind of kin available to an infant. The size of a lineage is determined by its growth rate, its mortality rate and by its propensities to fission. Growth rates appear to be influenced by the relative dominance status of the lineage (Sade *et al.* 1976); high-ranking lineages on Cayo Santiago grow faster than low-ranking lineages, primarily because infants in high-ranking lineages have lower mortality rates and females in high-ranking lineages tend to begin reproducing earlier than females in lower-ranking lineages (see also Drickamer 1974b).

In addition, some lineages are more cohesive socially than others in that their members engage in more positive interaction with one another and are more likely to defend one another than members of other lineages (Cheney 1977). Social cohesiveness tends to be lower in lower-ranking lineages (Cheney 1977; Seyfarth 1976) and in lineages in which the average degree of relatedness through maternal lines between pairs of members is low (Chepko-Sade & Olivier 1979). Average degrees of relatedness are likely to be low in large lineages and in lineages which have experienced high mortality among adult females, particularly among the females which link the sublineages. Such lineages (the latter of which will be referred to as having a fragmented kinship structure) are the most likely to undergo fission (Chepko-Sade & Sade 1979).

Although the specific effects of each of these factors on infant development are not fully understood, data from one social group on Cayo Santiago suggests that infants born into each of its three matrilines do indeed develop different patterns of social interaction and that these differences are probably due to differences in status and social cohesiveness between the lineages (Berman 1982a). Like their elders (McMillan 1982), infants in high-ranking lineages, for example, spend more time with their kin (lineage members) than do infants in lower-ranking lineages. This was shown by dividing each infant's lineage-mates into categories based on their age, sex and degree of relatedness to the infant through maternal lines. The median amounts of time infants in the top-ranking lineage spent near (within 5 m) each age–sex–kinship category was compared with the median amount of time infants in the bottom-ranking lineage spent near their kin in the same category. Within all age/sex classes of kin and during all age periods analysed from birth to 30 weeks, infants in the top-ranking lineage spent more time near their kin than did infants in the bottom-ranking lineage. This was partly because they had more kin but primarily because they spent more time near each of their kin. Although the bottom-ranking lineage was also smaller and more fragmented in structure than the top-ranking lineage, there was some evidence that differences in size and in degree of fragmentation were not critical; the two infants in the middle-ranking lineage, which was both the smallest and the most fragmented of the three, generally had scores which were intermediate between those of infants in the two other lineages.

Interestingly, no differences were found for measures of more intimate interaction between infants in each lineage and their kin. There was no evidence that infants in the top-ranking lineage engaged in more frequent approaches and departures from kin at 60 cm or had friendly contact with them (touching, embracing, grooming or play) more frequently than did infants in the bottom-ranking lineages. In fact, when infants in the bottom-ranking lineage were within 5 m of their kin, they tended to contact them at higher rates than did infants in the top-ranking lineage. Neither did infants in each lineage differ markedly in measures of mother–infant interaction (Berman 1978a). Hence, no infants appear to have been deprived of intimate social interaction with mothers or with other kin, and infants in the bottom-ranking lineage may have been at a disadvantage only because their kin were not so consistently available to them as were the kin of infants in other lineages.

The availability of kin, and particularly of close female kin, has been shown to be closely associated with the amount and quality of protection an infant receives during agonistic encounters (see sections 6.7, 6.8, 9.2 and 9.3). Infants in high-ranking lineages received fewer threats from other group members than infants in lower-ranking and were less likely to be threatened by unfamiliar individuals. When infants were threatened, mothers and other close female kin were likely to intervene on their behalf. Although infants in each lineage were equally likely to be protected by their mothers, infants in high-ranking lineages were more likely to be protected by close female kin than were infants in lower-ranking. Moreover, the protectors of infants in high-ranking lineages were more likely to be successful in that they were less likely to give a submissive gesture to the infant's threatener. Interestingly, the infants who had more close female kin and who spent more time near them were those who were threatened less, were more likely to be protected and were more likely to be protected successfully. Partial correlation tests, in which maternal (and close female relative's) ranks were held constant suggest that the presence of close female relatives inhibited threats and increased the chances of protection regardless of the mothers (or close female relative's) rank. Hence, infants in high-ranking lineages appear to receive fewer threats and more protection because they are near their protectors more and not simply because their protectors are high-ranking. On the other hand, the probability that the protector would give a submissive gesture was directly associated with rank. No strong associations were found between measures of protection and time spent near the mother or time spent near other close kin (brothers, aunts, uncles and cousins).

8.4 Effects of Parturition on the Mother's Relationship with Older Offspring

PHYLLIS C. LEE

The nature of the mother–infant relationship changes most dramatically at the time of weaning, when the mother alters her patterns of care for the infant and it becomes nutritionally independent. In many primate species (reviewed in Altmann 1980), weaning occurs several months to several years before the birth of a subsequent offspring.

At the time of the birth of a new infant, the relationship may undergo a second transition, often more abrupt than that of weaning, in which the amount of attention the previous offspring receives from the mother may change radically. This transition is marked by a shift in the focus of the mother's interactions from the older to the younger offspring. While Trivers (1972, 1974) has pointed to the importance of the mother's attempts to conserve resources for subsequent offspring and has focused therefore on weaning conflict, the transition in the relationship consequent upon a new birth may also reflect important aspects of the way in which maternal care is divided between offspring of different ages with different social needs.

After a new birth, the behaviour of both the mother and older offspring towards each other changes. Some offspring appear to be jealous, in that they demand increased levels of attention from the mother: these may be rebuffed (e.g. Dunn & Kendrick 1980; Hooley & Simpson in prep.). Other offspring may become depressed and withdrawn, moving away from the mother and becoming reluctant to interact with her or the new infant (e.g. Bolwig 1980). Generally, among the primates, this period after the birth has been described as one where young females, and particularly siblings, direct increased attention and either friendly behaviour or aggression to the mother and the new baby (reviewed in Berman 1982b).

When the author began to observe older offspring at the time of a new birth in vervet monkeys, she was struck by what appeared to be a major change in the behaviour of one yearling. In a subjective sense, he appeared to be depressed just after the birth: he moved away from his mother, spent time feeding and foraging on his own and rarely groomed with his mother. Relative to the other yearlings in his group, he played much less. No such changes were observed in offspring older than two at the time of birth. On the contrary, the young females seemed fascinated by their infant siblings. As a result of these impressions, changes in the relationship with the mother were examined by categorizing offspring into one and two year olds who had never had a younger sibling (the no-sibling group), two and three year olds with at least one previous sibling and four year olds who, again, had had experience of a previous sibling.

The majority of the analyses presented here were carried out as a *post hoc* examination of this suggestive trend towards depression on the part of offspring who had previously received uninterrupted attention from the mother. The sample sizes involved are tiny. Only one yearling and four two year olds without sibling experience were sampled throughout the three-month period after the birth. Only four two and three year olds (with a previous sibling) and only four four year olds were sampled. Data for the first month after birth were available for eight other offspring but the majority of these were restricted to the first month.

The behaviour of older offspring in the periods before and after the birth of a new sibling has been compared, using data for each month of the first 12 weeks of a new infant's life and those collected before the birth. Samples from the three-month period before birth for all behaviour which was not affected by seasonality have been used (Lee 1981). For play and proximity, which were affected by seasonality, the behaviour only in the month before birth has been compared with that in each month after the birth.

Changes in behaviour after the birth

In Fig. 8.2, the depression in yearling play is shown for a small sample of infants in the early months after birth. For infants who had no sibling born, the general seasonal increase in play rates began in the tenth and eleventh months. The yearlings with a younger sibling showed an initial seasonal rise in the eleventh month, which dropped until three months after birth. The play rates for the one infant sampled then rose to the normal wet-seasonal high.

The amount of time spent near the mother and other group members was compared before and after the birth to see if the depression in yearling play rates was a general trend in other types of interactions. Seasonal changes in the time spent near other group members suggested that during the birth season (the early wet season) immatures of all ages were spending more time close to other group members than they were in the preceding dry season. In Fig. 8.3, the time an offspring spent more than 2 m from other group members is shown for each individual in the three-month period before the birth and in relation to the early wet-season median for each age group in the period after the birth. While the majority (nine of eleven) of the older offspring showed the typical seasonal trend to spend more time close to others and less time away in the first period after the birth, the no-sibling group did not show this change so clearly. Half of them spent more time at a distance in the first month after birth than they had previously,

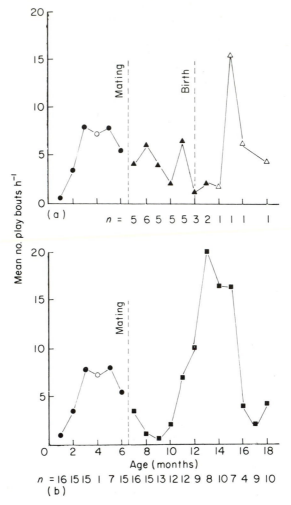

Fig. 8.2. Mean number of play bouts in each of the first 18 months of life for infants whose mothers gave birth (a) after mating and for those infants whose mothers did not give birth (b). These groups had been separated from the time of mating onwards. Solid circle = all infants; solid triangle = infants with sibs; solid square = infants without sibs. Open symbols indicate the sample size of 1.

contrary to the seasonal predictions. In addition, eight of the ten of the no-sibling group were above the median for their age group in the month after birth. The no-sibling group appeared to be responding to the births by moving further away from others than was common for that time of year.

At the same time, the percentage of time that the animals in this group spent close (within 2 m) to the mother declined for seven of ten of the offspring. The percentage of time spent close to the mother rose in the first month for all five two

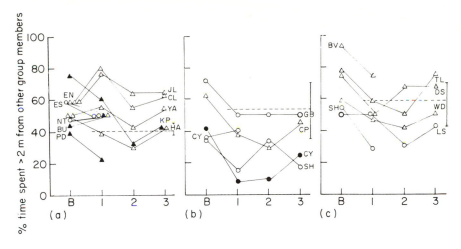

Fig. 8.3. Percentage of time spent at greater than 2 m from other group members for the month before birth (B) and in each of the three months after birth. Immatures are separated into one and two year olds with no previous sibling (a), two and three year olds with a previous sibling (b), and four year olds with a previous sibling (c). The median and interquartile range for each age group during that three-month period is shown by the broken line. Each immature is shown separately and the solid symbols are the youngest of the age group. Triangle = males; circle = females.

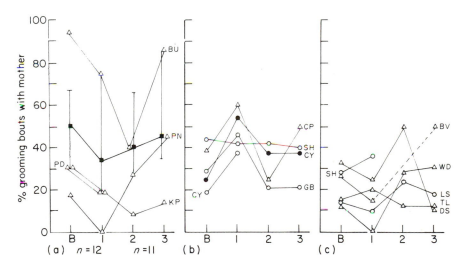

Fig. 8.4. Percentage of grooming bouts that were with the mother in the months before birth (B) and in each month after birth. The median and interquartile range is shown for all one and two year olds, and yearlings are plotted individually (a). Two and three year olds (b) and four year olds (c) are all plotted individually.

and three year olds and was variable for the four year olds. The no-sibling group were spending more time on their own and less time with their mothers than they had before the births. The two and three year olds were spending slightly less time alone, consistent with the seasonal changes, and an increased proportion of their time was spent close to their mothers in the first month after birth.

The hourly rate of grooming showed variable changes. The most consistent trend was seen in six of the eight females, who increased the time spent grooming with others during this period. This may be related to the high levels of interest in and contacts with infants by the young females (see section 8.7). However, the proportion of grooming with the mother in this period declined in the first month for eight of the 12 yearlings (Fig. 8.4). Among the two and three year olds, along with an increase in grooming rates, was an increase in the proportion of grooming that was with the mother during this period.

Summary of changes in mother–offspring interactions

The changes in associations and grooming with the mother in the period after a birth suggested that the two and three year olds interacted more with their mothers at this time than they had before the birth. Older offspring showed fewer consistent changes, while the yearlings and two year olds with no previous sibling tended to reduce the frequency of interactions with the mother in the period after the birth. These younger immatures moved away from the group and spent more time at a distance from the mother than they had previously. They groomed proportionately less with the mother after the birth than they had before. The small sample of data on yearling play also suggested that they did not show the seasonal rise in play rates along with the rest of the members of their group.

Five of the no-sibling group declined on both the measures of time spent near the mother and in the proportion of their grooming that was with the mother from the months before the birth to the first month after birth. Two rose on both measures. Two showed a decline in time spent near but a rise in the proportion of grooming and one showed a decline in grooming but a rise in time near. These offspring have been subjectively categorized into three groups on the basis of the changes in their relationships with their mothers (Table 8.1).

All three of the yearlings in the sample fell into the depressed category, suggesting that the relationship between the mother and her offspring changed most for this age group. However, even some two year olds without a previous infant sibling also showed some depression and two were consistently more demanding of the mother's attention. These categories correspond with those from observations on children (Dunn & Kendrick 1980) where some children became more withdrawn or depressed, while others became more demanding of the mother's attention after the birth of a new sibling.

Table 8.1. Identity of yearlings and two year olds who fell into descriptive categories reflecting changes in the mother–offspring relationship in the first month after birth.

	Yearlings		Two year olds	
	Males	Females	Males	Females
Depressed: decline in grooming and time spent near the mother	BU KP PD		JL	ES
Demanding: increase in grooming and time spent near the mother			HA EN	
Ambivalent: one or other measure decreases			YA CL	NT

These suggestive changes in the mother–infant relationship at the time of the new birth were not related to gradual age-dependent changes in rates of interaction but reflected abruptly altered levels of interaction between the mother and her previous offspring. Among the two and three year olds, the changes in the relationship tended to be temporary and reflected a strengthening of the relationship through increased grooming and interactions with the mother and the new sibling. The relationships of the four year olds with their mothers appeared to be relatively unaffected by the new birth.

Thus, there did appear to be a period after weaning when the relationship between a mother and her offspring went through another major transition, unrelated to conflict over access to the nipple. Yearlings and some two year olds with no previous sibling showed a tendency for their interactions with their mother to be less frequent in the first month after the birth of a new sibling. However, some of this age group interacted more intensively with the mother, and some were ambivalent. Older offspring either were attracted to the new infant or showed little change in any of their interactions with the mother.

8.5 Influence of Siblings on the Infant's Relationship with the Mother and Others
JILL M. HOOLEY and MICHAEL J.A. SIMPSON

With the exception of firstborns (see section 8.6) most infants are born into social networks containing siblings. Not only can siblings provide positive social stimulation for the infant in terms of play and learning experiences but they may also present threats to the infant arising from jealously or from competition for the mother's attention. In many macaque groups a new infant is a rival in the very direct sense that it will eventually displace its elder sibling and rank above it

(Datta 1981; Holman *et al.* 1982; Missakian 1972; Sade 1972b).

To examine sibling influences, Hooley and Simpson (in prep.) used data on the first 16 weeks for 33 rhesus monkey mother–infant dyads from the Madingley groups. At the time of their birth, 14 infants had siblings aged approximately one year (yearling siblings). The remaining 19 infants had nearest siblings who were aged two years or older (older siblings). Since there were no significant differences between the mothers of the two groups in age, parity or dominance rank, it can be assumed that differences between the two groups were due to the different ages of the nearest siblings rather than to differences between the mothers.

The most striking difference between the two groups was the precociousness of the infants with yearlings. Such infants showed high scores in the second and fourth weeks on behaviours usually associated with a more advanced developmental stage: even at two weeks they tended not to be in the close ventro-ventral position on their mothers but more loosely in contact with them. They also spent less time in physical contact with their mothers, more time out of arm's reach of her and more time alone. They were also more responsible for staying in proximity to their mothers than were the infants with older siblings.

Not only were the infants with yearlings precocious in relation to their behaviour with their mothers but they were also more socially advanced. In comparison with the infants with older sibs they were more likely to initiate social contact with other pen members and they also received more play. However, by the tenth week this was no longer true and the infants with older sibs became relatively more socially active.

In addition, there were certain effects which seemed to depend on the infants' sex. As well as being behaviourally precocious, male infants with yearlings had

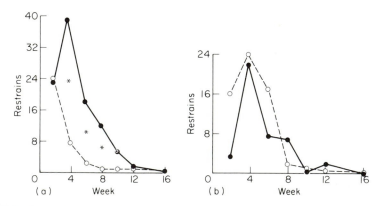

Fig. 8.5. Median frequency of restrains by mothers of sons (a) and daughters (b) when yearling (broken line) or older (solid line) siblings were present. Statistically significant differences at the 0.05 level or greater are denoted by asterisks.

mothers who were much more permissive than the mothers of the males with older sibs. The mothers of the males with yearlings physically restrained their infants much less frequently (see Fig. 8.5), and were also less likely to approach, initiate close contact or put their arm around the infant. The mothers of the females, however, behaved quite differently. Indeed it appeared that the mothers of females who had yearlings, rather than being more permissive, were actually more apprehensive about their daughters and they were more likely to approach them or initiate contact with them than were the mothers whose daughters had older siblings.

What is it about having an older sibling as opposed to a yearling sibling that discourages infants from being away from their mothers and makes mothers of males more restrictive? Observers who had taken data on the monkeys had the subjective impression that daughters whose younger sibling arrived when they were two years old were more disturbed than those who were only one year old. This was checked by examination of the watch notebooks in which were noted comments about events which occurred during a watch and which could not be recorded on the behaviour keyboard. The data concerning episodes which involved aggression between sibling and mother or sibling and infant are presented in Table 8.2.

Table 8.2. Median rate of incidents of aggression directed by siblings to mother and/or infant 12 h^{-1} of observation when a yearling was or was not present.

	Yearling sib present		No yearling sib
	Aggression by yearling	Aggression by older sib	Aggression by older sib
Male sib	0.22	0.00	0.47
Female sib	0.58	0.48	1.90

The following are typical examples of the type of episodes recorded:

Josie (two-year-old sister) takes Viv's nipple out of Celeste's (infant) mouth. Viv yanks Celeste away from Josie. Celeste squeaks. Josie has a tantrum. Viv displays, Josie grooms Viv.

Faye (three-year-old sister) pulls CT's tail *very hard*. CT (infant) squeaks.

As can be seen from Table 8.2, older siblings were more aggressive to the mother–infant dyad when there was no yearling between themselves and the infant. It seems that the arrival of a new infant is more distressing for the siblings and results in the sibling being more unpleasant to the mother and infant when it is a two year old encountering a new infant for the first time. If there has already been a birth to the mother before the sibling is two, the two year old is less aggressive, perhaps because it has not had such a long period of attachment to the mother without competition or because the yearling buffers the aggression more directly. Moreover, older sisters were more aggressive to the infant and mother than older brothers and older sisters were more aggressive than yearling sisters. Spencer-Booth (1968) found that two to three year olds interact more with infants than do individuals younger or older and that most interest comes from the females in this age group.

The early behavioural advancement of the infants with yearlings and the permissiveness of the mothers of the males may, therefore, be due to the fact that the yearling disturbs the mother–infant dyad much less than an older sibling. The older sibling on the other hand, by being much more interested in the baby and also by being more jealous at its arrival, may be more of a threat to the small infant and the mother therefore responds to the sibling's influence by being more protective and restrictive of her infant, at least in its early weeks. By the tenth week, however, when the infant is presumably less vulnerable to upset or injury, the mother becomes less restrictive and this may explain why the infant with an older sib becomes more socially active around this time.

The importance of the birth interval for sibling jealousy is of course also a focus for discussion in the human case, though there is no agreement as to the optimum interval. But we must not expect the issues to be simple in either monkey or human: a sibling who is a potential threat to the safety of a four-week-old rhesus infant may be a valuable play partner some weeks later. At present, there is no simple explanation for the differences between male and female infants and suspicion that interactions between a number of issues may be involved.

8.6 Primiparous and Multiparous Mothers and Their Infants
JILL M. HOOLEY

Firstborn infants differ from their younger siblings in two important ways. First, they have inexperienced mothers. Second, they have no siblings, so that the social context of the mother–infant relationship is quite different. Hooley and Simpson (1981) studied the relationships of primiparous and multiparous mothers and their infants living in the Madingley social groups during the first 16 weeks of the infant's life. Mothers were classified as primiparous ($n = 9$) if their infant was their first live-born offspring. All infants whose mothers were multiparous ($n = 33$) had at least one sibling living with them.

Differences between the mothers

There were several important differences between the primiparous and multiparous mothers. The primiparous mothers were much younger, their median age being only five as opposed to a median age of 13 years for the multiparous mothers. The primiparous mothers also had fewer relatives living with them in their social group. Although there was no statistically significant difference between the primiparous and multiparous mothers in dominance rank, no mother was rated by observers as being of high rank (relative to other adult females living in the group) at the time she had her first infant. Some of the primiparous mothers were however the daughters of dominant females. Perhaps as a consequence of their age, rank or lack of relatives to provide support in agonistic encounters, the primiparous mothers received more aggression from other adult females living in the group than did the multiparous mothers.

The primiparous and multiparous mothers also differed in personality measures (Stevenson-Hinde *et al.* 1980; see section 4.2) assessed in the November after the mothers gave birth. Although there were no differences in the Sociability of the mothers in the two groups, the primiparous mothers were found to be more Excitable and less Confident than the multiparous mothers. The following November, the primiparous mothers were still more Excitable than the multiparous mothers although they were no longer less Confident, possibly because many had by then had another infant.

Differences between the mother–infant dyads

These young and inexperienced mothers approached and left their infants more frequently than did the multiparous mothers (Fig. 8.6). This could have been due to general anxiety about the infant but could also have been associated with the fact that the primiparous mothers were much younger and perhaps just more active generally. Evidence in favour of the former view came from the finding that primiparous mothers who had daughters spent longer with their arm round the infant and were more likely to restrain her physically than were the multiparous mothers with daughters.

However, the issue was complicated by an interaction with the sex of the infant. This can best be demonstrated with an index of protectiveness, obtained by summing the frequencies with which a mother approached her infant, made contact with it, restrained it and put an arm around it. By this criterion primiparous mothers were much more protective of their daughters than of their sons but multiparous mothers protected sons more than daughters. Furthermore, multiparous mothers tended to protect sons more than did primiparous mothers (Fig. 8.7). Thus, both the sex of the infant and the parity of the mother influence how the mother behaves to her infant.

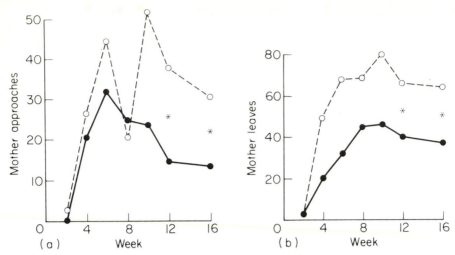

Fig. 8.6. Median frequency of approaches (a) and leaves (b) made to infants by primiparous (broken line) and multiparous (solid line) mothers. Approaches and leaves were all occasions when the mother's distance from the infant changed from less than 60 cm to more than 60 cm or vice versa. Statistically significant differences at the 0.05 level or greater are denoted by asterisks.

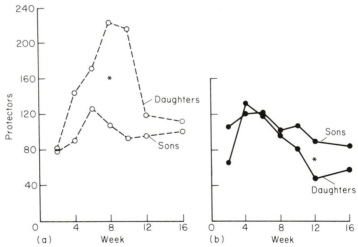

Fig. 8.7. Median frequency of protects made by primiparous mothers (a) and multiparous mothers (b) to sons and daughters. Statistically significant differences at the 0.05 level or greater are denoted by asterisks.

The explanation of this reversal of the protected sex according to parity must be sought in the social context. It has been shown that elder siblings may behave aggressively to new infants (section 8.5), so multiparous mothers may need to protect male infants from their elder siblings. The lack of siblings may also explain why firstborn males of primipares spend more time away from their mothers than do firstborn males of multiparous mothers. With daughters, however, non-relatives may pose a much greater threat to the infant than do elder siblings. Data from bonnet macaques (Silk *et al.* 1981b) suggest that most aggression by adult females is directed to female rather than male infants. A primiparous mother, with few kin to support her or her infant in agonistic encounters, may, therefore, need to protect a daughter more than a son.

The lack of direct sibling influences on infants in the primiparous group may also be the reason why they received less social contact (a measure which includes cuddles, carrying and pulling of the infant) from others. Since mothers frequently sit with their families, many of the social contacts that infants receive will come from siblings or other relatives who happen to be sitting near the mother. Since the primiparous mothers had fewer relatives living with them, most of the opportunities for other pen members to contact the infants came when they were away from their mothers.

The presence of siblings also seems to affect the Excitability of the infants, as assessed by the observers (section 4.2). Despite the fact that they had more Excitable mothers, male infants of primiparous mothers were less Excitable than the males of multiparous mothers, although there were no differences in either Confidence or Sociability. Since males with yearling siblings were also less Excitable than males whose nearest sibling was two, three or four years old, this suggests, for male infants at least, that the presence of siblings can increase the Excitability of an infant—all the more so if there is an age gap of two or more years between the infant and the sibling.

Results such as these emphasize the need to interpret relationships with reference not only to the individual characteristics of the participants but also to the social context within which the relationship is embedded. Parity differences in the mother–infant relationship are due not only to differences between the mothers but to the presence of siblings, and the influence of the latter can be both direct and mediated through the mothers. There would now seem to be a need for a study in which infants born to multiparous mothers but reared socially without siblings were compared both with infants with primiparous mothers and infants with experienced mothers and siblings.

8.7 Caretaking of Infants and Mother–Infant Relationships
PHYLLIS C. LEE

Mother–infant relationships are developed and maintained within the context of the social group and particularly within the extended family. Recent studies have emphasized the importance of other individuals in affecting the interactions between two animals (Berman 1978b; Datta 1981; Kummer 1967; sections 6.7

and 6.8 and Chapters 8, 9 and 12). As such, infant development and mother–infant relationships are affected not only by maturation and by changes in the interactions between mothers and infants but also by other individuals in the groups. The other group members who contact infants are varied and range from siblings (Berman 1978b; Small & Smith 1981; section 8.5), to adolescent and adult females, sometimes called 'aunts' (Rowell *et al.* 1964), and to adult males (Deag & Crook 1971; Packer 1980; Ransom & Ransom 1971). These other individuals who contact and care for infants are thus an important part of the social context for the mother–infant relationship.

Among vervets, this handling and caretaking of young infants is characterized by relaxed and friendly contacts, and infant handlers will respond to distressed or endangered infants in a protective fashion (Gartlan 1969; Johnson *et al.* 1980; Lancaster 1972; Struhsaker 1971). Protective caretaking of infants is also seen in langurs and patas (Chism 1978; Hrdy 1976; Jay 1968; Poirier 1968) and has been called allomothering by Hrdy (1976). In contrast to the type of infant caretaking seen in vervets and langurs, in macaques other animals can be abusive of infants and aggressive to the mothers (Hooley & Simpson in prep.; Rowell *et al.* 1964; Silk 1980). Spencer-Booth (1970) reviewed a range of different types of contacts to infants and showed that they vary with the age and dominance of the mother. In other species, the attractiveness of the mother (e.g. Altmann 1980; Cheney 1978b), the age and sex of the infant (e.g. Berman 1982; Ransom & Rowell 1972) and the age and sex of the partner (e.g. Busse & Hamilton 1981) appear to affect how much attention an infant will receive and whether it will be friendly or abusive.

In this section, the nature of the friendly contacts to infants in free-ranging vervet monkeys studied in Amboseli will be described and an attempt made to assess the ways in which these friendly contacts affected mother–infant relationships.

From focal samples of the infants in the first 12 weeks of life, the frequency with which infants were approached to within 2 m by others and how frequently they were contacted and groomed by others were analysed. Contacts to infants consisted of touching the body with the nose or kissing, touching, cuddling and putting an arm around the infant. Holding the infant ventral and carrying the infant ventral were also considered as friendly infant contacts and all these categories have been combined in the analysis of infant contacts as 'infant handling'.

Timing of infant handling

Handling of infants by other group members began within a few hours of birth. One infant, a son of a dominant female, spent 28 minutes out of a 30 minute sample on his first day of life sitting with or being carried in ventral contact by a three-year-old female. The mother stayed within arm's reach throughout and attempted to retrieve the infant only gently. Although she was dominant to the juvenile, she made no aggressive attempts to regain her infant. She finally retrieved him when the juvenile's attention was distracted by a predator alarm. At this time he was no more than 16 hours old and could have been less than four hours old.

Nine out of 15 infants in this sample received more contacts from others in their first month than during the second or third month of life (month 1: median = 41.6%; month 2: median = 28.6%; month 3: median = 21.3% of the total contacts in the first three months). The hourly rates with which infants were handled by both adult and immature females were highest in the first month for ten of the 15 infants (Fig. 8.8).

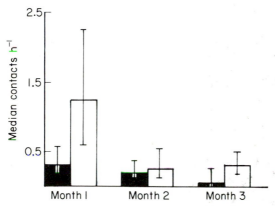

Fig. 8.8. Median and interquartile range of the number of contacts to infants made per hour by adult (solid column) and juvenile (open column) females in each of the first three months of an infant's life.

One factor which could affect how frequently a handler could make contact with an infant was how frequently she was near the infant or approached to within 2 m of the infant during this early period. Since mothers may attempt to control the access of others to her infant, only the approaches made to the infant while it was in ventral contact with the mother were considered. The mean number of approaches per hour by a group member while the infant was in ventral contact with its mother was divided by the proportion of time that the infant spent ventral, in order to correct for a decline in time spent ventral with age (Lee in press).

There were no significant differences between any of the first three months in the rate that each infant was approached while it was ventral. Only four infants were approached most frequently in the first month, four in the second month and seven in the third month. There were also no differences between infants in the different study groups. Thus, irrespective of differences between the groups in the number of potential infant handlers and in maternal behaviour, infants were sought at similar rates throughout the first three months.

Contacts to infants which consisted of grooming by others were also seen throughout the first three months. Grooming of the infants did not take place only in the first month and only five of the 15 infants were groomed the most in their first month. Three of these were the infants of high-dominance mothers and two were firstborn infants. Grooming appeared to be less restricted to the period of early attractiveness than did handling of the infant.

Age/sex class of infant handlers

Juvenile females handled infants at higher rates than did the adults (Lee 1981; Fig. 8.8). Juvenile males, particularly siblings, also contacted and cared for infants in their first three months. Unrelated juvenile males would respond to distress calls ('whirring') and would retrieve and care for infants who had been left behind by the group. However, juvenile females handled infants at significantly higher rates than did the males (median for males: 0.03; median for females: 0.72; $P < 0.01$) and the two- and three-year-old females tended to handle infants at higher rates than did the one- and four-year-old females (median for two and three year olds: 1.71 h^{-1}; median for one and four year olds: 0.27 h^{-1}).

Siblings, particularly sisters, handled their infant siblings at higher rates than other unrelated members of the sibling's age/sex class (Table 8.3).

Attractiveness of the mother

There was a weak trend for adult females to handle infants in the first month more when their mothers were of high dominance than when the mothers were lower in dominance ($r_s = 0.37$, $n = 16$, NS). The handling of infants by juvenile females in the first month of the infant's life was not significantly related to maternal dominance ($r_s = 0.27$, $n = 16$, NS). However, juveniles tended to handle firstborn infants, irrespective of the mother's dominance, above the median hourly rate, and thereby diluted the effect of maternal dominance on rates of handling. When these infants were excluded, juvenile and adult females handled infants of high-dominance mothers significantly more than those of lower-dominance mothers ($r_s = 0.68$, $n = 11$, $P < 0.05$).

In spite of this, there was no correlation between maternal dominance and the hourly rate that the infant was approached by another group member while it was ventral on the mother in its first three months ($r_s = 0.08$, $n = 16$, NS). Approaches initiated by the mother to others while the infant was ventral were

Table 8.3. Comparisons between the rates of a sibling handling its infant sibling and each member of the sibling's age/sex class handling that infant during its first three months. Higher = number of siblings who handled more than others of their age/sex class; lower = number of siblings who handled less than others of their age/sex class.

	Males	Females
Higher	16	11
Lower	4	4
Equal	17	1

positively related to her dominance but not significantly so ($r_s = 0.39$, $n = 16$, NS). Thus, the higher rates at which the infants of high-dominance mothers were handled did not appear to be related to how frequently they were approached nor to how frequently their mothers approached others while their infants were in ventral contact.

In addition to the friendly contacts and caretaking of infants, other group members also groomed with the mothers but less frequently with the infants. These grooming bouts with the mother may have been a means of contacting the infant, since the infant was in easy reach of the mother's grooming partner. Those infants who were handled the most by others were those who were groomed the most ($r_s = 0.58$, $n = 16$, $P < 0.05$).

Since the amount of grooming an infant received was related to the amount of handling and handling was also related to maternal dominance, there was a trend for infants of the higher-dominance mothers to be groomed by other group members more than were infants of low-dominance mothers (Fig. 8.9). When the firstborn infants were excluded from the correlation, it was significant ($r_s = 0.67$, $n = 11$, $P < 0.05$). However, the trend was reversed for the firstborn infants ($r_s = 0.67$, $n = 5$, NS). All but two of these grooming bouts were with juvenile or adult females, both of which appeared to be highly attracted to the higher-dominance and firstborn infants as grooming partners.

However, there was no correlation between the rate at which a mother groomed or was groomed by all others and her dominance rank ($r_s = 0.04$, $n = 16$, NS). Mothers did not receive or give grooming more in any one of the first three months than the others. Mothers may have maintained consistent relationships among themselves and this may have minimized the effect of rank on how frequently they were involved in grooming.

In studies on captive monkeys, many of the contacts to infants have been characterized as abusive of the infant (see p. 146). Among these vervets, infant handlers would use the maternal rejection techniques (pushing or gentle nipping) when they wanted to dislodge a tenacious infant but no threats or attacks were directed to an infant by an immature or an adult who was handling it at the time. However, mothers did receive aggression from other group members, even though they had a young infant. Harassments of the mother were scored when another animal caused her to retreat or abandon a resource and when she was threatened, chased or attacked. Harassments were separated into those initiated and those received in each of the first three months of an infant's life. In the second month of an infant's life, low-ranking mothers received proportionately more harassments than did the high-ranking mothers (Lee 1981). If this early period was one when mothers passed on their dominance to their infants, they may have been especially affected by aggressive interactions in the second month. Although harassments might be expected to be highest in the first month when

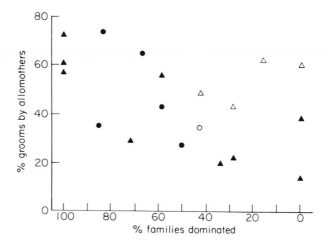

Fig. 8.9. Percentage of each infant's grooming bouts over the whole of the first three months of life that were with other group members (the ordinate) plotted against maternal dominance (the abscissa). Solid triangle = male infant; solid circle = female infant; open symbol = firstborn.

the infants were the most vulnerable (e.g. Silk 1980), mothers and their infants then received attention of a friendly sort rather than aggression. Some of the lower-dominance mothers may also have moved away from the group during this time to avoid both the aggressive harassments and the contacts of others with their infants.

Summary

Throughout the first 12 weeks of life, infants were approached, contacted and cared for by other group members. The friendly contacts were non-randomly distributed among the infants, infants of higher-ranking mothers receiving more attention and grooming. Firstborn infants also received more attention and this was unrelated to their mother's rank. Lancaster (1972) also found that juvenile females cared for infants more frequently when they were born early in the birth season than when they were born late.

The caretakers were also non-randomly drawn from the group. Juvenile females, especially the two and three year olds, handled infants most. Siblings of the infant were also more likely to contact the infant than were unrelated immatures of the sibling's age/sex class. However, adult females and some unrelated juvenile males also cared for infants, groomed with the mothers and approached both the mother and the infant. Infants were handled most in the first month but approaches and grooming of the infant and the mother did not appear to be concentrated in this early period. Thoughout the first three months of life, infants became familiar with those other animals who both avoided and were avoided by their mothers.

Chapter 9
Triadic Interactions and Relationships

9.1 Triadic Interactions and Social Sophistication
ROBERT A. HINDE

In the main, it has so far been assumed that the dyadic relationship is the basic element of social structure, e.g. relationships between mother, infant and sibling can to some extent be understood in terms of the dyadic interactions and relationships between the three participants. However, interactions frequently involve three animals, each directing its behaviour simultaneously with respect to one or both of the others. The most common form involves triadic agonistic interactions in which two individuals join forces against a third (e.g. Altmann 1962; van Hooff & de Waal 1975; de Waal *et al.* 1976; Watanabe 1979).

A related type of encounter has been labelled by Kummer (1967) 'tripartite'. Here, three individuals '*simultaneously* interact in three *essentially different roles* and *each of them aims its behaviour at both* of its partners' (Kummer 1967, p. 69). Kummer was primarily concerned with interactions between a harem-owning, male hamadryas baboon and two of his females. One female, sitting close to the male, may scream at or threaten the other female, while at the same time presenting to the male and attempting to remain between him and the second female. The latter attempts to get near the male and threatens the first female in return but because of the relative positions of the three animals her threat tends to be directed also at the male. While he may merely watch the proceedings, any intervention on his part is likely to involve an attack on the second animal. Kummer thus describes such interactions as involving a protected, a protector and an antagonist.

152

However, the difficulty of making precise distinctions here is illustrated by the phenomenon known as 'agonistic buffering' (Deag & Crook 1971). If a male picks up and carries an infant, this may render other males less likely to attack him (Deag 1980; Itani 1963; Kummer 1967; Packer 1980) but here the infant is merely being used by the male. Yet such instances intergrade with tripartite interactions as defined by Kummer.

Interaction involving three individuals are of special interest in revealing the social sophistication of non-human primates. Here another study of hamadryas baboons may be cited. In this species, the male normally herds a harem of several females. Under experimental conditions, a male without a mate caged with a male who already 'owns' a female may attempt to appropriate the female. Bachmann and Kummer (1980) investigated the circumstances under which this would happen. Using six males and six females, they carried out choice tests which yielded a preference score of each female for each male and of each male for each female. They also tested the extent to which each male would respect or interfere with various male–female pairs. The latter was not related to the preference of the owner or of the rival for the female. However middle- and low-ranking rivals were less likely to attempt to interfere with the pair, the higher the female's preference score for the male she was with. The more dominant males were not so inhibited. Apparently, a female with a strong preference for her present owner would be too costly a prize for a middle- or low-ranking male, for even if she were won she would have to be herded closely and there would be considerable risk of losing her again. In any case, these data show that baboons are able to assess certain aspects of the relationship between two other individuals (see also Kummer *et al.* 1974).

Another example has been provided by Cheney and Seyfarth (1980) in their recent experiments on vocal recognition in vervet monkeys. They found that when a mother was played a recording of her offspring's screams, she responded by looking towards the offspring. Other females in the group responded by orienting towards the mother as soon as they heard the scream, thus displaying some understanding of either the relatedness of the individuals or of their persisting relationship.

The complexity of tripartite interactions and the understanding of the relationships existing between others on the part of a third has led Kummer (1981) to suggest that knowledge about interactions—their predictability and outcome—has led to the use of other individuals as tools in the attainment of individual goals (see also Humphrey 1980; de Waal 1982).

The contributions in this Chapter are concerned primarily with the consequences rather than the mechanics of triadic interactions and relationships. Sections 9.2 and 9.3 are concerned with rank acquisition in female monkeys and thus provide further data on issues raised in sections 6.7 and 6.8. Interference by third parties in agonistic encounters plays a large part in determining rank. However, full understanding of the nature of interferences is facilitated by simultaneous consideration both of the proximal factors determining the behaviour of the individuals concerned and of the ultimate costs and benefits that result from their behaviour: further discussion is thus postponed to sections 12.6–12.8.

The other two contributions (sections 9.4 and 9.5) are concerned with the factors determining male emigration from the troop. In the three species principally considered in this book, males usually leave their natal troop. Section 9.4 reviews evidence and provides further data on the factors determining whether and when a male rhesus leaves his troop and where he emigrates to. The final section is a case study showing how occasional males who remain in their natal troop can nevertheless reach high rank with the aid of female relatives. Further discussion of both proximate and ultimate factors influencing male emigration is given in section 11.3.

9.2 Early Differences in Relationships between Infants and other Group Members based on the Mother's Status: their Possible Relationships to Peer–Peer Rank Acquisition
CAROL M. BERMAN

Rank acquisition among rhesus monkeys occurs in at least two stages. Peer–peer rank is acquired early in life and correlates with maternal rank. Adult rank is established generally in adolescence, at least among females. A number of mechanisms by which non-human primates acquire rank at each stage have been suggested. The most frequently cited involve the direct participation of the mother. For instance offspring may learn their mother's rank through her active intervention in agonistic encounters (e.g. Cheney 1977) and they may learn by observing their mother's interactions with group members before they engage in agonistic interactions themselves (e.g. Altmann 1980; see section 6.8).

At the same time, evidence is accumulating that rank acquisition may also occur in the mother's absence (Loy & Loy 1974; Walters 1980) and that offspring can learn their mother's rank through interaction with group members without the direct involvement of the mother: thus there is evidence that infants of high-ranking mothers are more attractive play partners than infants of low-ranking mothers (Cheney 1978a; Fady 1969; Tartabini & Dienske 1979). In addition, data from 20 rhesus infants on Cayo Santiago suggest that infants of high- and low-ranking mothers have different agonistic experiences during their first 30 weeks of life and that these differences are due more directly to differences in their relationships with group members than with their mothers (Berman 1980b; section 8.3).

On Cayo Santiago, the earliest indications that infants interact with one another according to their mother's ranks were observed between 27 and 30 weeks of age. The direction of fearful gestures between pairs of infants (cowers, fear-grins, screams and runnings away) were consistent and in harmony with the mother's ranks in 27 of 35 (77%) dyads. By one year of age a nearly perfect hierarchy could be constructed (Sade 1967).

Threats and threateners

Infants were rarely threatened during their first few months but were threatened with increasing frequency over time. Although absolute frequencies of threat directed toward infants were not at first strongly related to maternal rank, by 27 to 30 weeks of age infants in the top-ranking lineage received fewer threats than infants in the two other lineages (Mann-Whitney U-test, $P < 0.05$). When frequencies of threat were corrected for differences in opportunities of access to infants (by expressing them as ratios of the amount of time threateners spent within 5 m of the infant), lineage rank also appeared to be an important factor between 11 and 14 weeks and between 19 and 22 weeks of age.

Although the most common threateners of all infants were adult males and adult females, infants in the top-ranking lineage were more likely to be related to, and hence familiar with, their threateners than were infants in the lower-ranking lineages. This was shown by comparing: (1) the proportions of threateners who were not relatives of the infant they threatened for infants in each lineage; and (2) the median percentages of point time samples during which infants in each lineage were recorded within 5 m of their threateners. During all age periods for which the analyses were possible (11–14, 19–22 and 27–30 weeks), infants in the top-ranking lineage were threatened by smaller proportions of non-kin and spent more time near their threateners than did infants in other lineages. These differences could be important if infants are less easily frightened by threats from familiar animals or if threats from related and familiar individuals (who may also interact with infants in friendly ways) are less intense or less likely to have serious consequences.

Protectors

Infants were protected primarily by their mothers and by other close female relatives. Of a total of 29 interventions, 19 were by mothers, six by sisters, three by less closely related females and one by an uncle. An intervention or protection was defined as a threat which was followed within 15 s by one or more of the following reactions by a third individual (the protector): (1) approaching the infant, making contact and/or carrying it away; (2) screaming at, threatening or chasing the threatener; and (3) neither avoiding nor breaking contact with an infant who had approached and made contact with it for at least 5 s. During the first 22 weeks, infants in each lineage were equally likely to be protected by a third individual after being threatened. However, between 27 and 30 weeks, infants in the top-ranking lineage received protection after a larger proportion of threats than did infants in other lineages ($P < 0.07$, Mann-Whitney U-test).

Interestingly, these differences were not due to differences between patterns of protection by high- and low-ranking mothers; high- and low-ranking mothers protected their infants at similar rates ($P < 0.22$, Mann-Whitney U-test). However, in cases where infants were not protected by mothers, infants in the top-ranking lineage were more likely to be protected by other group members than were infants in the other lineages. ($P < 0.014$, Mann-Whitney U-test). Indeed, between 27 and 30 weeks, no infants in the middle- or bottom-ranking lineage were observed to be protected by individuals other than the mother, whereas six out of seven infants in the top-ranking lineage received this sort of protection.

Although almost no case of intervention was followed by continued threatening, protectors of infants in the top-ranking lineage were less likely to gesture fearfully while attempting to intervene than were protectors of infants in the two lower-ranking lineages. The protectors of six out of seven infants in the top-ranking lineage were never observed to fear-grin, scream, cower or run away, whereas the protectors of all five infants in the other lineages did so in at least half the cases. In this sense, infants in the top-ranking lineage had more effective protection and possibly more successful models from which to learn. Perhaps not surprisingly, infants in the top-ranking lineage were also less likely to be frightened when threats were not followed by interventions. These infants emitted fearful gestures in a smaller proportion of such cases than did infants in lower-ranking lineages ($P < 0.05$, Mann-Whitney U-test).

In summary, several aspects of the infants' experiences with agonistic interaction, including those which do not directly involve the mother, tend to ensure that they will take on ranks similar to their mothers', even before they are fully integrated into a peer–peer dominance hierarchy. These aspects are: (1) infants in high-ranking lineages are threatened at lower rates than other infants and are less likely to be threatened by unrelated or unfamiliar individuals; (2) by 27 to 30 weeks of age, just as infants are beginning to interact with each other according to rank, infants of high-ranking mothers are more likely to be protected than infants of lower-ranking mothers; and (3) when protected, their protectors are less likely to react with fear. Group members other than the mother, particularly close female relatives, appear to play a large role in the protection of infants and may be more directly responsible for the differences discussed than the mother.

9.3 Influence of Close Female Relatives on Peer–Peer Rank Acquistion

CAROL M. BERMAN

The outcome of aggressive interactions among infant and yearling rhesus monkeys is usually predictable from their mothers' rank. One way for infants to acquire their mother's rank is through the active intervention of group members, especially close female relatives, in their agonistic encounters. At about the time that infants are beginning to interact with each other according to maternal ranks, infants of high-ranking mothers are more likely to receive protection after being threatened than infants of low-ranking mothers. When protected, their protectors are more successful in that they are less likely to emit fearful gestures to the infant's threatener. The possibility that the interventions of close female relatives on behalf of infants contribute more towards the establishment of infant ranks than do those of the mother will be explored here. An examination is made of the extent to which individual differences in the frequencies of threats infants receive, the proportions of threats followed by intervention and the success rates of intervention are related to variables concerning the kinds of relationships infants have with their mother, close female relatives, other relatives and with the highest-ranking members of the group. Does maternal rank influence infants indirectly (as well as directly) by influencing their relationships with their principal protectors and, if so, how? It may be, for example, that infants of high-ranking mothers have more protectors. It is also possible that they have closer relationships with their protectors, spending more time near them where they can be protected by them more easily. The fact that infants in high-ranking lineages have high-ranking relatives may also contribute to differences between infants, because high-ranking protectors may be better at inhibiting aggression and winning fights. The data were obtained on Cayo Santiago.

Protection by close female relatives other than the mother (i.e. sisters and grandmother)

Infants of high-ranking mothers did indeed spend more time near (within 5 m) their close female relatives than infants of lower-ranking mothers, partly because they tended to have more of them ($P < 0.06$, Kendall rank-order correlation test) and partly because they spent more time near each of them ($P < 0.004$, Mann-Whitney U-test). This probably reflects both the greater social cohesiveness of high-ranking lineages (see Cheney 1978b) and their faster reproductive rates (Sade *et al.* 1976).

In addition, infants who had more close female relatives and spent more time near them were those who were threatened less frequently ($P < 0.05$ in both cases, Kendall rank-order correlation test). When the mother's rank was controlled through partial correlation tests, these correlation coefficients were virtually unchanged, suggesting that the presence of close female relatives may inhibit others from threatening an infant regardless of its mother's rank. Nevertheless, infants of high-ranking mothers are likely to be at an advantage because they tend to associate with close female relatives more than infants of low-ranking mothers.

Infants who had more close female relatives and who spent more time near them were more likely to be protected from threateners by group members other than the mother ($P < 0.05$ in both cases, Kendall rank-order correlation test). However, these correlations were partly dependent on variations in mothers' (and close female relatives') ranks, suggesting that differences in proportion of protection were due to the relative amounts of time spent with close female relatives and to differences in their ranks and/or the mother's ranks.

Finally, infants who had more close female relatives and who spent more time near them tended to be those whose protectors were less likely to emit fearful gestures when intervening on their behalf. However, these correlations were highly dependent on variations in mothers' and/or close female relatives' ranks, suggesting that differences in the success of intervention were due largely to differences in mothers' and/or close female relatives' ranks.

Protection by high-ranking group members

There was also evidence that infants who spend more time near high-ranking individuals may receive better protection regardless of their mothers' ranks. There were modest tendencies for infants who spent more time near the top-ranking adult males of the group to be threatened less and to have slightly higher success rates for intervention than infants who spent less time near them. Time spent near these males was not strongly related to maternal rank. However, infants of high-ranking mothers did spend more time near members of the top-ranking lineage than did infants of lower-ranking mothers ($P < 0.01$, Kendall rank-order correlation coefficient $= 0.60$). Hence, they were probably in better positions to enjoy the benefits of associating with high-ranking individuals than were infants of low-ranking mothers.

Protection by mothers

In contrast, differences between infants of high- and low-ranking mothers were not due directly to their relative proximity to their mothers. The amount of time infants spent near (within 5 m) their mothers was not associated strongly with the mother's rank, the frequency of threat, the proportion of interventions or the success rate of intervention.

Protection by other close relatives

The number of other close relatives infants possessed (brothers, aunts, uncles, nephews or nieces) and the amount of time infants spent near them were not strongly associated with mother's rank, frequencies of threat, proportion of intervention or success rates of intervention. Hence, differences between infants of high- and low-ranking mothers were probably not due directly to differences in their relationships with these relatives.

These findings strengthen the suggestion that close female relatives play an important role in the infants' early agonistic experiences in ways which are likely to contribute to their acquisition of rank which is similar to that of their mothers. Maternal rank appears to influence directly the number of protectors an infant has, their ranks and the amount of time spent near them. These factors in turn influence the frequency with which infants are threatened, the likelihood of their protection, the effectiveness of their protection and possibly the quality of their models (see Fig. 9.1). This is not to imply that the mother may not also influence rank acquisition more directly (e.g. Altmann 1980) but rather to emphasize that the mother's rank can also have indirect and subtle consequences for an infant. By influencing its relationships with other companions the mother can influence her infant's development without necessarily being directly involved or even present (e.g. Loy & Loy 1974; Walters 1980).

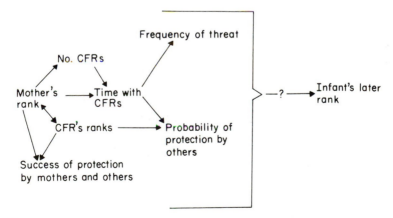

Fig. 9.1 Hypothetical causal relationships linking variations in the mother's rank with variations in the infant's early agonistic experiences. CFR = close female relative. (After Berman 1980b.)

9.4 Influences of the Social Situation on Male Emigration
JOHN COLVIN

In many animal species there is dispersal of the adult phase away from the natal site (Baker 1978). In a number of species of birds and higher mammals this dispersal entails emigration from the natal social group to another group, in which breeding may occur. In primates, as in other mammals, it is males that emigrate, while females remain with their natal group throughout their lives (Greenwood 1980). Thus, in 25 of 29 species studied, almost all males born within the group transfer to neighbouring groups before full adulthood (Wrangham 1980).

Male emigration, and natal emigration in particular*, is of interest in the present context because it leads to a radical restructuring of all the emigrant's relationships. To what is this restructuring due? Is it due simply to a decreased interest in all relationships in the natal group, accompanied by an increased interest in potential relationships in a neighbouring group? Alternatively, does it arise in response to particular cues, with the result that most bonds are broken and then rebuilt elsewhere because of some overriding but unrelated propensity? If this latter possibility were found to be the case, then it is necessary to ask whether the particular cues involved are ecological or social. If they are social, it should also be asked whether they are related to a single dyadic relationship or to general aspects of the social situation. Finally, if it is the social situation which is important, is it intra or extragroup factors or some combination of these that have the greatest effect?

While many authors have focused on the ultimate causes of emigration (Greenwood 1980; Harcourt 1978b; Marsh 1979; Packer 1979a; section 11.3), the question of proximate causes will be discussed here. The proximate causes of natal emigration have also been studied in several species of birds and mammals and the comparative evidence available points to the importance, in a highly social primate species, of negative extratroop factors on the one hand and positive intratroop factors on the other.

One method of determining which social factors are associated with male emigration in primates is to compare the social relationships of immature males who do emigrate with those of immature females who do not (Seyfarth *et al.* 1978). A more precise method is to take advantage of individual variation between males in the age of emigration. Since males emigrate at a variety of ages through adolescence and early adulthood (Drickamer & Vessey 1973; Kawanaka

*The term 'natal emigration' is used here to distinguish the first emigration from the natal troop from later emigrations during adulthood from non-natal troops (cf. 'natal dispersal' versus 'breeding dispersal', Greenwood *et al.* 1979).

1977; Packer 1977; Sugiyama 1976), the social situations of males who stay in their troops can be compared with those who leave at each particular age.

This latter method was used in a detailed study of the natal emigration of immature male rhesus monkeys (Colvin 1982). In this study, factors associated with emigration at different ages were analysed in two ways. In the first part of the study, patterns of emigration shown by 141 males from all six troops of the Cayo Santiago population over an eight-year period were examined. At this level of analysis, based upon data taken from censuses and genealogical records over a long period, only general aspects of troop structure could be examined. These included troop size and lineage structure but more importantly rank relations both within and between troops. The second part of the study, by contrast, focused on much more detailed differences in the social conditions of the immature males of one troop—Group I.

For the Cayo Santiago population censused between 1973 and 1980 the median age of natal emigration was four years old, with most males emigrating between three and six years of age. Occasionally, males emigrated while still juvenile (two-and-a-half years old) or remained in their natal troops into adulthood (six-and-a-half years old) (Fig. 9.2). Of the latter, one is currently in his twelfth year as a natal male, another in his fifteenth and both are alpha-males of their respective troops.

Given the wide range of ages at which natal emigration can occur, it is likely that the relevant causal factors will vary with age. Concern here is mainly with

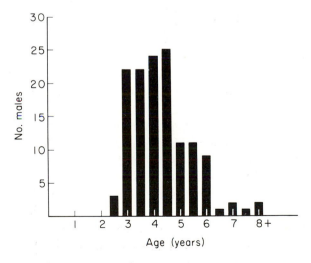

Fig. 9.2. Frequency distribution of natal emigration ages for rhesus monkey males born on Cayo Santiago between 1970 and 1975. The median age of natal emigration is four years old.

differences between those males who stay and those who emigrate at three to three-and-a-half years old or around the age of puberty. Intertroop factors shall be considered first and then intratroop factors in more detail.

In some primate species, there is good evidence that negative intertroop factors may delay male emigration. In vervets and Japanese macaques, female aggression during intertroop encounters may be the primary factor acting to limit male movement (Cheney 1981; Packer & Pusey 1979), while in baboons competition with the resident males of the opposing troop may exert the major inhibitory influence (Packer 1979a). However, these negative factors must be offset against more positive attractions. A number of authors have noted an attraction during intertroop encounters between unfamiliar males and females, especially young females in a sexually receptive state (Cheney & Seyfarth 1977; Hrdy 1977b; Marsden 1968) and both Packer (1979a) and Pusey (1980) have suggested that the immediate motivation to transfer may be a strong sexual attraction to unfamiliar individuals of the opposite sex. Affiliative interactions with subadult males from other troops may also facilitate transfer into those troops (Boelkins & Wilson 1972; Cheney & Seyfarth 1977; Packer 1979a), particularly for younger males who initially integrate into peripheral all-male subgroups (Colvin unpublished data).

In some populations, however, intertroop encounters are either rare and non-aggressive (Anderson 1981) or, as in the present case, frequent yet seldom aggressive. As a result, those transfers which have been observed are accomplished with little or no opposition (Anderson 1981; Colvin unpublished data). Thus, in some populations it appears that negative intertroop factors do not influence emigration. The absence of any data relating individual differences in age of emigration to individual differences in involvement in aggressive intertroop encounters supports this interpretation. In addition, in view of the fact that Cheney (1981) found no strong individual differences in rates of aggression received during intertroop encounters in vervets, this argument could apply also to those populations and/or species in which encounters are both frequent and aggressive. In these cases, aggression from females appears to play a role in determining to *which* troop a young male transfers.

In a similar fashion, the positive intertroop factors of attraction to unfamiliar individuals and affiliative interactions between strangers probably influence an emigrating male's choice of troop (and perhaps, as a consequence, the exact timing of his emigration) but not his age of emigration. Attraction to unfamiliar but proceptive females or familiarity with subadult males is likely to offset the effects of aggressive encounters with these individuals' troops. As a result, the pattern of male intertroop movements may not be random but may be affected by the distribution of previous transfers. There is some evidence to support this in vervet (Cheney 1981) and rhesus monkeys. In the latter case, the predominant tendency was for males to transfer to troops of adjacent but higher rank (Colvin 1982) (Fig. 9.3). In so doing, they joined potential previous acquaintances from their natal troops (either brothers or non-fraternal relatives) significantly more often than expected by chance (see also Meikle & Vessey 1981).

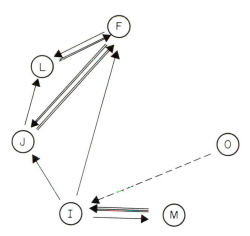

Fig. 9.3. Intertroop preferences based on male transfers among the six Cayo Santiago troops, arranged in order of intertroop dominance (F [highest] to O[lowest]). The percentage of emigrations from each troop are as follows: 20–40% (single arrow); 40–60% (double arrow); 60–80% (treble arrow); and 75% (broken arrow [$n = 4$]).

The fact that intertroop factors are unlikely to influence directly the age of emigration, apart from affecting the immediate opportunity to transfer troops, suggests that *intra*troop conditions might be of primary importance. Such conditions, to which the male would be exposed more or less continuously, could determine both the onset and/or the strengthening of a propensity to emigrate, with intertroop factors acting only as the final stimulus to emigrate. As argued above, an overall balance of positive intratroop factors should delay emigration, whereas an overall balance of negative intratroop factors should stimulate emigration.

A major clue to the nature of these factors comes from the analysis of population data. A number of studies have looked at the effect of maternal rank on the age of male natal emigration (Cheney 1978; Drickamer & Vessey 1973; Itoigawa 1975; Kawanaka 1977; Sugiyama 1976). None of these studies has demonstrated particularly striking rank effects, although reanalysis of Kawanaka's (1977) data, taken from observations of a single troop of Japanese macaques, reveals a significantly positive correlation between lineage rank and the mean age of male emigration. Seventeen out of 20 males were from medium- or low-ranking lineages and all but one of them had emigrated by six years of age.

The three males from the high-ranking lineage, on the other hand, emigrated at five, eight and eleven years old respectively.

In the present rhesus monkey study, the effects of lineage rank showed much more strongly. Within each of five troops, lineages could be classified as either high or low ranking, with intermediate-ranking lineages grouped into a third, medium-ranking class (a single lineage troop, Group O, was excluded from this analysis). Males from low-ranking lineages emigrated at a mean age of 3.45 years, significantly earlier than males from medium- or high-ranking lineages (where the mean ages of emigration were 4.11 years and 4.87 years respectively). The only males to emigrate at two-and-a-half years were from low-ranking lineages, while it was only high-ranking males who emigrated later later than six years of age. Males from the single-lineage troop behaved as high-ranking males, with only two out of eight males emigrating before six years of age.

The effect of lineage rank on the age of natal emigration could be transmitted through any set of interactions or relationships whose patterning is influenced by rank. At one extreme, this could involve a single, key relationship of a male. At the other extreme, it could involve general aspects of a male's interactions with all troop members, e.g. overall time spent in proximity to others or overall rates of aggression received from others, i.e. situational factors. Between these two extremes a male could be affected by various subsets of his relationships. These subsets may be divided into two types: (1) those in which all partners also have relationships with each other (closed networks), these relationships being likely to exert a direct effect on those of the male within the network; and (2) those in which the other individuals concerned do not necessarily interact with each other (open networks).

From the study of focal males in Group I there is little evidence to support the notion that rank-related effects might be transmitted through the most general aspects of a male's interactions. Thus, low-ranking males between two and three-and-a-half years of age spend no less time than do high-ranking males in proximity to others or being groomed by others, nor do they receive aggression from others at a higher rate. However, low-ranking males are less likely to be related to their aggressors than are high-ranking males and this may affect the psychological value of the aggression received.

The two main open networks of the immature male comprise the subset of his relationships with mature females and the subset of his relationships with adult males. Silk *et al.* (1981b) have shown that among captive bonnet macaques *(Macaca radiata)* low-ranking immature males receive aggression from adult females at higher rates than do high-ranking immature males, although immature females receive considerably more aggression than immature males of similar ranks (see also Dittus 1977; Silk *et al.* 1981a). There is no evidence, however, that in rhesus monkeys a similar rank-related effect is operating.

Indeed, of those immature males who remain with their natal troop at three-and-a-half years old, high-ranking males actually receive aggression from adult females at a higher rate* than do low-ranking males.

In relationships with adult females, there may on the other hand be advantages to be gained by high-ranking males who delay emigration, which are not available to low-ranking males. In particular, while all males who remain with their natal troop at three-and-a-half years old spend considerable time during the breeding season observing at close range consorting females, high-ranking males devote more time to this activity than do low-ranking males. The benefits to be gained from this could include the observational learning of a wide range of social skills relevant to mating. In addition, at both three-and-a-half and four-and-a-half years, only high-ranking immature males copulate with receptive females. If, as a result, these males were to father offspring in the natal troop, this might then delay their emigration even further. In chacma baboons *(Papio ursinus)* it is known that males having infant offspring in the troop emigrate significantly less than expected if paternity were not taken into account (Hamilton & Busse 1980).

By contrast, lack of mating success in the natal troop, considered by a number of authors as a key factor in causing emigration (Altmann & Altmann 1970; Marsh 1979; Packer 1979a), is likely to take on increasing importance as males grow older. Males may be affected either because access to receptive females is frustrated or, in non-seasonal breeders, by finding most females impregnable; this latter factor is known to stimulate 'breeding emigration' by adult males in yellow baboons (Rasmussen 1979). In addition, lack of mating success may lead to increased intrasexual competition, in turn causing emigration (Dunbar 1979a).

The effects of relationships with adult males on natal emigration have been stressed by Dittus (1979). In langurs and patas monkeys, immature males are rejected by the adult male from the natal troop at an early age (Hall 1967; Poirier 1969; but see Gutstein 1978). In other species, however, it is the consequences of weak or inhibited relationships with adult males which appear to be more important. Harcourt and Stewart (1981) have shown that for immature male gorillas nearing maturity, lack of a close relationship with the current leading male when they were younger leads to gradual peripheralization and dispersal, rather than to continued integration and potential take-over of group leadership. In rare cases in bonnet macaques, lack of strong bonds with adult males may also induce peripheralization and emigration in subadult males (Simonds 1973).

For the rhesus macaques on Cayo Santiago, relationships with adult males may have less importance in relation to emigration than relationships with male peers (see below), at least at younger ages. Thus, immature males do not begin seriously to establish a position for themselves in the adult male hierarchy until

---

*Rate per female higher ranking than the male's mother.

they reach the age of four years although, when they do, previous relationships with adult males may contribute to the new rank they achieve. In this respect, high-ranking immature males may be at an advantage, since there is a positive correlation between immature male rank and the rank of the 'central' adult males with whom strong bonds are formed. However, with regard to emigration at three-and-a-half years, relationships with adult males probably play a minor role. On the one hand, all immature males regardless of rank have at least one strong bond with an adult male, who grooms them frequently and on occasion may provide aid. On the other hand, while immature males receive the most aggression from peers and older males, there are no rank differences between the immature males in the amounts of aggression received. Furthermore, for males emigrating at puberty (three to three-and-a-half years) there is no direct competition with adult males for access to receptive females; the degree of male–male competition at later ages remains to be assessed.

The two, main closed networks of the immature male comprise the male peer network, which will cover a varying breadth of ages depending on demographic factors (cf. Altmann & Altmann 1979) and the family network (mother and close male and female kin), extending into the lineage network. Brothers may occupy a key position in both networks. On Cayo Santiago male age-cohorts are large (with a mean size of 3.2 males in 1970 increasing to 7.2 males by 1976); accordingly, consideration of peer networks was restricted to males within these cohorts only (Colvin 1982). Since the ranks of immature males follow those of their mothers until puberty (Loy & Loy 1974; Sade 1967), males within each cohort rank in a stable hierarchy, reflecting maternal and lineage ranks. A relative rank can be assigned to each male on the basis of the percentage of other males within his cohort to whom he is dominant. The relative rank of the individual within his cohort (RRC) thus reflects a male's relative dominance or subordinance in his peer network, with values greater than 50% indicating dominance over most individuals and values less than 50%, subordinance. From the population data there emerges a strong relation between the RRC at one year of age and the age of emigration (Fig. 9.4). The mean age of natal emigration for males with an RRC at one year of 100% is 5.20 years. Such males delay emigration for longer than those with an RRC of less than 100% but greater than 50% (their mean age of natal emigration being 4.26 years). Males with an RRC less than 50% but greater than 0% delay natal emigration on average by 3.61 years, whereas males with an RRC of 0% delay the least (their mean age of natal emigration being 3.26 years). This effect is both clearer than that described for lineage rank alone and also more precise since it is related specifically to relationships within the peer network.

How might relative overall dominance or subordinance in peer relationships influence emigration? The study of the focal males in Group I points to a number

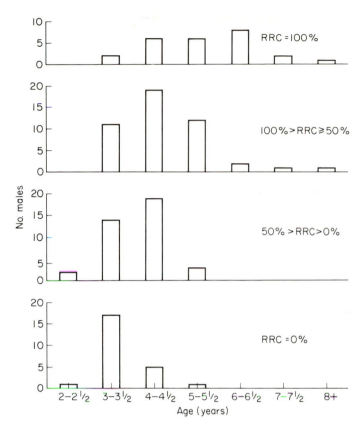

Fig. 9.4. Frequency distribution of rhesus monkey male natal emigration ages in relation to their relative rank in the cohort (RRC) at one year. Males with the lowest RRC emigrate at the youngest ages and the age of emigration increases with the RRC.

of factors which might be involved. Considering only those peer relationships with an aggressive component, at three and three-and-a-half years low-ranking males receive higher rates of aggression from their peers than do high-ranking males (Fig. 9.5). Considering all peer relationships within a network, low-ranking males at three years show the least responsibility for maintaining close proximity to their peers out of deference to their higher-ranking partners. Related to this, these males are the most inhibited from interfering between males of other peer dyads and thus least able to protect their most-valued peer relationships. Finally, taking a role often as a dominant rather than as a subordinate partner in the most important peer relationships (the 'strong' and 'weak' peer relationships described in section 3.2) may also be associated with early emigration. Taken together, these four factors might create in the low-ranking male a weakened sense of affiliation towards his peers, encouraging him to emigrate early.

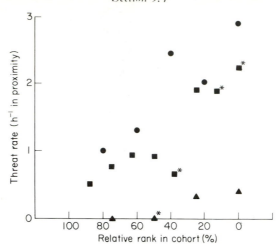

Fig. 9.5. Aggression received in rhesus monkey male peer relationships in relation to their relative rank within a cohort. The rate of aggression received increases as the relative rank decreases. Data for two male cohorts in Group I are shown: the 1974 cohort at three years (solid triangle) and the 1975 cohort at three years (solid square) and three-and-a-half years of age (solid circle). Some low-ranking males who emigrate at three-and-a-half years of age (asterisked) received particularly high rates of aggression at three years old.

Conversely, high-ranking males might be expected to delay emigration due to the absence of weakening or inhibiting influences in their peer relationships.

Concerning the effects of the other type of closed network—the family/ lineage network—on age at natal dispersal, there is as yet less evidence available. However, the unusually low rate of emigration from one troop on Cayo Santiago—Group O (see p. 164)—may be due to lineage structure in this troop. Group O comprises a single lineage and as the troop has grown in size there has been an increasing proportion of adult males who are natal and thus also members of this lineage. The network of relationships in this lineage might therefore show greater 'cohesiveness'* than that of other lineages for at least two reasons. First, adult females of a number of Old World species are known to harass those of lower rank than themselves, especially those of other lineages (Dunbar 1980b; Silk *et al.* 1981a). As a result, subgroups of females from low-ranking lineages are more likely to be broken up than those of females from high-ranking lineages, leading to decreased cohesiveness in those lineages. Second, the presence of related rather than unrelated males in association with a lineage would tend to increase rather than (possibly) decrease the cohesiveness of that lineage. Furthermore, as a whole, in a single-lineage troop this would tend to

*As measured, for example, by the ratio of strong to weak bonds in the network or by the percentage of all possible relationships which are realized (network 'density', Barnes 1972; see also Hanby 1980a).

increase overall troop cohesiveness due to greater male–male affiliation and more widespread adult male–juvenile bonding. This latter effect has been described for both Barbary and bonnet macaques (e.g. Taub 1980a; Wade 1979) and is probably an important cause of the delayed male emigration in these species (Ali 1981; Simonds 1973; Taub pers. comm.). However, it is not known whether this delayed male emigration is a direct consequence of overall troop cohesiveness *per se* or due simply to stronger and more widespread bonds between immature and adult males. If it is overall troop cohesiveness that is important, then its direct effect on immature male relationships has yet to be assessed. One possibility is that troop cohesiveness influences the 'compatibility' of an individual's relationships, i.e. the extent to which relationships can, when the situation arises, overlap without conflict.

Finally, we turn to the possibility that the effects of lineage rank on age of male emigration could be transmitted through a single, key relationship. Alternatively, such a relationship could influence natal emigration in other ways not directly related to lineage rank. Since immature males have potentially key relationships with their mother, with brothers, with one or two adult males and with one or two male peers, each of these types of relationship should in principle be considered. The key peer relationships, the sibling and the mother–son relationship, will be briefly considered here.

While the relative degree to which a male is a dominant rather than a subordinate partner in both strong and weak peer relationships (section 3.2) taken together may affect the age of emigration (see p. 168), there is only weak evidence for a similar effect when only the key (strong) peer relationship(s) is considered by itself. There is little evidence either for any connection between the quality of the sibling relationship and the age of emigration. Probably this is because, irrespective of rank, all immature males who have brothers have strong friendships with at least one of them (see section 3.2). However, when they emigrate, there is no evidence that on Cayo Santiago males do so in the company of their brothers. Thus, only 19% of males who could have accompanied older or younger brothers did so, while almost half the males who emigrated alone had brothers who were potential companions (Colvin 1982).

Finally, the relation between the nature of the mother–son relationship and emigration shall be considered. Data on orphaning in vervet monkeys suggest that this may lead to early emigration (Cheney *et al.* 1981), although there is no evidence to support this in Japanese monkeys (Hasegawa & Hirawa 1980) or rhesus monkeys (Colvin unpublished data; Drickamer & Vessey 1973). On the other hand, Itoigawa (1975) found that often it was possible to predict when young male Japanese monkeys would leave their natal troop on the basis of the degree of association between mother and son at one year of age and changes in association thereafter.

In the study of focal rhesus males, a similar relation was found (Colvin 1982). Data from this study suggest that the nature of the changes in the mother–son relationship between two and three years of age is predictive of emigration at three-and-a-half years old, although, thereafter, differences in the relationship with the mother between those who stay and those who emigrate are less clear. There is also some evidence that the emigration of males is delayed if their mothers are alpha-ranking in their lineages; this effect is strongest in middle-ranking lineages. Finally, while the expression of mother–son bonding declines from four years of age onwards, some very high-ranking females may continue to influence their sons' social integration by assisting their integration into the adult male hierarchy (section 9.5; Colvin unpublished data; Koford 1963).

To summarize, it is clearly not the case that a uniform decrease of interest across all relationships in the natal troop precedes male emigration. It would appear instead that certain key social factors influence emigration, though we are not yet able to specify precise causal links or to say whether emigration is preceded by psychological changes only or by physiological or physical intermediaries, such as stress or weight loss. These key social factors include, for the young male, a change in the relationship with his mother but more importantly, the social situation *vis-à-vis* his male peer network. Greater overall dominance within this network delays emigration, whereas greater overall subordinance encourages emigration. In addition, various other aspects of troop structure may influence the age of male emigration, including lineage structure and factors affecting relationships with mature females. By contrast, the available evidence suggests that *inter*troop factors are less likely to influence directly the age of emigration but may influence the emigrating male's choice of troop.

This survey of the wide range of factors which may influence the age of male natal emigration has shown that the influence of the social situation on the individual is not only diverse but can also be highly complex. Nonetheless, certain factors, such as the closed networks comprising peer relationships, would seem to hold more promise for further investigation than others. It is fortunate in the present case in that it is only a single outcome being dealt with — the choice either to emigrate or to stay at any one age. However, it should be borne in mind that the key factors which influence this outcome are likely to vary with age. They may even vary across individuals, although here the simplifying assumption that all individuals are affected by a similar set of factors has been made. The inability to provide adequate descriptions of social situations and the social structures, such as network structures, to which these give rise perhaps most hampers investigation at present. To return to the peer network for a moment, primatologists are now used to demonstrating without too much difficulty how certain dyadic relationships can be constrained by others, either actively through interference or passively through inhibition based upon fear or respect. However, there has as yet been little thought given to the problem of describing and explaining how contextual factors enhance dyadic relationships and, more importantly, under what conditions the dyad is more properly to be considered as a component of an undifferentiated triad or higher-order structure.

9.5 Matriline Membership and Male Rhesus reaching High Ranks in the Natal Troops

BERNARD CHAPAIS

As a rule, male rhesus leave their natal troop around puberty. While residing in it, they usually rank among themselves according to their mother's rank and cannot be assigned an individual (basic) rank in relation to transferred, adult males (Drickamer & Vessey 1973; Kaufmann 1967). The purpose of this section is to illustrate how the long-term social situation (here matriline membership) can differentially affect the social relationships of natal males, through a description of the process by which the two mature sons of the alpha female of group F at Cayo Santiago achieved and maintained top-ranking positions in the central male hierarchy of their natal troop (contrary to the above mentioned rule). The phenomenon described here is not idiosyncratic. The achievement of high rank in their natal troop by males related to high-ranking females has been observed frequently in groups of rhesus and Japanese macaques (e.g. Koford 1963; other references in Chapais 1981).

In January 1978, group F included four matrilines totalling 72 individuals (Fig. 9.6) and a central (i.e. non-peripheral) male hierarchy of 14 adult males (Table 9.1). Of the 11 natal males aged four years or older, two were members of the central hierarchy (seven-year-old 415, the alpha-male and the son of the highest-ranking female of the group, FB, and 436, also aged seven, who had been out of group F before January 1978). Six natal males belonging to lower-ranking matrilines left group F during the birth season, while two others stayed with the group but did not become part of the central male hierarchy. The remaining natal male, four-year-old 580, the other mature son of the alpha-female FB, eventually rose in rank. His story will be reported below.

FB's status was unique among females. She was subordinate to the alpha-male (her son 415) but dominant over second-ranking male EE and fourth-ranking male 1J and she was never observed to direct any submissive gestures to third-ranking male TJ and fifth-ranking male Z2. Furthermore, all these males seemed desirous to avoid any circumstances which might have triggered an agonistic interaction between them and FB. On the other hand, FB was clearly

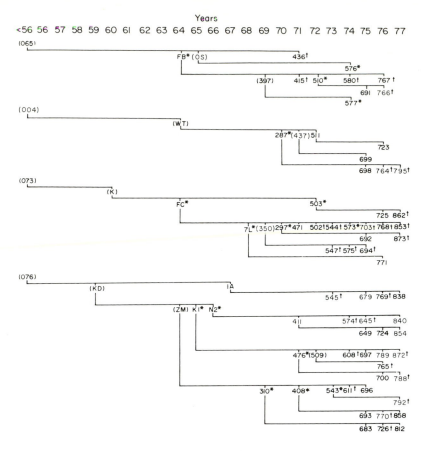

Fig. 9.6. Date of birth and genealogical relations through the maternal line of the natal members of group F in January 1978. Monkeys who died before this date are not included except for connector females (in brackets). † = male; * = female who gave birth to a live offspring during the 1978 birth season.

subordinate to other *lower-ranking* non-natal males. Various lines of evidence suggest that this triangular dominance structure is best explained in terms of alliances between FB and high-ranking males.

Between May 13th and 18th 1978, 580 was observed for the first time to displace passively or actively high-ranking males EE and TJ. His status in relation to lower-ranking males was to be ascertained over the following days. Two aspects of 580's behaviour seem relevant to the understanding of his rapid ascent to the second-highest rank, below his brother 415. First, 580 regularly approached and groomed 415, often just after attacking other males and females. In doing so he might have demonstrated the existence of a potential alliance between himself and his brother. Second, and probably more important, 580

Table 9.1. Distribution of submissive behaviour among central males (all sampling sources combined). The rank order presented is the one that produced the smallest frequency of behaviours above the diagonal. N = a male born in group F; the figures in brackets give the age of the males in January 1978.

Targets†

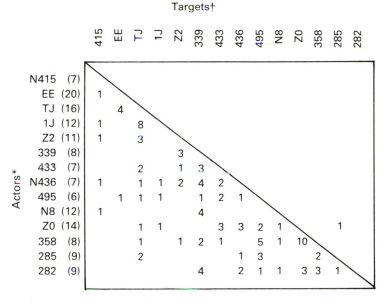

Actor (age)	415	EE	TJ	1J	Z2	339	433	436	495	N8	Z0	358	285	282
N415 (7)														
EE (20)	1													
TJ (16)		4												
1J (12)	1	8												
Z2 (11)	1	3												
339 (8)					3									
433 (7)						2		1	3					
N436 (7)	1			1	1	2	4		2					
495 (6)			1	1	1	1	2	1						
N8 (12)	1					4								
Z0 (14)						1	1		3	3		2	1	1
358 (8)					1	1	2	1	5	1	10			
285 (9)						2			1	3		2		
282 (9)				4		2	1	1			3	3	1	

Actors*

*The males approached, threatened or attacked the targets.
†The males fear-grinned, cowered, fled or were supplanted.

associated closely with his mother (FB) and his sister (510). These three individuals groomed each other on a regular basis and 580 used to displace actively males EE, TJ and 1J at times when these were near FB. The latter never opposed such interferences and was even observed to support her son against adult males. FB, 580, 510 and, to a lesser extent, 577 and 576 (see Fig. 9.6) formed a monolithic and persistent coalition. High-ranking males would literally jump away from FB when 580 approached her.

The foregoing suggests that as soon as 580 began to challenge higher-ranking males, he could take advantage of support from his matriline. It was then only a matter of time before 580 could assert his status independently of the actual presence of his allies. By July, 580 was fully integrated in the central hierarchy, being deferred to by all males, except 415, independently of the presence of his relatives.

On September 10th, 415 lost one testicle in the course of a fight with an extragroup male. Twelve days later, he exhibited the first signs of tetanus (limb and body stiffness) and was observed for the first time to be challenged by adult males. On the following days, 580 was seen threatening his brother 415, who could hardly move and kept fear-grimacing. After an absence of three days, during which he was thought to be dead, 415 was back with group F, showing subtle signs of recovery and obviously ranking at the bottom of the male hierarchy.

From that time on, 415 was involved with males and females in a large number of social interactions which differed strikingly from previous ones. One aspect of 415's efforts to regain a central position in group F came out clearly. In sharp contrast to his previous indifference towards adult males, 415 systematically undertook to threaten (staring at, hoarse-grunting at and even chasing) males N8, 433, 339, Z2, TJ and EE (1J had emigrated) who were just passing by. Most of these responded at first by counter-threatening and attacking 415. A series of spectacular and highly meaningful interactions took place in these circumstances: 415's relatives (FB, 580, 510, 576 and 691) as well as unrelated adult females were observed to defend 415 successfully against retaliating males. They would immediately and vigorously chase and even bite males who threatened or attacked 415. Such interactions expressed both the readiness of 415's relatives to defend him as often as he would be challenged and the strong dependence of 415's status on such support. Over the following weeks, 415 rose to the second rank, below his brother 580 who continued to express his dominance over his older brother.

Necessary conditions for males to achieve high ranks in their natal troop can now be summarized on the basis of the present data. 415's high status at the beginning of the study period indicates that support from a higher-ranking brother is not such a necessary condition. However, support from high-ranking female relatives appears to be crucial. First, female support against other *natal* males is ensured by a strong asymmetry of power among females, expressed as a well-defined dominance hierarchy (Missakian 1972; Sade 1967, 1972b): the rank of natal males corresponds to their mother's rank (Kaufmann 1967). Second, female support against *non-natal* males is facilitated by the weakness of the sexual dimorphism (= 1.09, based on calculations made from data on the weight of adults in Napier & Napier 1967) and expressed through coalitions of both related and unrelated females (Chapais 1981).

The fact that only sons of high-ranking females reportedly achieve high ranks in their natal troop might depend not so much on the existence of rank-related differences in the capacity of females to support male relatives against non-natal males as on rank-related differences in the willingness of males to stay in their natal troop. This possibility is discussed elsewhere (Chapais 1981) in the context of a functional model concerned with the timing of male transfer. According to this hypothesis, only males from high-ranking matrilines would be in a position to secure precociously a rank (and an associated reproductive success) high enough to offset the costs of postponed emigration. (Given this perspective, FB's status in relation to high-ranking males, while undoubtedly facilitating the process of status assertion of her sons, might not have been a necessary condition to it.)

Chapter 10
Description of and Proximate Factors
Influencing Social Structure

10.1 Description
ROBERT A. HINDE

Species differences in group structure may be revealed in data about group composition, e.g. generalizations can be made about species which usually live in male–female family groups, one-male groups, multimale groups, etc. Within such categories, data on interindividual spacing may indicate more subtle differences between species or between the relationships of particular age/sex classes (Kummer 1971). However, groups of a particular demographic type may depend on very different types of interindividual relationships: for instance the one- (or multi-) male groups of mountain gorilla are maintained by the attraction of the females for the male (Harcourt 1979a), while in patas troops the situation is reversed, the male being a rather peripheral hanger-on (Rowell 1978). Furthermore, while data on interindividual distances can indicate something about the nature of interindividual relationships (Sigg 1980; Vogt 1978), the extent to which the spatial arrangement of individuals in a troop is consistent is a matter of debate (cf. Rhine *et al.* 1979 with Altmann 1979) and more precise behavioural data are necessary to describe social structure.

Description of the 'surface structure' of a troop, i.e. the structure observed in a specific instance, requires description both of the constituent dyadic (and higher-order) relationships and of how they are patterned (see e.g. Hinde 1976; Strayer & Harris 1979). Since relationships involve a time dimension, the description of surface structure must have reference to a specified time period.

176

Indeed, the patterning may vary with time, for instance between reproductive and non-reproductive seasons (Rowell 1972a; sections 6.9, 6.10 and 12.3), or with food availability (Aldrich-Blake *et al.* 1971; Crook 1966) and it may change as a small group matures into a larger one (Dunbar & Dunbar 1976; Eisenberg *et al.* 1972). Each relationship is likely to affect and be affected by a number of others and therefore any stability of structure is likely to be dynamic: a change at one point may have far-reaching repercussions.

In practice, of course, we must generalize from the 'surface structure' of particular troops to aspects of structure which are common to a number. However, description of all the relationships and of how they were patterned within even one primate troop would be a formidable task. As always, description requires selection from among the mass of detail presented by nature. And it must be admitted that there is as yet no generally agreed method for abstracting the essentials from data on the nature and patterning of relationships. Of course, as with relationships, to some extent the method must be suited to the problem under study: what is important for one issue will not be so for another. Even the patterning poses formidable problems however: while in small groups patterning can sometimes be described in terms of relatively few paradigms, such as linear, circular, radial, hierarchical and so on, larger groups may embrace all these and other possibilities. Sociograms (e.g. Kummer 1971), networks (Deag 1974), first-order zones (Hinde 1974) and dendograms (e.g. Bygott 1979) have all been used to illustrate aspects of social structure. Since structure is dynamic and changes with time, a succession of structures may be necessary (e.g. Kummer 1971). Another more general solution to this problem is presented by Schulman in section 10.7).

In addition, it may be profitable to describe social structure at a number of levels. In the Hamadryas baboon, for example, the basic unit is the family, consisting of a male and his females and their offspring. Families are grouped in clans, which may also contain older males, who no longer have females, and subadult male 'followers'. These clans may be grouped into bands, each containing about three clans or 60 individuals. A number of clans may form a troop, tolerating each other on the sleeping cliffs but otherwise interacting little (Kummer 1971, pers. comm.). Similarly, macaque troops contain females and their immature offspring grouped into matrilines, themselves associating in a troop.

Description is of course but a means to an end; we also need principles which will help the understanding of the processes which underlie the structures that we observe (cf. Chapter 1). Since the surface structures observed tend to have some consistency over time (e.g. Dunbar 1979a) and since the surface structures of different troops tend to have features in common, it seems likely that most interspecies differences have a genetic basis. This view receives some support from the finding in the one case which has been studied that interspecies hybrids showed an intermediate type of structure (Nagel 1971). The principles sought must therefore first concern characteristics of the species or of subgroups within the species. Those found useful so far fall into the following categories.

1. Principles concerned with relationships within or between age/sex classes

Within every species there are propensities for relationships of particular kinds within or between particular age/sex classes. Young males may be attracted to their age-mates, older males to females, females to their female relatives, etc. A number of examples appeared in Chapters 5 and 8.

2. Status

The most widely used concept in the study of social systems is that of dominance. As discussed in Chapter 3, when applied to a dyadic relationship, the concept of dominance refers to the pattern of the directions of complementary interactions within the relationship and can have some of the properties of an intervening variable. In a group situation, dominance refers to the patterning of relationships: if A is dominant to B and B is dominant to C, it usually happens that A is dominant to C. The individuals can thus be arranged in a 'dominance hierarchy'. This may be merely descriptive if it refers only to one aspect of the relationships or in some degree explanatory if it refers to a number of correlated ones. (In practice, opinions differ on exactly how the concept of dominance should be used: some recent discussions are given by Bernstein [1981]; Deag [1977]; Hinde [1978]; Hinde and Datta [1981]; Jones [1980, 1981]; and Rowell [1974].)

Dominance is usually assessed in terms of the direction of agonistic responses or precedence to resources, a precedence which, it is reasonable to suppose, depends on past or potential aggressive responses. However, dominance rank may be predictive of the nature and direction of interactions other than those associated with agonistic behaviour or competition. A number of lines of evidence indicate that subordinate individuals are attracted to more dominant ones. For example Sade (1972a) found that rhesus females high in the dominance order tended also to have a high 'grooming status' (defined as the sum of the number of animals grooming the focal animals, the number grooming the groomers, etc.). In chimpanzees the more dominant individuals tend to be involved in more grooming bouts and to be groomed for longer than their inferiors (Simpson 1973). Similarly, the frequency with which Barbary macaques occur as nearest neighbours to others in focal observations is correlated with their dominance rank (Deag 1974), and subordinate, individual macaques tend to watch more dominant ones (Haude *et al.* 1976).

While there is a general tendency for dominant individuals to receive more grooming than subordinate members of a hierarchy (though *Alouatta palliata* seems to be an exception [Jones 1979]), most grooming goes on between animals that are adjacent in rank (e.g. Oki & Maeda 1973). Seyfarth (1976, 1980) explained this in terms of a tendency to groom individuals of high rank coupled with competition for access to such individuals. Thus, while high-ranking individuals can groom others of similar status, middle-ranking individuals cannot compete with them and are limited to other middle-ranking monkeys, and so on (see also Fairbanks 1980).

As discussed in section 6.1, grooming can be seen as a form of 'social approval'—a reward given in the (perhaps non-conscious) 'expectation' of future returns (e.g. Hinde & Stevenson-Hinde 1976). Seyfarth (1976) suggested that high-ranking partners are attractive because frequent grooming between two individuals increases the probability that they will subsequently support each other in agonistic encounters with third parties, and high-ranking individuals are more likely to win their disputes.

Rank in a group cannot always be predicted from success in dyadic encounters (Maslow 1936) and coalitions may play an important part both in determining dominance in the first place and in facilitating changes in dominance (e.g. Dunbar 1980b; Chapters 6, 9 and 12). While relative rank among juvenile male baboons depends on their size or age, that of females is related to their mothers' ranks (Lee & Oliver 1979) and depends primarily on their mothers' support (Cheney 1977) (see also Chapters 6, 9 and 12). Although dominance relationships usually show considerable stability, dominants are always liable to be challenged. In langurs, young females rise in dominance over the older ones, who remain active in troop defence (Hrdy & Hrdy 1976). Sudden changes in dominance often depend on coalitions or changes in coalitions (e.g. Chance *et al.* 1977; de Waal 1975, 1982; Gouzoules 1980; Nash 1974; Weisbard & Goy 1976). In rhesus monkeys, daughters tend to rise in rank above their older sisters and this also depends on support from other related individuals (see Datta 1981; sections 12.6 and 12.8).

Not surprisingly, subordinate animals pay a good deal of attention to more dominant ones and it has been suggested that the 'attention structure' of a troop is basic to the patterning of dominance relations (Chance 1967). However, the issue should certainly be considered the other way round: the characteristics of individuals that attract attention are just those that are indicative of aggressive potentiality. In any case, the ubiquity of a relation between attention structure and dominance structure has been questioned (de Assumpcao & Deag 1979).

Exchange theorists attempt to understand human interactions on the supposition that individuals tend to maximize their profits in interactions and minimize their costs. Homans (1961) suggested that distributive justice is realized when the profit (i.e. rewards minus costs) of each individual is proportional to his investments (i.e. that with which he has been invested and/or which he has acquired). What counts as an investment is to a large degree culture-specific but such things as age, wealth, maleness, strength and wisdom are commonly recognized in human societies.

In non-human primates, dominance, as determined by potential for aggression (with or without support), is clearly the most important form of investment. However, there is increasing evidence that other characteristics may also count as investments and contribute to status. For instance Bygott (1974), studying chimpanzees, found that age was correlated more significantly than any agonistic measure with the frequency of grooming and of being groomed, and in hamadryas baboons older males may play an important part in deciding where to move, while younger males take the initiative when to move (Kummer 1968; see also de Waal 1982).

In some species the highest-ranking male (or males) seems to contribute to troop organization in a number of special ways, acting as a policeman and protector, tending to support weaker animals against aggressors and so on (e.g. Buskirk *et al.* 1974; de Waal 1977; Estrada *et al.* 1977; Watanabe 1979). In Japanese macaques, it has been claimed that the alpha-males have an important influence on the structure of the groups (e.g. Grewal 1980a; Yamada 1971; see also Izawa 1980 on *Cebus apella*), though it is also argued that female aggressiveness can limit the numbers of males in the troop (Packer & Pussey 1979). High-ranking males have been reported as threatening troop members away from traps (Maxim & Buettner-Janisch 1963) and drugged baits (Fletemeyer 1978). However, the leadership activities of high-ranking animals are not necessarily matched by a high learning ability when faced with complex problems in artificial situations (Bunnell & Perkins 1980; Strayer 1976).

3. Blood relatedness

Among many species, individuals tend to associate preferentially with and aid their kin (e.g. Kaplan 1977; Kurland 1977; Massey 1977; Suomi 1977). This usually involves associations along matrilines. Thus, troops of macaques usually involve a number of matrilines, each with a matriarch, her daughters ranked in inverse order of age and their daughters similarly ranked, etc. (see Fig. 2.1). This may lead to a degree of genetic differentiation (Olivier *et al.* 1981). These ties between related females permeate all aspects of their behaviour—some illustrative examples are given in Chapters 6, 8, 9 and 12—and adult males form liaisons with particular kinship groups (Grewal 1980a). There is even evidence that mothers and daughters tend to give birth within a shorter interval than non-related females (Anderson & Simpson 1979). The probability of a macaque troop splitting is influenced by the average degree of relatedness between its members (Chepko-Sade & Olivier 1979). Furthermore, species differences in the intensity of matrilineal ties have been related to differences between species in the degree of relatedness between individuals in the troop (Defler 1978; Wade 1979). Similarly, in the quite different social system of chimpanzees, long-term preferential associations along matrilines are well known (Goodall 1978; Pusey 1978). Associations between related males are less common but have been recorded in chimpanzees (Bygott 1979) and it is known that related males sometimes move together between troops (Boelkins & Wilson 1972; sections 9.4 and 11.3).

The fact that individuals tend to associate with kin does not of course necessarily indicate that they recognize kin as such, though there is some evidence that they do (Small & Smith 1981; Wu *et al.* 1980). In general, it is more probable that the direction of social behaviour is influenced by some correlate of blood relatedness, such as the relative familiarity of other individuals (e.g. de Waal 1978).

The principles discussed so far concern characteristics of the species or of age/sex classes within the species. Variations in demographic structure between troops may permit these principles differential expression and thus lead to differences in group structure (Altmann & Altmann 1979; Dunbar 1979a). Variations in the sex ratio of births may also have important consequences on group structure (Rowell 1972a). Furthermore, troops may grow in size and then divide (e.g. Eisenberg *et al.* 1972; Furuya 1968) or change cyclically with seasons of mating, birth and infant development.

In addition, local conditions that affect interindividual relationships may in consequence affect group structure. For example food availability may affect the incidence of aggression in captive rhesus monkeys (Bernstein 1969) and free-living chimpanzees (Wrangham 1974) and this will affect group structure. Several species of baboons split up into smaller troops when food is scarce (Aldrich-Blake *et al.* 1971; Anderson 1981; Crook 1966; Sharman 1981) and the extent to which adult chimpanzees and vervet monkeys feed in company with each other depends on food availability (Fairbanks & Bird 1978; Wrangham & Smuts 1980). Scarcity of sleeping sites may also affect the size of baboon troops (Crook & Aldrich-Blake 1968; Kummer 1968).

The next four contributions in this chapter concern the manner in which individuals distribute their affiliative behaviour among other members of their group. In the first (section 10.2), Seyfarth reviews his heuristically valuable suggestion that three main principles determine the direction of an individual's grooming behaviour—the attractiveness of high-ranking individuals, competition for access to the more attractive individuals and the importance of relationships with genetic relatives. His arguments involve consideration of the benefits and costs to be derived by individuals from the establishment and maintenance of relationships with particular others and thus foreshadow the discussions in Chapter 12. He also emphasizes the cognitive complexity implied by some of the data.

Another possibility, raised in Section 10.3, is that under some circumstances individuals of similar rank may be especially attractive. This is discussed with reference to the peer–peer relationships of young male rhesus and the importance of familiarity in relationships is stressed. In rhesus male–female relationships during the birth season, the attractiveness of high rank is a major issue (section 10.4), while considerations of both rank and relatedness are necessary to account for the structure of relationships between females (section 10.5).

Section 10.6 provides an example of the existence of a subunit within a group. In section 10.7 a new method for describing social structure is introduced. This is potentially able to take account of the relative frequencies of different types of interactions within the dyads in the group. While the mathematical techniques it employs will be unfamiliar to most primatologists, it offers considerable hope for the future. Finally, it must be emphasized that social structures are not rigid and may be influenced by local ecological conditions (section 10.8).

10.2 Grooming and Social Competition in Primates
ROBERT M. SEYFARTH

Introduction

A recent study of feeding behaviour in juncos *(Junco hyemalis)* clearly illustrates some of the costs and benefits of group life for dominant and subordinate individuals. In a series of laboratory experiments, Baker *et al.* (1981) measured the foraging success of individuals that were housed alone. Birds were then housed together in flocks, where each group formed a dominance hierarchy. Within each hierarchy, high-ranking birds acquired significantly more food when feeding in the flock than when foraging alone. In contrast, because of feeding competition, low-ranking birds acquired roughly the same or less food in flocks than when feeding alone. When a low-ranking bird found food, it was frequently approached by a dominant individual who supplanted the subordinate and then ate. Low-ranking birds did benefit from joining flocks, however, because within flocks there was less chance that they would find no food at all.

In this section it is argued that the function of grooming in primates is best understood if grooming is studied in the same way as feeding competition. The argument may be summarized as follows. In addition to its short-term function of removing ectoparasites, grooming in primates has a number of long-term consequences, since grooming seems to establish or strengthen the social bond between the individuals involved. On account of these consequences, it is possible to identify, for each individual, a theoretically optimum set of grooming partners, interaction with whom maximizes the long-term benefits of grooming. Moreover, individuals will attempt to maximize the benefit they derive from social interactions like grooming, in much the same way that they attempt to maximize the benefits derived from foraging or feeding.

Assuming that grooming strengthens or establishes social bonds between individuals, the distribution of grooming among individuals should vary, depending upon the benefits to be derived from different kinds of grooming partners. If individuals derive maximum benefit by interacting with kin, for example, each individual will direct all of its grooming to its close genetic relatives and there will be minimal conflict between the interests of the individuals involved. In other cases, however, many individuals in a group will prefer the *same* grooming partners and competition for access to these individuals will result. In species with stable dominance hierarchies, one should first imagine in theory, how an 'optimal groomer' would distribute its behaviour among others, and then compare this theoretical optimum with the behaviour of dominant and subordinate individuals. The grooming distributions of dominant animals will be relatively unconstrained by competition and should therefore approach the hypothetical optimum grooming strategy for a particular species. The grooming of subordinate animals, on the other hand, will show how this strategy is constrained by competition.

A model of grooming based on these assumptions was originally formulated (Seyfarth 1976, 1977) to explain the distribution of grooming among adult females in four species of Old World monkeys: rhesus macaques (Sade 1972a), geladas (Bramblett 1970; Kummer 1975), baboons (Seyfarth 1976) and stump-tailed macaques (Rhine 1972). Since distributions of grooming in these species exhibited a number of similar features despite wide variation in degrees of relatedness among individuals, the original model attempted to explain the function of grooming in terms of both preference among kin and other factors. Subsequently, the assumptions and predictions of the model have been tested on adult female vervet monkeys, both in the wild (Seyfarth 1980) and in captivity (Fairbanks 1980), hamadryas baboons (Stammbach 1978), geladas (Dunbar 1980b) and bonnet macaques (Silk *et al.* 1981b). The model is briefly reviewed below as it was originally formulated and the results of subsequent tests are considered. The model's relation to theories of grooming based on kin selection (Hamilton 1964) and theoretical approaches to social structure in human and non-human primates are then discussed.

A model of grooming in primates

To begin, assume that grooming in non-human primates has at least two beneficial consequences for the individuals involved—the removal of their ectoparasites and the establishment or maintenance of a close social relationship with their grooming partner, regardless of who gives or receives. Following Hinde (1974), 'social relationship' is used as a short-hand term to describe the frequency, quality and patterning of different interactions between the same individuals over time.

Second, assume that while all individuals are equally skilled at removing each other's ectoparasites, the benefits to be derived from social relationships with different individuals are not equal: for example some individuals will be more successful than others in forming alliances or in finding food and as a result will be more desirable partners.

Third, assume that individuals are selected to maximize the benefit they receive from social relationships with others.

Given these three assumptions, the following may be predicted. First, an individual's attractiveness as a grooming partner will be derived from the potential benefit it can offer others. Thus, if close genetic relatives are most likely to reciprocate each other's grooming with tolerance at feeding sites and support in alliances, etc., then grooming will occur most often between these individuals. In addition, however, in species with stable, long-term dominance hierarchies, high-ranking individuals may also offer great potential benefit from grooming interactions. High-ranking animals are most able to support others in aggressive alliances (e.g. Cheney 1977), they have freest access to scarce food and water (e.g. Wrangham 1981) and they are most likely to give alarm calls in response to predators (Cheney & Seyfarth 1981).

Second, the model predicts that *both* frequent grooming of high-ranking animals and frequent grooming between individuals of adjacent dominance ranks will result when high-ranking females are preferred grooming partners and when females compete for the opportunity to groom with those of high rank. This is because high-ranking females will be least constrained by competition and will distribute the majority of their grooming to others of high rank. Middle-ranking females, however, will meet competition for access to those of high rank and will compromise by grooming those of middle rank. Finally, low-ranking females will meet competition for access to all individuals and will groom those of low rank.

As originally formulated, the model's predictions were consistent with the two major features that characterized grooming among adult females in four different species of Old World monkeys: (1) high-ranking individuals received more total grooming than others; and (2) the majority of grooming occurred between females of adjacent dominance ranks (Seyfarth 1977). It was argued that where adjacently ranked females were close genetic relatives (e.g. Sade 1972a; Kurland 1977), *both* preference among kin and competition for access to high-ranking individuals were likely to occur. During adulthood, competition for access to high-ranking individuals would reinforce bonds formed during ontogeny between members of the same maternal lineage.

Tests of the model using groups of females where genetic relatedness was unknown

Stammbach (1978), Fairbanks (1980) and Seyfarth (1980) have tested the model of rank-related attractiveness by observing grooming among adult females in groups where genetic relatedness was unknown. In Stammbach's and Fairbanks' caged studies, the adult females had been captured separately and were assumed not to be genetic relatives. In the author's study, genetic relatedness was unknown. Long-term data on vervets, however, suggests that some of the adjacently ranked females in the groups observed were close genetic relatives.

All three observers found that high-ranking females were the most attractive grooming partners and that individuals competed for the opportunity to groom with those of high rank. In addition, Stammbach found that frequent grooming was also related to 'mutual preference (among individuals) in choice tests, i.e. to an aspect of attraction or friendship that was itself not correlated with dominance.'

In Fairbanks' study, which involved three groups with three adult female vervets in each, competition for access to the highest-ranking female caused almost no grooming to occur between third- and first-ranking individuals. In the author's study, which involved three groups of eight, seven and eight adult female vervets, competition was less strong and many middle- or low-ranking individuals were able to direct considerable grooming to the highest-ranking female in their group. At the same time, ten of 23 females in these three groups gave significantly more grooming than expected to females of adjacent rank.

The author found considerable evidence that high rates of grooming, alliance formation and proximity were correlated among adult female vervets. While grooming of high-ranking females did not cause changes or instability in the ranks of low- or middle-ranking females, there was nevertheless some suggestion that these individuals could minimize the cost of their status by grooming more dominant individuals.

At the same time, Fairbanks argued that there may be a cost involved in maintaining a close relationship with high-ranking females. Animals who approach high-ranking females most often are most likely to receive aggression from them and they must also devote a considerable amount of time and aggression to keeping others away. Against this cost, Fairbanks found no evidence that grooming with the alpha-female increased the likelihood of her support in an aggressive coalition. All of the coalitions formed by the alpha-female were formed with immature kin. However, both Fairbanks (1980) and Kummer (1978) list a number of additional benefits, such as access to food and water, predator defense or information sharing, that might accrue to those who form bonds with high-ranking individuals. In captive macaques, low-ranking females may be able to minimize the amount of harrassment they receive from high-ranking females by grooming these individuals (Silk *et al.* 1981a). Among free-ranging vervet monkeys, high-ranking animals have freest access to limited supplies of water during the dry season (Wrangham 1981) and they occasionally tolerate drinking by low-ranking females who sit near and/or groom them (pers. obs.). High-ranking females are also those most likely to give alarm calls in response to predators (Cheney & Seyfarth 1981) and thus most likely to protect those nearby. This benefit, however, also involves a cost, since high-ranking vervets are more likely than others to be taken by a predator (Cheney *et al.* 1981).

In summary, a number of studies indicate that an adult female's attractiveness as a grooming partner is directly related to her rank and that competition for access to high-ranking animals has an important effect on the distribution of grooming in primates. Less is known about exactly *why* high-ranking females should be so attractive. In some studies, high-ranking animals do seem to reciprocate the grooming they receive, while in other studies they do not. Returning to the analogy between grooming and feeding competition, what is the 'giving-up time' (e.g. Krebs 1979) for a middle- or low-ranking female? How long will she persist, without receiving any benefit, in her attempts to form a close relationship with a high-ranking individual? What alternative strategies are open to her? The most likely alternative, of course, involves the establishment of mutually beneficial relations among kin. Bonds between genetic relatives and their relation to the competition described above are considered below.

Grooming and kin selection

In many species of Old World monkeys, genetically related females occupy similar or adjacent dominance ranks. Frequent grooming among these animals develops at an early age and persists into adulthood (e.g. Sade 1968). In one respect, therefore, a grooming model that assumes rank-based attractiveness is difficult to test against a model that assumes kin-based attractiveness, because both predict frequent grooming between females of adjacent ranks.

The model of rank-based attractiveness was not, however, originally presented as an exclusive alternative to grooming among kin. Instead, its aim was to suggest that there may be at least two behavioural tendencies that determine how a monkey distributes its grooming. Taken together, the author argued (Seyfarth

1980), kin-based and rank-based attractiveness may allow a greater proportion of grooming to be explained than either hypothesis could explain on its own. In Sade's (1972a,b) study, for example, 32% of all grooming occurred between close genetic relatives and thus could be explained as the result of attraction between kin. In addition, however, the total amount of grooming received by each female was correlated with rank, not family size, and frequent grooming was also observed between adjacently ranked members of different matrilines. These data seemed best explained as the result of competition for attractive, high-ranking individuals.

Dunbar (1980b) and Silk *et al.* (1981a) have tested the model of rank-related attractiveness in two cases where genetic relatedness among adult females was known. Dunbar found little support for the model, since most grooming occurred between kin and he could find no evidence of competition for high-ranking individuals. Silk *et al.* also found that the great majority of grooming occurred among kin. However, when they examined grooming among non-kin there was clear evidence that 'adult and immature females are attracted toward females from high-ranking lineages'.

The precise manner in which a monkey distributes its grooming will depend on: (1) the frequencies with which kin or other individuals reciprocate grooming; and (2) the benefits to be gained if reciprocation occurs. Dunbar's (1980b) data indicate that, in some cases, benefits to be gained from grooming kin may be so great that they outweigh the benefits to be gained from grooming any other individuals. Thus Dunbar found that the great majority of grooming and coalitions were formed between female geladas and their offspring and he argued that perhaps 'the parent–offspring bond is the only relationship of sufficient 'strength' to provide an adequate foundation for a bond that must last a life-time to show a profit'. In contrast, Silk *et al.*'s data suggest a more 'mixed strategy', in which a large proportion of grooming was allocated among kin and the remainder was directed at high-ranking animals. Earlier (1977), the author had shown that a wide variety of such mixed strategies, in which animals attempt to distribute varying proportions of grooming among kin and high-ranking animals, *all* lead to grooming networks with similar general features: high-ranking animals receiving the most total grooming and the majority of grooming occuring between females of adjacent ranks.

There is, however, at least one way to test whether individuals are behaving solely according to attraction among kin or whether they are attracted both to kin and to those of high rank. If attraction among kin were the only factor affecting the distribution of grooming, and if all other variables such as age and sex were equal, it would be predicted that grooming relations would be similar in all matrilines, regardless of their dominance rank. In contrast, if both kin based and rank-based attractiveness are at work, it would be predicted that grooming relations between members of high-ranking kin groups would be stronger than grooming relations between members of low-ranking kin groups. This is because, within high-ranking kin groups, individuals will be attracted to each other both because of bonds formed during ontogeny and because of the benefits associated with high status. In contrast, within low-ranking kin groups, individuals will be attracted to each other because of bonds formed during ontogeny and because they are forced to compromise their attraction for high-ranking individuals. Berman (1980b), Cheney (1977), Fady (1969), Guiatt (1966) and Yamada (1963) all present data suggesting that high-ranking kin groups may indeed be more 'cohesive' than others but at present this prediction remains to be thoroughly tested.

Finally, demographic factors may affect the relative importance of kin-based and rank-based attractiveness in primate grooming. As Altmann and Altmann (1979) and Dunbar (1979a) have pointed out, rates of mortality among free-ranging monkeys and apes are high and in many cases it will be unlikely that an adult animal will have many close adult kin within its group, e.g. among free-ranging vervet monkeys, infant mortality is roughly 60% within the first year (Cheney *et al.* 1981). Assuming a 50:50 sex ratio at birth, this means that there is less than a 10% chance that two sisters will be born in successive years and will both survive their first 12 months. A grooming 'strategy' based on rank-related attractiveness is likely to be particularly evident in these cases, where the number of close genetic relatives is small.

Grooming networks and social structure

In the author's original model of grooming among adult females, analysis began by abstracting a number of features that were common to grooming net-works in a variety of species. An attempt was then made to show how these relatively complex features could be produced by animals whose behaviour followed two simple rules: (1) direct some grooming to close genetic relatives; and (2) direct the remaining grooming to the highest-ranking animal available. This approach, which uses the relatively simple 'tendencies' of individuals to generate more complex social structures, was not original but owed a great deal to the prior work of Hinde (1974), as well as to that of Barth (1966), Homans (1961), Kummer (1975) and Vaitl (1978).

Implicit in the original discussion of grooming, however, were a number of assumptions about the animals' cognitive abilities that were not based on empirical data. The credence of the argument rested, among other things, on the assumption that monkeys could rank one another and that they knew which individuals were blood relatives. It was argued that monkeys behaved *as if* they knew such things about others but no evidence was provided that they did so.

In this respect, a long-standing dichotomy between the way social structure is studied in humans and the way it is studied in non-human species was skirted. In humans (anthropologists tell us), abstractions like rank hierarchies and kinship relations are known to the participants, who communicate with each other about them and adjust their behaviour accordingly. Monkeys and apes, on the other hand, are assumed to know nothing about these concepts (e.g. Altmann 1981): they just *behave* and the researcher provides the abstraction.

However, do monkeys really know nothing about the network of social relationships within their group? The author believes that they know more than we realize and thinks this can be demonstrated empirically by considering how monkeys classify each other.

From a participant's point of view, social structure is essentially a matter of classification. An individual distinguishes others according, for example, to kinship, marriage, status or residence and he lumps or splits others into different categories that may overlap to varying degrees. One supposedly unique feature of human knowledge about social structure is that human classifications are non-egocentric: an individual knows not only how others stand in relation to him- or herself but also knows how they stand in relation to each other.

Constructing a rank order of individuals is a simple form of non-egocentric classification because an individual must know not only its own rank but also the relative ranks of others. The grooming model discussed above assumes that monkeys can rank each other and there is evidence from at least two species that they do so (Seyfarth 1981). If these observations are borne out by further tests, the results would be important, because they would indicate that concepts like dominance hierarchies are *not* just abstractions employed by humans to explain non-human social structure. Instead, they are concepts that exist, in some form or another, for the animals themselves.

A similar argument can be applied to two other aspects of non-human primate social structure: group membership and kinship. In this instance, human systems of classification are assumed to be unique because they are hierarchical: at a comparatively simple level, humans distinguish different individuals, while at a more complex level, these individuals are lumped or split into broader categories based on group membership, kinship or some other factor.

Playback experiments on free-ranging vervet monkeys, however, question the uniqueness of such human taxonomies because they have shown the following. First, individuals can recognize the vocalizations of those outside their own group and they can associate the calls of particular individuals with particular groups (Cheney & Seyfarth 1982). Second, within each group, animals can distinguish kin from non-kin by vocal cues alone (Cheney & Seyfarth 1980). Third, within each group animals go beyond this relatively simple discrimination and can 'arrange' individuals into higher-order units, apparently on the basis of matrilineal kinship (Cheney & Seyfarth 1980). Here again, there is evidence that the abstractions used by human observers to describe social structure also exist in the minds of the monkeys themselves. An experimental approach like vocal playbacks thus allows us to test whether our inferences about social structure, obtained from observations and correlational data, are in fact accurate representations of what is socially important from the monkey's own perspective.

Conclusion

In summary, it has been argued that group life requires animals to balance the costs and benefits of different social relationships, much as they balance the costs and benefits of different feeding strategies. Grooming in primates is a social tool, used by animals in their attempts to establish and maintain those social bonds that return the greatest benefit.

From an observer's point of view, data on grooming thus offer an insight into the benefits and constraints of group life. In addition, however, such data can also be used to study the complex ways in which monkeys perceive one another. To explain efficient foraging, it must be assumed that an animal can distinguish among different sorts of food but to explain efficient grooming, as has been attempted here, it must be assumed that a monkey can make complex classifications of its fellow group members. A monkey needs to rank the others in its group, to know who its own kin are and to know which individuals make up the other kin groups nearby, if it is to distribute its grooming to best advantage. Thus, because grooming reflects both the benefits to be gained from social relationships and the constraints imposed on individuals by others, there is a direct link between the study of grooming and the study of social structure in primates. If grooming has the function of establishing and maintaining social bonds and if particular social bonds are more beneficial than others, then animals must have specific knowledge about thir fellow group members. In addition, if a situation can be created in which the animals reveal this knowledge, understanding of social structure (from the animals' point of view) can be sharpened.

10.3 Familiarity, Rank and the Structure of Rhesus Male Peer Networks
JOHN COLVIN

In section 5.4, some of the ways in which rank affects the relationships of immature male rhesus monkeys with their male peers were considered. However, the interactions of rank and relationships were considered *in vacuo* rather than by asking how they might be influenced by other causal factors, such as familiarity. Similarly, the way in which rank affects the patterning of social relationships within a social network was not considered and no attempt was made to provide an explanation of the interactions of rank and relationships, for instance in terms of their possible functional consequences.

In this section an attempt will therefore be made to broaden our perspective in order to explore the multiple effects of two presumed causal factors— rank and familiarity—on male peer relationships and network structure. The functional significance of these effects will also be touched upon. As in section 5.4 the studies by Colvin (1982) and Tissier *et al.* (in prep.) of four male cohorts on Cayo Santiago will be drawn upon. Excluding the effects of age and sex allows a focus to be placed on the effects of familiarity and rank.

Among the first attempts to examine the nature of proximate influences on social structure was a series of papers by Seyfarth (1976, 1977, 1980; see also section 10.2). Seyfarth developed a model to explain how a number of interacting causal factors, including the attractiveness of high rank, could determine the social structure of adult female relationships. This model was tested on a number of species and in the first part of this section some of its strengths and weaknesses will be considered briefly. Seyfarth's model will then be contrasted with that developed by Colvin (1982), which emphasized the attractiveness of similar rank over that of high rank. Through a critique of both models, in the last part of this section there will be a move to a broader perspective incorporating additional factors.

In his review of six studies on the grooming behaviour of adult female monkeys in four species, Seyfarth (1977) noticed that there were two major consistencies across all the grooming networks that had been described. The first of these was that high-ranking individuals receive more grooming than others and the second was that the majority of grooming occurs between females of adjacent rank. Working on the theoretical premises that attractiveness is derived from the benefits that an individual can offer to others but that the complete expression of that attraction may be curtailed by other social factors, Seyfarth put forward a basic model of female grooming. This proposed that frequent grooming of high-ranking females and between females of adjacent rank resulted because high-ranking females are preferred grooming partners and females compete for the opportunity to groom those of high rank.

In his basic model, Seyfarth relied primarily on one central assumption about female grooming behaviour. This is that every female has an identical goal, that of attempting to form as intense a grooming relationship as possible with the highest-ranking female available. On account of the fulfilment of one individual's goal necessarily precluding the fulfilment of another's, a corollary to this assumption is that competition between individuals will be a major factor determining the distribution of grooming partners.

Seyfarth recognized that there were other causal factors which could influence grooming relationships, in particular preferences for close genetic relatives and changes in reproductive state. These he considered in modifications to his basic model and, while he could not specify their precise influence, he demonstrated their importance and concluded that female grooming networks must be considered in terms of a number of interacting causal factors.

The model of grooming relationships proposed by Seyfarth has a number of important strengths. First, it is constructed, with sound theoretical logic, from premises derived from and compatible with established research. The model is able to reproduce accurately critical features of female grooming networks and to explain these parsimoniously as resulting from the interaction of basic principles. It has provided a hypothesis whose explanatory and predictive power has been testable by future research. Many of the predictions generated by the model have also received support from more recent research (Fairbanks 1980; Seyfarth 1980; Stammbach 1978), including data on grooming in immature, male rhesus monkeys (Colvin 1982).

On the other hand, Seyfarth's model also possesses a number of weaknesses, which strongly suggest that attraction to high rank is not necessarily a primary factor underlying either adult female grooming interactions or primate interindividual associations in general.

In the first case, two important predictions generated by the model have not been supported by Seyfarth's more recent study (1980). One of these is the expected positive correlation between rank and the ratio of grooming received to grooming given. The other is the expected greater reciprocity in grooming between females of adjacent rank. Further lack of support for predictions of the model also comes from studies of adult female grooming in gelada baboons (Dunbar 1979b) and bonnet macaques (Silk 1982).

More importantly, a number of the basic assumptions of the model are perhaps too great an over-simplification. For example it focuses on the benefits which might accrue within relationships but, as Fairbanks (1980) points out, ignores costs, both within the relationship and associated with defending that relationship against others. Since such costs should increase with partner rank, individuals might well minimize costs by preferring those of lower rank. In addition, it is unlikely that the goals of each individual in associating with the other are identical. With a unitary principle of attraction, differences between individuals' goals would not, in practice, affect the distribution of associations. However, with more than one principle of attraction, a possibility which Seyfarth allows for in his discussion of the influences of kinship and of attraction to infants, the interaction of these principles would determine different goals for different individuals.

Finally, it must be recognized that the focus of the model is upon grooming, although this is related also to agonistic behaviour, alliance formation and other components of adult relationships. While Seyfarth never argued that the model could be widely applied to other types of behaviour, this nevertheless becomes a limitation if the examination of networks of relationships whose functional significance rests on very different types of interaction, such as play, is required.

Such was the case in the study of Colvin (1982). While confirming the predictions of Seyfarth's model in their grooming interactions, the peer relationships of immature, male rhesus macaques did not in most instances support the predictions over a range of other interactions, including alliance formation, mutual feeding, passive association and especially play. Instead, in many cases, Colvin's findings (see section 5.4) are explained better by a basic model of attraction to similar rank than of attraction to high rank. The model of attraction to similar rank (Colvin 1982) was proposed as an alternative and was also based on assumptions concerning the costs and benefits of relationships. However, these assumptions were more general and as such were also more applicable to relationships as a whole.

The assumptions of this second model were designed to take into account the problem both that costs and benefits will vary between relationships and that in any one relationship these costs and benefits may be not only extensive (cf. Kummer 1978) but also difficult to measure. It was assumed that individuals select partners so as to maximize their profits and minimize their costs. One way to maximize profits is to choose a single partner with whom the *greatest range* and/or value of benefits can be exchanged. At the same time, dependency on a single relationship for a range of outcomes can also minimize costs. First, a single relationship allows for more efficient investment of time. Second, it is more dependable in the short term because the behaviour of a single partner becomes easier to predict than that of several. Likewise, in the longer term, the potential costs of desertion are minimized because changes in behaviour of a single partner are more easily monitored than changes in several partners.

These assumptions suggest several reasons why partners of similar rank should be the most attractive. First, their needs are most likely to coincide and thus be reciprocal, allowing the greatest range of benefits to be exchanged. Second, each can expect similar costs to be incurred by the other. These will include both the costs of direct aggression and those associated with the possibility of desertion. Given the relation between rank and kinship structure in many primate species (i.e. that the most closely ranking individuals are also likely to be the most highly related), there is also a third reason for the attractiveness of similar rank. This is based on the relation between kinship, familiarity and interpersonal perception. Familiarity, which is presumed to be related to prior exposure, provides an estimate of the degree to which each partner in a relationship can predict the behaviour of the other and is thus an important basis for more effective monitoring of the partner. Since, for ontogenetic reasons, familiarity is associated with kinship ties, those of similar rank will be attractive also by virtue of their familiarity. Not only can a relationship with a familiar individual provide the individual with the maximum opportunity for making social comparisons and for testing the other's abilities (Fady 1969; Owens 1975 a,b) but it may also provide an opportunity to learn about the nature of a relationship under the widest variety of extradyadic influences and in the greatest range of social contexts.

The basic assumption of Colvin's model is therefore that males are attracted to others primarily by virtue of their closeness in rank. A second assumption is that in a choice between two partners of equal closeness in rank, the higher ranking of the two is chosen.

One of the attractive features of this model is that the partnerships predicted on the basis of attraction to similar rank normally coincide with those predicted on the basis of attraction to kin-based familiarity. However, to the extent that these attractive forces do not coincide, the model is weakened, e.g. on the basis of familiarity the highest-ranking male of a medium- or low-ranking lineage should be more attracted to his relative who ranks beneath him than to the unrelated male above. In this model however, he is more attracted to the male above, for the reasons given above. In addition, the model has further weaknesses. While in some cohorts (the 1974 spring and summer cohorts and the 1975 summer cohort) its predictions were better supported than those of Seyfarth's model, in others (the 1975 spring cohort), the reverse was the case. In addition, considering the members of all cohorts individually, in a number of cases the predictions of neither model were well supported, suggesting that a more complex set of factors influences partner choice than either model could elucidate.

The requirements for a multifactorial model have recently been explored by Tissier *et al.* (in prep.) and permit both a resolution of the differences between the models of Seyfarth and Colvin and a correction of their weaknesses. Tissier *et al.* recognize that preferences are determined by a small set of causal factors which take effect through the operation of several principles. It is the interactions between these principles which determine the rate at which male peers approach, leave and interfere with each other and thus determine the patterning of peer associations. These causal factors and the principles to which they are related are listed in Fig. 10.1.

The two main causal factors are familiarity and rank but others include time budgets and other relationships. These factors then operate through principles which both encourage and discourage association.

Familiarity increases attractiveness between peers primarily because it enhances predictability and permits monitoring within a relationship (principle 1). In this respect, kin will be preferred because for ontogenetic reasons they are

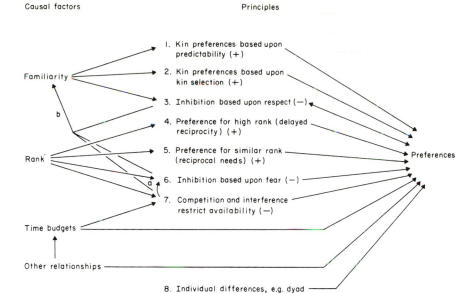

Fig. 10.1. Relations between causal factors and principles which influence partner preferences in immature rhesus male peer networks. Some principles promote associations (+); others inhibit them (−). The two feedback effects (a and b) are shown.

the most familiar. Kin may also be attractive by virtue of their familiarity since this is a proximate mechanism through which the ultimate forces of kin selection may act (principle 2); this is unlikely to be a potent selective force in primate social evolution but its possible effect here is acknowledged. Familiarity may also discourage association (principle 3). Studies of inhibition in adult hamadryas baboons have shown that a 'rival' male respects the bond between a male 'owner' and his 'target' female if he has previously seen them interact (Kummer *et al.* 1974). However, this respect is modified by the degree of preference shown by the female for the owner (Bachmann & Kummer 1980) and also by the degree of familiarity between rival and owner (inhibition is less, the more familiar the rival is with the owner [Kummer *et al.* 1978]). Similar processes influence the degree of inhibition experienced by immature rhesus males when attempting to interfere with dyads of lower-ranking peers (Colvin 1982; Tissier *et al.* in prep.; section 5.4).

   Rank influences attractiveness in the two ways outlined by Seyfarth (1977) and Colvin (1982) (see above). On the one hand, there is a preference for high rank, rewarded through a system of delayed reciprocity (principle 4) and on the other hand, there is a preference for similar rank, rewarded through the fulfilment of similar needs (principle 5). Rank effects can also discourage association. They can both lead to an inhibition which is based upon fear (principle 6), e.g. rhesus males are inhibited in this way from interfering between two more highly ranking peers (section 5.4), and, through competition for preferred partners, they can lead to interference between individuals and so to their restricted availability and decreased association (principle 7). While such competition is likely to be intense in relation to partners of high rank, it is likely to be far less strong in relation to partners of similar rank. In the main, therefore, competition will be over access to high-ranking individuals.

The pattern of associations in a peer network is therefore the outcome of interactions between all males. On account of the processes of competition and inhibition it is necessary to take a dynamic view of network structure. Competition has the effect of excluding certain preferred relationships for some males, as Seyfarth (1977) originally argued. Over time, these exclusions will then become established within a more stable structure. This occurs as a consequence of two effects. First, interferences will reinforce inhibitions based upon fear (Fig. 10.1a). Second, the combined effects of interference and inhibitions will reduce familiarity (Fig. 10.1b), which in turn will reduce familiarity-related preferences and increase inhibition based upon respect.

It may be noted that the opposite process will take effect in cases where three (or more) males are attracted to associate as a triad (or any other higher-order subgroup)*. Here competition will be reduced and familiarity increased, so increasing further their mutual attraction.

From a dynamic viewpoint, therefore, over time a stable and sharply delineated pattern of peer relationships should develop, through the facilitation of certain relationships and the inhibition of others, from situations in which competition is high and relationships are not well established. Changes in peer relationships, for instance due to deaths, emigrations or rank reversals, or in other relationships, for instance due to increased interest in receptive females during the breeding season, will on the other hand act to destabilize network structure.

Given some understanding of the operation of the various principles outlined above, Tissier *et al.* (in prep.) outline two, more comprehensive, models of immature male peer associations. The first is a predictive model, while the second is a descriptive model. Both models attempt to take into account differences between cohorts in their lineage structures.

The predictive model of Tissier *et al.* is a relatively simple combination and elaboration of the models of Seyfarth (1977) and Colvin (1982). It assumes that preferences are influenced both by kinship (familiarity) and rank effects and

---

*Within the scheme of Fig. 10.1, such principles as those governing the formation of higher-order relationships are included under 'individual differences'. These are, in essence, the 'error factor' of the model, in terms of which all undiscovered principles are accounted for.

that these interact to determine network structure. Since it is unable to specify the nature of their interaction, it assumes a hierarchy of effects, namely that kinship is the primary determinant of preferences and that high and similar rank are secondary determinants. In one variation of the model, high rank exerts a stronger influence than similar rank, in the other variation, similar rank exerts a stronger influence than high rank. It is also predicted that, even in the former case, in very large cohorts the influence of similar rank will be stronger among low-ranking individuals while the influence of high rank will remain stronger among high-ranking individuals. This is because as cohort size increases a point will be reached where high-ranking males are prepared to invest less in low-ranking males than are males of intermediate rank. As a result, among low-ranking males, the influence of similar rank will override that of high rank. A final assumption of the model is that if a male is competitively excluded from a given relationship, in all cases he will attempt, if possible, to find an alternative partner rather than to accept a diminished association with his excluded partner.

Tests of the two variations of the predictive model on association data for the four rhesus male cohorts revealed that each was more powerful than the Seyfarth or Colvin model alone but that for some cohorts (the 1978 and 1979 cohorts) the data were better predicted by the first variation of the model, whereas for others (the 1974 and 1975 cohorts) the data were better predicted by the second variation. As expected, in the large 1975 cohort (spring), low-ranking males showed patterns of partner preference that correlated more closely with those predicted by the second variation of the model than by the first.

The results suggest that both variations of the model on their own have a certain limited explanatory power. Taken together, they provide an adequate approximation to the patterns of association found among immature male rhesus macaques, since they are able to describe general rules which lead to the observed distribution of partner preferences. However, the model still fails to specify the precise conditions under which one variation will provide a better prediction than the other.

Tissier *et al.* (in prep.) propose a second, descriptive model, which avoids some of the problems of oversimplification encountered by the predictive models when faced with such complex phenomena as male peer networks. In dealing with the considerable variations in network structure, where all the factors and interactions which give rise to that variation are not fully understood, it is useful to go beyond a few simple rules of intrinsic generality to postulate *post hoc* more complex explanations which are supported by the available evidence. The heuristic value of such interpretations remains of course to be assessed by future research.

A descriptive model can only be understood fully by running it against the data. While there is insufficient space to do this here, an understanding of some of its features can at least be gained by looking at some of the problems it is designed to overcome. These are of several types. First, the two versions of the predictive model need to be integrated so as to account for the interaction between the influences of high and similar rank. Second, how these influences interact with those of familiarity must be determined, since the effect of the latter is unlikely to be an absolute one (as proposed in the predictive model). Third, how these sometimes conflicting attractions are resolved, both directly by the individual and indirectly through the processes of competition needs to be reconsidered. Thus, it is not always the case that when excluded from one peer relationship, an individual will attempt to associate to the same degree with another, less preferable male. Fourth, it must be expected that the degree of competitive exclusion that is apparent will vary between cohorts, most probably as a result of variations in the stability of relationships within each cohort. Finally, to contend with individual differences problems of idiosyncracy must still be expected.

Here, no more can be done than to consider briefly certain aspects of two of these problems in further detail. The first concerns the conflict between attraction to a closely ranking non-relative of higher rank and to a closely ranking relative of lower rank. There were test cases in all four cohorts. In seven out of eight cases of a choice between partners of adjacent rank, the lower-ranking relative was preferred to the higher-ranking non-relative, thus supporting Silk's and Dunbar's finding that preference for kin is of crucial importance: in the eighth case both partners were preferred at an equally low level. However, in one out of two cases where the choice was between partners of rank difference ± 2, the higher-ranking non-relative was preferred to the lower-ranking relative, thus supporting the notion that, in some cases, the partner's rank is crucially important. There were also cases in which higher, adjacent ranking non-relatives were preferred to lower-ranking relatives of rank difference -2. Most of these cases were among low-ranking males and here it appears that males may defer relationships with lower, closely ranking partners in favour of those with partners of higher but close rank.

The second aspect touched upon here concerns the problem of idiosyncracy. One 'idiosyncratic' feature of peer association which appears susceptible to analytic interpretation concerns the strength of relationships between very closely related peers ($r \geq 1/8$). There was at least one such relationship in each cohort (Fig. 10.2). These closely related individuals were particularly attractive to their other peers in a way that did not relate to their rank or kinship. In all but one case, they were the most 'popular' individuals in their cohorts. In the 1979 cohort, 998, the higher ranking of the two closely related males, was considerably more attractive than 992, the highest-ranking male of the cohort. Indeed, this appears to account for much of the difference in the structure of associations between the

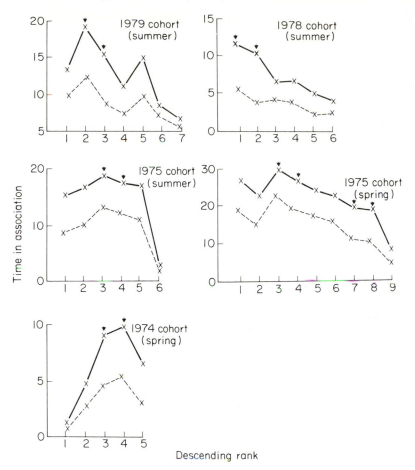

Fig. 10.2. Relation between individual rank and total association time with peers in four cohorts in spring and/or summer. In each cohort the two males who belonged to the most closely related dyad (arrows) spent the most time in association with their peers (solid line). In general this was also the case when the time spent with the most preferred peer was deducted from the time spent with all peers (broken line).

1979 and 1978 cohorts, which otherwise show a superficially identical internal structuring of lineages. A possible interpretation of this effect can be found in Tissier *et al.* (in prep).

In conclusion, it may be noted how the proposed causal factors governing the peer network cause it to exhibit a number of dynamic properties which allow for its effective maintenance as an 'open system' (in the sense of von Bertalanffy 1968). Thus attraction to familiar individuals is a mechanism which, by virtue of being self-reinforcing, acts as a conservative force maintaining stability within the system. Stability is also achieved through intermale competition, for this acts to reinforce pre-established rank and familiarity differences, such that stronger relationships are strengthened, weaker relationships are weakened and conflicting interests are minimized. However, the conflicting tendencies to associate with high-ranking, closely ranking and familiar peers, while providing the potential for alternative associations and thus reducing active conflict, also maintain a tension between individuals which allows for adaptation to changing circumstances. In this way, a shift in the distribution of partner preferences within a cohort can be achieved following rank changes between individuals or male emigration.

10.4 Structure of the Birth Season Relationships among Adult Male and Female Rhesus Monkeys

BERNARD CHAPAIS

Since most studies on adult male–female interactions in multimale primate groups have focused on sexual behaviour, comparatively little is known of the interactions between mature males and non-cycling (i.e. pregnant or lactating) females (but see Grewal 1980a,b; Seyfarth 1978a,b; Smuts section 6.9; Strum in press; Takahata 1982). The assumption that multimale primate groups originated from the addition of a number of males to an original core of a group of females and one resident male (e.g. Eisenberg *et al.* 1972) implies that a better understanding of the nature of non-sexual, male–female interactions in such groups might throw light on the issue of their adaptive significance (e.g. Clutton-Brock & Harvey 1977; Wrangham 1980).

This principle governed the analysis of data on the birth-season interactions between the 19 adult females and 15 non-peripheral adult males of group F on Cayo Santiago. Each adult female was the object of 12 five-minute focal samples (Altmann 1974) every week. Male–female interactions occurring during focal periods but not involving the focal female were also recorded and scan samples were used to estimate the relative observability of animals during such concurrent sampling. Finally, male–female interactions occurring outside focal periods were recorded *ad libitum*. The date of birth and genealogical relations through the maternal line of all natal members of group F in January 1978 are given in Fig. 9.6. The rank order of the 15, non-peripheral (i.e. central), adult males is given in Table 10.2 (for data see Table 9.1). The rank order of the 19 adult females (data in Chapais 1981) is also indicated in Table 10.1.

Table 10.1. Distribution of grooming relations among central males and adult females. Males and females are arranged in descending rank order. A = long-term, high-frequency grooming relations extending throughout the birth season or over half of it*; B = long-term, low-frequency grooming relations (as for A but grooming occurring at longer intervals); C = long-term, low-frequency grooming relations interrupted over the first weeks of lactation; D = medium-term, low-frequency grooming relations (between two weeks and two months); E = short-term grooming relations (one week); F = very occasional grooming bouts.

Females

Males	FB	510	577	576	511	287	503	7L	FC	573	297	471	N2	411	K1	476	543	408	310
415	D	F	D	F	A	F	B	A	C	C	F		C	A	D	C			
580	A	B	E	B*	A*	F	D			C									
EE	A		D	A*	D	A	B	C	C	A	F		C	A	C	F	B	A	C
TJ	A												B						
1J	A	A*					F				F	A							D
Z2	F																		
339								D				D							
433							E	D											
436	F			B			F						D		F			F	
495														F					
N8					B*							F	F	F	F				
Z0																			D
358	F		F		F					D	F	F							
285								F											D
282																			

Proximity

A 'maximum spanning tree' (see Morgan *et al.* 1976) was constructed on the basis of a matrix of comparable proximity scores for all male–female dyads (Fig. 10.3). Its main features are as follows. First, three males are clearly central: the alpha-male 415 (connected to six females), third-ranking male EE (connected to 11) and fifth-ranking male 1J (connected to four). Second, the alpha-female FB, connected to four males ranking second to fifth, is also highly central. Third, females tend to gather around males according to their degrees of relatedness to each other: sisters 543, 408 and 310 connected to EE; three of the four females of the subunit (see section 10.6) linked to 415; FC and her daughters 573 and 297 joined EE.

Whether it was the female or the male that was mainly responsible for the observed proximity was assessed on the basis of the frequency of approaches (the distance decreasing from more to less than 1.5 m) and leavings executed by members of each dyad. These figures and the method used to assess responsibility are given elsewhere (Chapais 1981). An examination of the direction of arrows in Fig. 10.3 reveals that each of the males connected to the alpha-female FB contributed more than she did to maintaining mutual proximity. Excluding these four dyads and considering only the 19 dyads involving the

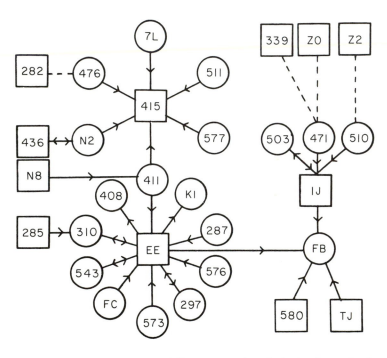

Fig. 10.3. Maximum spanning tree (MST) constructed on the basis of a matrix of comparable proximity (≤1.5 m) scores for the 15 males forming the central hierarchy and the 19 adult females. The characteristics of MSTs are as follows: (1) all individuals are included in the tree, i.e. visited by at least one line; (2) no closed loop occurs; and (3) the sum of the similarities (proximity scores) composing the tree is the maximum possible sum. The spatial distribution of individuals is arbitrary. In the present case, each female is connected to a male but three males are not connected to any female: 433 was never recorded within 1.5 m of the focal females, while 495 and 358 had tied and, in any case, had very low proximity scores. The broken lines correspond to weak and probably non-significant links. The arrows originate from the individuals who played a greater role in the maintenance of mutual proximity. Square = male; circle = female.

highest-ranking males (415, EE and 1J), one finds that for four dyads the responsibilities were approximately equal, for 13, the female was more responsible than the male, and in the remaining two, the male was more responsible than the female.

In summary, the five highest-ranking males (415, 580, EE, TJ and 1J) were highly central (i.e. close to the core of adult females) compared to lower-ranking males and this resulted from two factors: the attractiveness of the alpha-female FB and the attractiveness of high-ranking males 415, EE (and 1J to a lesser extent).

Grooming

On the basis of data on the frequency, direction and temporal patterning of grooming bouts (data and definitions in Chapais 1981), six types of male–female grooming relations were defined (Table 10.1).

Although all males (except lowest-ranking 282) groomed and/or were groomed by females, only the five highest-ranking were involved in long-term, high-frequency grooming relations (type A). Males 415 and EE in particular groomed and were groomed by the majority of females (15 and 17 respectively). Most of the ten lower-ranking males had either type D or type F grooming relations. For their part, all females, except 577 and 297, maintained long-term grooming relations (types A, B or C) with males, the alpha-female FB being involved in four type A relations (with four of the five highest-ranking males).

Males groomed the alpha-female FB more often than they were groomed by her (Wilcoxon test, two-tailed $n = 7$, $P < 0.02$). FB excluded (in view of her special status that will be further evidenced below), females groomed first-ranking male 415 more often than he groomed them ($n = 14$, $P < 0.01$) and third-ranking male EE also received from females more grooming than he gave ($n = 15$, $P < 0.05$). Similarly, second-ranking male 580 tended to direct less grooming than he received but this difference was not significant ($n = 7$, $P > 0.05$). Thus, the three highest-ranking males appeared to be highly attractive for adult females. The three lowest-ranking males groomed females more often than they were groomed in all cases (nine dyads, ten if FB is included) while middle-ranking males directed more grooming than they received in ten cases and less in seven cases (11 and seven if FB is included).

In summary, adult females were attracted to the highest-ranking males, who were themselves attracted to the alpha-female. Lower-ranking males affiliated comparatively little with adult females but appeared to be attracted to them.

Table 10.2. Distribution of agonistic acts among central males and adult females (all sampling sources combined). The upper left figure in each cell = the frequency of mild aggressive acts (threats and lunges) directed by males to females; the upper right figure in each cell = the frequency of serious aggressive acts (chases and bites) directed by males to females; the lower figure in each cell = the frequency with which females were passively displaced by males.

In each cell the upper figure is the aggressive-act count (mild – serious) and the figure in parentheses is the lower (passive displacement) count.

Females

Males	FB	510	577	576	511	287	503	7L	FC	573	297	471	N2	411	K1	476	543	408	310
415	1-0			2-0					1-0			1-0	1-0		0-1 (1)	1-0 (1)			0-2 (1)
580	3-2 (1)	1-0	0-3 (7)	4-0 (2)	1-4 (11)	3-4 (6)			3-1 (10)	2-2 (13)	0-3 (8)	2-2 (6)	6-2 (1)	6-2 (6)	3-1 (6)	1-4 (13)		1-2 (13)	4-3 (3)
EE				2-0	2-0	1-0 (1)		1-0	1-0	3-0		0-1							1-0 (1)
TJ	5-1 (7)	6-1 (3)	2-0 (2)	7-1 (3)	5-0 (9)	5-2 (5)	2-1 (10)	4-4 (5)	5-1 (5)	2-0 (3)	2-3 (1)	1-0 (2)	2-1 (3)	2-2 (4)	2-1 (8)	2-0 (2)	0-1		5-1 (13)
1J	2-0 (1)	2-2 (1)	6-0 (1)	8-0 (2)	7-0 (1)	3-0 (1)	0-1 (1)	5-0 (4)	3-0 (2)	3-1 (2)	2-1 (3)	2-0 (1)	3-1 (1)	2-0 (1)	4-0 (1)	2-2 (1)	2-0		3-0 (1)
Z2	2-0	2-0 (1)	1-0 (1)	2-0 (3)			1-0 (2)		1-0	2-0	1-0	1-0 (1)			2-0 (1)			2-2 (1)	1-0 (2)
339	1-0	1-0	0-1					0-1					1-0 (1)		1-0 (1)	0-1 (2)			
433	1-0	0-1							0-1		0-1		(1)						
436				1-0															

	3-0	1-1	1-1	3-0	3-0	3-0	1-0	2-0	3-0	2-0	2-0	2-2	2-0	4-0
495	5	5	2	3	3	5	3	1	6	1	1	2	4	2
N8										0-1 1	0-1 1-0 2			
ZO	1-0 1	5-0 1	1-0 1	3-0 2	1-0 2	2-0 1	3-0 1	1-0 3	2-0 1	1-0 1	3-0 2	1-0 1	4-0 2	5-0 1
358	1-0 1		2-0 2-0 1 1	1-1 1-0 1 1	2-0 1-0 1 1	3-0 1	1-0 2-0 1 1	2-0 1-1 1 1			1-0 1	2	1-0 2-0 2 1	1-1 1
285			1-0 1-0 1				1-0 1			1	1-0 1	1-0 1		1
282		1	1											

Agonism

Table 10.2 presents data on the frequencies of aggressive acts directed by central males to adult females and on the frequencies with which females were passively displaced by males. Adult females (FB excepted) were subordinate to all central males and were not observed to initiate aggressive interactions with them. FB's status was unique among females. She was the only female that did not receive aggression from any of the nine highest-ranking males but she was displaced and threatened by lower-ranking males (Table 10.2). FB was dominant over her son 580 (second-ranking), male EE (third-ranking) and IJ (fifth-ranking). Although her rank relations with some other high- or middle-ranking males (e.g. TJ and Z2) were never overtly expressed, various interactions suggested that these males were dominant to FB but highly attracted to her.

The distribution of agonistic interactions among central males and adult females (Table 10.2) is more spread out (136 dyads) than the patterning of grooming bouts (74 dyads). However, high-ranking males 415 and EE, the males most attractive to females, were among the least aggressive, ranking eleventh and twelfth respectively in the frequency of all aggressive acts and ninth and 12.5th in the frequency of bites (data in Chapais 1981); they were also little feared (Table 10.2). By far the most aggressive male, 580 systematically attacked females (except his mother FB) by the time he had risen to the second rank in May (see section 9.5). 580's behaviour is best interpreted as a status-assertion strategy which had deferred positive effects on his attractiveness during the following breeding season. Other aggressive and (though to a lesser extent) feared males were TJ (fourth-ranking), IJ (fifth-ranking), Z2 (sixth-ranking), 495 (tenth-ranking), Z0 (twelth-ranking) and 358 (thirteenth-ranking). Most of these affiliated very little with females.

To summarize, the most attractive males (415 and EE) were not often aggressive with females, whereas the majority of aggressive males were those who affiliated relatively little with females. Membership of the first matriline appeared to confer special status: FB was highly attractive to high-ranking males; 580 could assert his status with relative impunity; and 415 was the alpha-male (see section 9.5).

Assistance in agonistic interactions

Central males rarely interfered in fights involving adult females only but they often interfered in fights between adult males and females, intervening more often in the more severe attacks. Out of 63 recorded interventions, males supported other males against females on seven occasions only (11.1%). In the remaining 56 cases, one male (two males in two instances) defended one or more

Table 10.3. Frequency with which central males assisted adult females in their agonistic interactions with adult males (all sampling sources combined).

Males \ Females	FB	510	577	576	511	287	503	7L	FC	573	297	471	N2	411	K1	476	543	408	310
415														3	1				
580	5		2		1	1		1	1		1	1							
EE				2	2			2	3	2	4	2	1		3	1	2	2	2
TJ	1			1	1			2		1		1	1	1	2			1	1
1J	2	1		1				1		1				2					
Z2																			
339																			
433					1														
436																			
495																			
N8																			
Z0																			
358																			
285																			
282																			

females who were attacked by a male or supported one or more females against a male (Table 10.3).

All females received aid from central males but only the five highest-ranking males aided females (except for one instance of aid from male 433) and they did so only when dominant to the targets (except for two instances where EE aided females against 580). There were two consequences of males only entering fights against lower-ranking males. First, male intervention terminated the agonistic interaction in most cases. Second, the lower a male ranked, the smaller the number of males he defended females against and, as a result, the less valuable an ally he was for females.

Adult females were observed to participate in fights of opposing adult males. Although such fights were relatively infrequent they always elicited much excitement. In the majority of cases, females barked and hoarse-grunted at one of the two male combatants but they sometimes intervened more directly by chasing one of the males alongside the other. Fifteen of the 19 adult females supported 12 adult males in 29 recorded female interventions (birth and breeding seasons data combined). The dominant male was supported in 27 cases. Thus, for a male, challenging a higher-ranking male could mean facing a

coalition including adult females. Therefore, females may contribute to the
stabilization of male—male rank relations and this factor might account for the
low frequency of intermale fights taking place in sight of adult females.

In summary, the five highest-ranking males appeared to be useful to adult
females in their agonistic relations with lower-ranking males, while adult females
appeared useful to the highest-ranking males in their dominance interactions
with lower-ranking males.

Conclusion

The main structural features extracted from the above data are the attraction of
females to the highest-ranking males, the attractiveness of the alpha-female and
the patterning of male assistance to females. Such long-term, rank-related,
multidimensional male—female relationships are best interpreted in terms of a
system of mutual benefits, whereby both the adult females and the highest-
ranking males enhance their competitive ability in relation to lower-ranking
males (see section 12.7).

10.5 Dominance, Relatedness and the Structure of Female Relationships in Rhesus Monkeys
BERNARD CHAPAIS

In order to elucidate the social structure of primate groups, many authors have
studied the relations between three important structural dimensions—
demographic composition, genealogical structure and dominance structure—
and the distribution of various behaviours and interactions. In view of its
quantitative and social importance, grooming has been analysed in an attempt to
abstract principles governing the structure of social relationships among females
in particular. Two main tendencies arose. The first one emphasized the
importance of genetic relatedness: for example Sade (1965) found that rhesus
females preferentially groomed the members of their lineage.

The second tendency emphasized the importance of dominance. Seyfarth
(1976; section 10.2) presented a model which appeared to account for the
distribution of grooming among female baboons on the basis of the integration of
two principles—the attractiveness of high rank and the existence of competition
to groom high-ranking females. Both authors, however, did not restrict their
explanation to either factor. Sade (1972a) found that the dominance rank of a
female correlated positively with her grooming status, defined as the number of
individuals grooming her, and Seyfarth (1977) considered the effect of
relatedness on his basic model. However, neither author was specifically

concerned with assessing the relative weights given respectively to dominance and relatedness by individuals in allocating grooming among group members.

Such an attempt has been made here, not only for grooming but for other behaviours as well. The subjects are the 19 adult females of group F on Cayo Santiago and the period covered is the 1978 birth season. Data were recorded through focal sampling on the 19 females, concurrent sampling during focal periods and sequence sampling outside focal periods (Chapais 1981). The age and genealogical relations of adult females are given in Fig. 9.6. A linear and stable dominance hierarchy could easily be extracted from data on submissive gestures (see the order of females in Table 10.1).

Proximity

On the basis of data on dyadic proximity (≤ 1.5 m) a maximum spanning tree (see Morgan *et al.* 1976) was constructed (Fig. 10.4). Most females (16 out of 19) are connected to their relatives, a finding which points to relatedness as the main determinant of the structure of proximity.

The responsibility of females for the maintenance of mutual proximity was assessed on the basis of data on approaches (the distance decreasing from more to less than 1.5 m) and leavings using the proximity index (% AP − % L due to A) (see Chapter 2). The index was calculated for the 32 dyads which filled the criterion of a minimum of ten approaches and ten leavings per dyad. It was found that 18 dyads verified the hypothesis that the proximity between adult females was maintained to a greater extent by the subordinate female of the pair; six dyads neither supported nor contradicted the hypothesis and eight dyads ran counter to it. It may be concluded that if relatedness is needed to account for the structure of proximity (see above), dominance appears as a necessary variable to understand its dynamics, i.e. the role of each partner in maintaining proximity.

Grooming

Lactating females are more attractive than pregnant females in many species of primates (reviewed in Chapais 1981). As a result, the asynchrony of births can distort the assessment of the effects of dominance and relatedness on grooming preferences (e.g. A groomed B more often than C) in three different ways: (1) B might have lactated for longer than C over the period of time considered; (2) A might not have been in the same reproductive state when grooming B and C; and (3) the proportion of females pregnant and lactating (the competing recipients of A's grooming) might have differed for B and C, taken at comparable stages of the same reproductive state but at different times of the year. The analytical procedure which was adopted in the present study to minimize these distorting

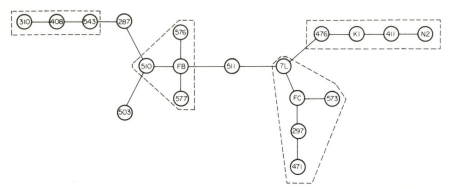

Fig. 10.4. Maximum spanning tree constructed from a matrix of proximity data (≤1.5 m) for the birth season and connecting the 19 adult females of group F. Relatives are enclosed within the broken contours (see Fig. 10.3). The spatial distribution of individuals in the tree is arbitrary.

effects is only briefly summarized here. It consisted of dividing the birth season into four subperiods of comparable duration, compiling a matrix of grooming frequencies for each subperiod and assessing for each subperiod the relative reproductive attractiveness of the 19 adult females taken two by two. Females were considered to be more reproductively attractive, the longer they had been lactating. Knowing for any of the 171 female dyads which female had been the more attractive on the basis of her reproductive state alone during a given subperiod, it was possible to verify if any of the 17 other females had directed significantly more grooming to the more reproductively attractive female. If this was the case, the effect of reproductive state could not be discarded as an explanation for the observed difference in grooming frequencies. If, on the other hand, it was found that a female groomed more often the less reproductively attractive female of the two considered or that she groomed significantly more often one of two equally reproductively attractive females, the effect of the asynchrony of births on grooming preferences could be discarded.

The analysis of the data on grooming preferences which could *not* be explained in terms of the asynchrony of births produced a set of ten rules which are now summarized. Among unrelated females, females groomed those of higher rank more often. They did not automatically prefer their relatives over non-relatives; rather they distributed their preferences among females of both categories if the non-relatives were dominant to their relatives. On the other hand, they preferred the more related rather than the more dominant of their relatives. Finally, they groomed their relatives, whether dominant or subordinate to themselves, more often than unrelated subordinates and showed little concern for females of the latter group.

Interestingly, the analysis of the data on grooming preferences which *could* be explained in terms of differences in reproductive attractiveness produced many exceptions to the above rules.

Seyfarth (1976) interpreted his finding that adjacent-ranking females groomed each other more often than expected as the expression of competition for access to high-ranking females. Although he acknowledged the possibility that a correlation between closeness in rank and degree of relatedness might account for the disproportionate amount of grooming between females of adjacent rank, the degree of relatedness of the females he studied was unknown.

In group F, 18 of 171 female–female dyads (10.5%) were pairs of adjacently ranking females and these 18 dyads accounted for 29.3% of all the grooming exchanged by adult females (573 bouts of 1958). Females of adjacent rank groomed each other significantly more than expected (one-sample chi² = 730.0, $P < .001$) as in Seyfarth's study. Females were related in 15 of the 18 dyads. The average coefficient of relatedness of the 18 dyads, assuming that sisters were half-sibs, was 0.27, while the average coefficient of relatedness of the remaining 153 dyads was 0.03. Adjacent-ranking females were thus, on average, nine times more related than non-adjacent-ranking females. This supports the view that the greater than expected proportion of grooming between females of adjacent rank reflects primarily their higher relatedness rather than the outcome of competition to groom high-ranking females. Moreover, an even stronger argument in favour of the present hypothesis concerns the form of each female's distribution of grooming among females ranking above herself. Seyfarth (1976) produced eight graphs illustrating the distribution of grooming of each of eight adult female baboons. These curves were presented in support of the model, as it appeared that each female tended to groom females ranking above herself in inverse relation to their rank and females ranking lower than herself in direct relation to their rank. The left part of such distribution curves are given for females of group F (Fig. 10.5). According to Seyfarth's model, these curves should be decreasing towards the left (/ shape), indicating that females encountered more and more competition the higher the rank of their grooming partner. Examining the curves, it can be seen that two females (577 and 576) groomed higher-ranking females inversely to the above prediction (\ shape); four females (511, 287, 503 and 7L) groomed, among females ranking above them, those ranking higher more than those ranking intermediately (U shape); six females (FC, 297, 471, N2, K1 and 543) groomed higher-ranking females in no apparent relation with rank (ᴡᴡ shape); and finally five females (573, 411, 476, 408 and 310) also groomed higher-ranking females in no apparent relation with rank except that they each directed more grooming to the female ranking immediately above themselves (their mother in three cases and a sister in two cases (ᴡ/shape).

The shape of all 17 curves clearly demonstrates that adult females did not conform to the above prediction (/ shape). One could conclude on this basis that competition was not an important factor affecting the way these females distributed their grooming. However, inasmuch as the existence of such competition and of preferences for dominant females has been established and will be further documented, a different explanation has to be sought. It is hypothesized that females were indeed more attracted to higher-ranking females but that they did not, as a result of competition, compromise by directing more grooming to intermediate-ranking females. The attractiveness of high rank is not

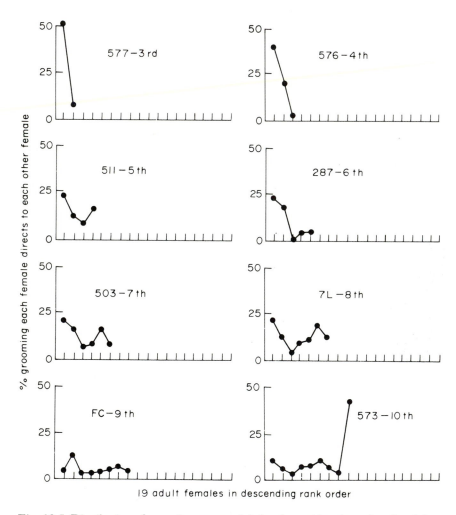

Fig. 10.5. Distribution of grooming among adult females ranking above them by adult females ranking third to nineteenth. The rank of the groomer is indicated after her name. The figure is based on grooming frequencies recorded by focal and concurrent sampling during the birth season. These figures were not corrected for interindividual differences in

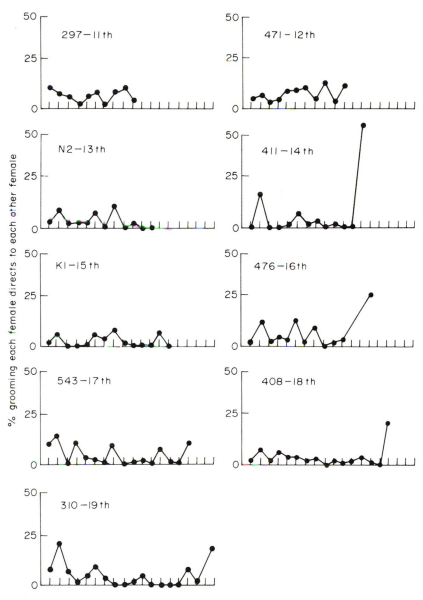

19 adult females in descending rank order

the time spent pregnant and lactating. However, it is believed that the order of females on the reproductive attractiveness scale during the birth season (Chapais 1981: Appendix A) cannot account for the similarities in the trends observed over the 17 distribution curves, a linear function of rank. This militates against the possibility that females groomed adjacent-ranking females at high rates as a result of competition, since females of adjacent rank are even less attractive than intermediate-ranking females on the basis of rank alone. Therefore, relatedness, and not competition, appears to be the proximate cause of high rates of grooming between females close in rank.

Affiliation subsequent to aggression

On 48 recorded occasions, an adult female, after interacting aggressively with another female or a male, approached and directed friendly behaviours to another adult female (Table 10.4). After such an aggressive episode, and independently of whether she was the aggressor or the aggressee, the female preferred to affiliate with a female ranking above herself: 41 cases out of 48 (83%) fall under the diagonal. Of particular interest are the 27 instances where the affiliation was subsequent to an attack on a subordinate female, since the immediate advantages of affiliation are not obvious. In 26 cases, affiliation was directed to a female ranking higher than the target female and in 23 cases, to a female ranking higher than both the aggressor and the target. Incorporating relatedness, affiliation was directed to unrelated, higher-ranking females in 12 cases versus related higher-ranking females in 14 cases; it followed an aggressive act towards an unrelated subordinate female in 14 cases versus a related subordinate in 12 cases. These data suggest that affiliation subsequent to aggression with lower-ranking females might function to demonstrate, establish or reinforce alliances with higher-ranking females against lower-ranking ones whether related or not.

Affiliation interferences

An active affiliation interference consisted of a female approaching two other females who were usually grooming and threatening one of the two females, who responded by leaving immediately. The interferer would then direct friendly behaviours to the competed-for female, who never opposed the interference. Of the recorded affiliation interferences, 87 of the 89 fall under the diagonal in Table 10.5, indicating that the competed-for females were the more dominant of the pair. If interferences were executed for reasons independent of dominance (e.g. for access to lactating females), one would expect the 89 instances to be distributed on each side of the diagonal. The observed asymmetry was due to 75 of 89 interferences (84.3%) being executed by females ranking intermediate between the interfered females. Considering relatedness, females were observed to prevent the access of subordinate relatives and non-relatives to both unrelated

Table 10.4. Distribution of affiliative acts (approaches, grooming, touching and huddling) subsequent to aggression. A female on the Y axis affiliated with a female on the X axis after aggressing a female subordinate to herself (*) or after being aggressed either by an adult male (†), a female subordinate to the female subsequently approached (‡) or a female dominant to the female subsequently approached (§). The table is based on data from all sampling sources collected during the birth season.

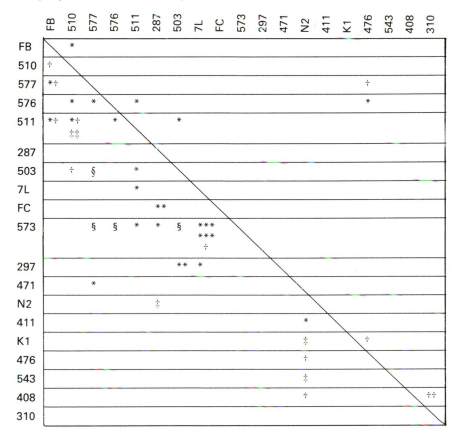

	FB	510	577	576	511	287	503	7L	FC	573	297	471	N2	411	K1	476	543	408	310
FB		*																	
510	†																		
577	*†														†				
576		*	*		*											*			
511	*†	*† ‡‡	*		*														
287																			
503		†	§	*															
7L				*															
FC						**													
573			§	§	*	*	§		*** *** †										
297								**	*										
471		*																	
N2						†													
411													*						
K1													†			†			
476													†						
543													†						
408													†						‡‡
310																			

and related higher-ranking females. If affiliation interferences constitute an additional component of the dynamics of alliance formation, the above data show that females sought to weaken alliances existing both within and outside their own matriline.

Fight interventions

A fight intervention occurred when a female (the interferer) entered a fight between other females and aided one of them (the beneficiary) against the other (the target). The present analysis excludes interventions where all the actors were relatives.

Table 10.5. Distribution of affiliation interferences among adult females. A female on the Y axis left a female on the X axis as a third female threatened the former. The table is based on data from all sampling sources collected during the birth season.

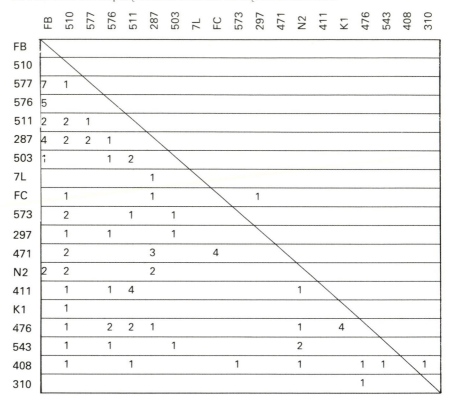

	FB	510	577	576	511	287	503	7L	FC	573	297	471	N2	411	K1	476	543	408	310
FB																			
510																			
577	7	1																	
576	5																		
511	2	2	1																
287	4	2	2	1															
503	1				1	2													
7L						1													
FC	1					1				1									
573	2				1	1													
297	1			1		1													
471	2					3	4												
N2	2	2				2													
411	1			1	4								1						
K1	1																		
476	1			2	2	1							1		4				
543	1			1		1							2						
408	1				1					1			1			1	1		1
310													1						

Consider first the aid given to female relatives against unrelated females. A total of 50 instances where females aged three years or more aided immature or adult female relatives against unrelated adult females were recorded. These 50 interactions involved 57 different interferer–beneficiary–target triads, which were partitioned according to the degree of relatedness between the interferer and the beneficiary and the rank relation between the interferer and the target. The important point for the present discussion is that females aided relatives *both* against higher- and lower-ranking, unrelated, adult females. In doing so they incurred some risk of retaliation from the more dominant target. Thus, an important characteristic of the patterning of aid given to female relatives against unrelated females is its bidirectionality in relation to the female dominance hierarchy.

Consider now the aid given to unrelated females. Table 10.6 presents the 31 instances where a female aided an unrelated female combatant aged three years or more. These interactions are of particular interest since they cannot be explained in terms of kin selection. First, the cells above the diagonal correspond to those occasions when a female aided a female subordinate to herself. By examining the identity of females in these cells (the targets), it can be seen that all of them are subordinate to the beneficiary. Thus, when a female aided a subordinate, she always supported her against an even more subordinate animal. Second, the cells under the diagonal correspond to those cases where a female aided a female dominant to herself. By examining the identity of females in these cells (the targets), it can be seen that all of them are subordinate to the interferer. Thus, when a female aided a higher-ranking female she never supported her against a female either dominant to the beneficiary or ranking between herself and the beneficiary. In other words, females supported unrelated females at no risk. Thus, a constant in the allocation of female aid among unrelated females is its unidirectionality in relation to the dominance hierarchy: females never supported unrelated females upwards. This is in sharp contrast with the bidirectionality of aid given to relatives. Some possible adaptive consequences of such a patterning are discussed in section 12.7.

An attempt has been made to assess the relative weight given to relatedness and dominance by adult females interacting among themselves. Results point to the inadequacy of unitary models (e.g. only relatedness or only dominance) and to the methodological advantages of combining a fine-grained level of data partitioning with a careful control of distorting factors: the abstracted principles suffer few or no exceptions. The pervasive role of dominance and the coalitionary nature of the social relationships of adult females are interpreted in functional terms in Chapter 12.

Table 10.6. Distribution of female aid among unrelated females. All participants were aged three years or more and are ordered according to their dominance rank. A female on the Y axis (the interferer) aided a female on the X axis (the beneficiary) against a female whose name is given in the corresponding cell (the target). The interferer and the

	FB	510	691	577	576	511	699	287	698	503	7L	FC	573	297	471	692	679	N2	411
FB						N2		471											
510						503						*411*							
691						287 *													
577						297	573						*411*						
576						503										310			
511		287																	
699																697 *476*			
287									297										
698																			
503			297			N2													K1
7L				411		N2													
FC	*411*							573											
573																			411
297				*411*													543		
471				679													N2		
692																			
679						408 310	*476*												
N2																			
411																			
649																			
K1																			
697																			
476																			
700																			
543																			
696																			
408																			
693																			
310																			
683																			

beneficiary could not be differentiated in three cases corresponding to the six italicized targets. * = 691 aided 699 against 503 on three occasions, against 573 once and against 297 once.

10.6 Autonomous, Bisexual Subgroups in a Troop of Rhesus Monkeys

BERNARD CHAPAIS

Multimale, female-bonded primate groups include a stable core of resident female relatives with their young and a number of adult, transferred males (Wrangham 1980). The spatial, social and temporal cohesiveness of females is a fundamental characteristic of these groups but one that is not static.

To illustrate the dynamic stability of social structure and the variability encompassed by the notion of the social group, the existence in one troop of rhesus monkeys (group F on Cayo Santiago) of two autonomous, bisexual subgroups (the main group and the subunit) is reported here. This subdivision persisted throughout the 1978 birth season and part of the breeding season and did not seem to correspond to a process of troop-fissioning.

The subunit comprised two adult sisters (N2 and K1) and their respective matriline, making a total of 21 individuals, including four adult females (see Fig. 9.6). The latter had a strong common interest in the alpha-male 415, who appeared as the *raison d'être* of the subunit. The four adult females and their offspring followed 415 and interacted with him on a regular basis. Other group F members generally did not follow 415. However, three adult females (577, 511 and 7L) showed a marked, though less regular and long-lasting, interest in 415. A non-natal adult male, N8, also associated with the subunit.

Even though the main group consisted of fairly well delimited subgroups defined along genealogical lines, its homogeneity was insured by rank-related, transmatrilineal interactions among adult females (see section 10.5) and by inter-actions between females and a few high-ranking adult males (see section 10.4). It is interesting that the three lowest-ranking females of group F, sisters 543, 408 and 310, preferred to stay with the main group even if they were related to members of the subunit through their mother ZM (dead), the sister of N2 and K1.

The main group and the subunit were very distinct spatially, ranging inde-pendently over the island about half of the time and acting as truly separate social groups. Their memberships were highly constant, especially during the birth season. However, the main group and the subunit reunited at least once every day and members of both groups mingled and interacted in such a way that a superficial analysis would not have revealed the existence of two subgroups.

In summary, two factors were responsible for the existence of a subdivision in group F: (1) the reciprocal independence between 415 and most members of the main group (415 would leave group F independently of whether the main group followed him or not and vice versa); and (2) the preference of females K1, 476, N2 and 411 for 415 over members of the main group.

Group F subdivisioning was salient throughout the birth season. By the end of this period, however, the subunit was seen a few times without 415 (who remained with the main group); the subunit would join later. This unusual pattern suggests that the subunit was becoming autonomous in relation both to the main group and to 415. However, three factors contributed to attenuate the distinctiveness of the two subgroups during the breeding season. First, females of the subunit were consorted by males of the main group and 415 consorted females of the main group. Second, in September, 415 contracted tetanus and was unable to keep pace with the group for a few days. During his recovery he associated closely with his matriline and with other females of the main group. Third, group F as a whole lost much of its cohesiveness, with receptive females dispersing widely and many temporary subgroups waxing and waning.

The persistent subgrouping observed in group F is not unique. Southwick *et al.* (1965) described the structure of a troop of rhesus monkeys in India that consisted of three persistent subdivisions, including a peripheral all-male subgroup. The two other subgroups were reminiscent of what was observed in group F: (1) a 'central-male subgroup' composed of two, old, dominant males, eight females, one 'subdominant male' and nine immatures; and (2) a 'dominant-male subgroup' composed of the most dominant male, four females and nine immatures.

Chepko-Sade and Sade (1979) studied five troop fissions that occurred on Cayo Santiago and identified a number of 'tendencies in intragenealogical splitting'. In light of their analysis, group F's long-lasting subdivisioning appears to differ in one important respect from a process of troop-fissioning. The autonomy of the subunit was clearly related to the attraction of its adult female members to the alpha-male and to the reciprocal independence between this male and the main group.

10.7 Analysis of Social Structure

STEVEN R. SCHULMAN

While much of this book, and a considerable amount of research, has centered on the study of relationships, it is not immediately apparent how the transition can be made to the study of social structure. The search for methods to describe how relationships are patterned, in a way which incorporates the possible effects of relationships upon other relationships, is not a trivial analytical problem. In a pioneering approach to the subject, Sade (1972a) introduced classical sociometry (Festinger 1949; Forsyth & Katz 1946; Katz 1947; Luce 1950; Luce & Perry 1949) to primatologists. Two significant difficulties were inherent in this approach. First, these techniques were primarily applicable to one class of interactions at a time (grooming in Sade's example) and hence the picture of social structure which emerges from such an analysis does not fully incorporate the effects of other kinds of interactions on the ultimate model of surface structure which an investigator is trying to construct. The second shortcoming of classical sociometric techniques is that they attempt to discern particular types of structures, most often cliques of individuals. In practice, social structure may take on many forms, e.g. cliques, linear hierarchies, and so on.

In recent years, mathematical sociologists have developed techniques which overcome both of these problems. The techniques permit analysis of the effects of relationships on other relationships and are not limited to searching for a predetermined type of structural pattern such as cliques. These techniques offer promise in the study of animal social structure. It is already customary to present agonistic interaction data in the form of a matrix, where each element represents the frequency with which row individual i threatened column individual k. If a linear dominance hierarchy is present, the rows and columns of the matrix can be simultaneously permuted to reveal an upper triangular structure, i.e. a matrix with mostly zeros below the main diagonal (see Sade 1967 for a typical example). Departures from linearity (e.g. triangular dominance relations) are depicted by non-zero elements below the diagonal. A similar permuting of the rows and columns of a matrix to reveal structural patterning in social-network data underlies the techniques advanced here. However, simple idealized patterns, such as a linear hierarchy producing an upper triangular matrix, are unlikely to be common when multiple ties in a socially complex species are being considered.

Such assumptions about underlying structure are not required by the block-model approach recently advanced by sociologists (Arabie *et al.* 1978; Boorman & White 1976; Breiger *et al.* 1975; White *et al.* 1976). In constructing block-models, the rows and columns of a data matrix are simultaneously permuted and partitioned to create blocks of dense ties and blocks of sparse ties between individuals. Individuals belonging to the same block are said to be *structurally equivalent* in that they show a similar pattern of relationships with other members of the entire group. For instance in a study of multiple social networks in a large group ($n = 110$) of rhesus monkeys on Cayo Santiago, a particularly robust structural equivalence group consisted of six subadult, peripheral, non-natal males attempting to gain access to the main group; they were distinguished by their absence of friendly ties with other group members.

Employing one of a variety of clustering algorithms we take our population of n individuals and assign each individual to one of c structural equivalence groups (in the literature also referred to as blocks). To be useful, c must be considerably smaller than n. Thus, a large $n \times n$ matrix of grooming interactions may be collapsed into a small and compact $c \times c$ matrix of grooming relations among structurally equivalent groups. This collapsed matrix is called an *image graph* or *blockmodel* as it represents a coarser image or model of the structure of grooming relations within the entire group.

Various algorithms are available for differentiating blocks of individuals from network data. A particularly useful one is the CONCOR algorithm which is distinguished by its capacity for simultaneously treating and incorporating contributions from multiple social networks into a blockmodel. Take a data matrix M_o and use it to construct a new matrix M_i. Each element, m_{jk}, in the new matrix is the Pearson product-moment correlation between columns j and k of the old matrix. By applying the same procedure of computing intercolumn-correlation coefficients to the new M_i matrix and repeating this procedure with each newly obtained correlation matrix, it will be found that all of the coefficients converge at $+1$ or -1 as the number of these iterations approaches infinity. The individuals of the original raw data matrix may then be permuted (reordered) and divided into the blocks A and B, possessing only positive correlations among members of the same block and only negative correlations between members of different blocks. The CONCOR procedure can next be applied to just the columns of A block or B block individuals to subdivide either or both of the blocks into smaller and smaller blocks. The beauty of CONCOR is that it can be applied to multiple networks (of similar dimension) merely by stacking them and applying CONCOR to the stacked raw data matrix. Thus if a researcher has four 25×25 matrices reporting grooming, mating, agonistic and play behaviour within a group of 25 monkeys, stacking these matrices will yield an M_o matrix of dimension 100×25*.

Such blockmodels, based upon multiple types of interaction data (e.g. grooming, mating, agonistic and aiding interactions), represent a quantifiable, concrete picture of surface structure. Sophisticated techniques are now available for comparing blockmodels of one group over time, e.g. during mating and non-mating seasons. The same techniques can be used to compare blockmodels of different groups or species. The groups being compared need not be constrained to contain the same number of individuals and the blockmodels being compared may contain different numbers of blocks (see below).

The procedure for comparing blockmodels of surface structure in different populations is more involved and readers should consult the original development in Boorman and White (1976) (see also Boorman & Arabie 1980;

---

*For more extended discussions of the use of CONCOR and particularly for advice on preliminary coding and weighting of data for input to the algorithm, see Arabie *et al.* (1978) and Breiger *et al.* (1975) and for primatological applications see Pearl and Schulman (1982) and Schulman and Boorman (in prep.). A new algorithm (ICON) is a combinatorial optimization procedure which seeks an optimal assignment of individuals to blocks by maximizing a measure of contrasts between blocks. ICON is described and applied to a rhesus social-structure example in Schulman and Boorman (in prep.).

Pearl & Schulman 1982; Schulman 1980)*.

The patterning of relations between relations is algebraically summarized in a 'role table' (Boorman & White 1976). Such a role table is an algebraic abstraction of the typically complex ways relations between blocks interweave. The principal value of a role table lies in its use for comparing structural patterning in blockmodels derived from different groups or the same groups at different points in time. The role table is an algebraic abstraction of the common denominator of the two surface structures (for details see Birkhoff 1967; Boorman & White 1976; Gericke 1966). A key attribute of this algebraic approach to comparing surface structures is that it is unaffected by differences in the sizes of social groups being compared and relatively unaffected by differences in number of blocks.

Blockmodelling techniques were used by Pearl and Schulman (1982) to compare the surface structures of a group of 25 rhesus macaques in northern Pakistan with a provisioned group of 110 rhesus macaques on Cayo Santiago based on six types of ties. Despite contrasts in group size, environments and number of blocks, several striking similarities in social structure were apparent. The influence of dominance and genetic relatedness were pervasive in determining block membership. Analysis of the role tables demonstrated that the surface structure of the two groups were similar in the integration and patterning of relations between blocks. When comparing CONCOR and ICON block assignments against a network of genetic relatedness Schulman and Boorman (in prep.) uncovered some interesting associations. While blocks tended toward a division along genealogical lines, the anomalies were illuminating. Specifically, individuals of low rank within their genealogy showed greater variability in their block assignment and often were assigned to a block containing members of the next lower-ranking genealogy. In other words, the interactions of low-ranking members within a genealogy were more similar to those members of the next lower-ranking genealogy than to those of their own genealogy. The converse appeared true of subadult females rising in rank near the uppermost positions within matrilines. Thus, the group fissioning pattern observed by Chepko-Sade and Sade (1979) and Chepko-Sade and Olivier (1979), as well as the dynamics of female social-rank acquisition (see Chapais & Schulman 1980; Schulman & Chapais 1980; sections 6.7, 6.8 and 12.6), were detectable in the blockmodels.

In conclusion, these powerful new techniques provide primatologists with analytical tools for modelling the surface structures of different social groups,

---

*The procedure can only briefly be sketched here. Image graphs for each type of tie (e.g. grooming, mating, etc.) are used to construct an algebra, mathematically a Boolean semigroup, describing the pattern of relations between relations ('role interlock' in the terminology of Boorman & White [1976]).

different species and changes within a group over time. The approach is inductive and does not presume that investigators are looking for a particular kind of structure. Blockmodelling allows the data to determine their own configuration. Aggregation is inferred at the end of the analysis rather than imposed at the beginning. Another advantage is that the investigator does not need to make assumptions about the underlying distribution of particular behaviours or interactions.

10.8 Ecological Influences on Relationships and Social Structure

PHYLLIS C. LEE

Previous work on squirrel monkeys (Baldwin & Baldwin 1973, 1976), lemurs (Richard 1974), mangabeys (Chalmers 1968) and toque macaques (Dittus 1979) has suggested that variations in the rates and nature of social interactions within a single social group of primates can be related to food availability and dispersion and in particular to seasonal variations in the nature of the food supply. Since the social structure of primate groups can be viewed as the content, quality and patterning of the relationships existing between the members of a group (see section 10.1), the social structure of groups of monkeys living in different habitats may be expected to differ.

Until recently, few data were available to examine the ways in which social groups of a single primate species respond to local variations in resource availability and quality, in order to assess whether differences in observed social structure from one group to the next could be related to local ecological variation. Studies of baboons (Altmann & Altmann 1979; Dunbar 1979a; Ohsawa & Dunbar in press) have shown that differences in the nature and number of social relationships between individuals living in troops in different habitats can be explained by the demography of the group, i.e. by differences in sex ratios, survivorship and the age/sex classes present in the troop. Since certain types of relationships can exist only when specific age/sex classes are present, e.g. alliances between adult males (cf. Packer 1977) or competitive hierarchies among males for access to mates (cf. Hausfater 1975; Rasmussen 1980), demographic variation is important in restricting or providing opportunities for relationships to exist.

In addition, ecological differences between groups can affect primate social structure in several ways. First, as a result of food stress or abundance, both birth rates and mortality and ultimately group composition can be affected. Second, under conditions of food stress or widely dispersed resources, certain spatial and energetically costly relationships may be difficult to maintain. It is primarily this second case, where resource availability can directly constrain or alternatively provide opportunities for the expression of relationships that is considered here.

During a study of olive and yellow baboon troops in two habitats (Oliver & Lee 1978) and a study of three adjacent groups of vervet monkeys (Lee 1981), the effects of differences in food availability and distribution on rates of immature social interactions and general group structure were assessed.

All of the studies were carried out in highly seasonal environments, where rainfall and correlated food biomass peaked during a short period of the year. The effect of seasonal changes in food availability and distribution on the time spent in different maintenance activities, such as feeding or travelling, was pronounced for all study groups (Fig. 10.6). Both the Gombe baboons and all three groups of Amboseli vervets spent more time feeding during the dry season, when food availability was generally lower, while the Ruaha baboons spent more time feeding during the wet season periods of abundance. Travelling and resting time showed changes related to those of feeding. The time spent feeding appears to be related to the distribution of food as well as to its overall availability (Iwamoto & Dunbar 1983) and this may explain why groups that forage on widely dispersed or low quality foods have to spend more time feeding, even during the periods of abundance.

The frequency of different types of social interactions, both competitive and friendly, was also affected by seasonality. Individuals in both baboon troops had significantly higher rates of competitive interactions over food during the dry season than in the wet (Oliver & Lee 1978), while the reverse was true for the three vervet groups (Lee 1981). Again this may result from differences in the distribution of the food resources, in that baboons foraged together on larger-clumped resources which provided opportunities for increased competition, while the vervets fed singly on small-clumped resources during the dry season (Lee 1981). There were no significant differences between the seasons in the rate of grooming for any of the study groups, while the frequency of social play was highest, and significantly so, during the wet season for all five groups (see also section 6.5).

The habitats of each of the five groups were assessed for their relative quality (Table 10.7). In comparisons between the groups in different habitats, a consistent trend could be observed. Those groups with the lowest food availability spent the most time feeding and the least time in social interactions, irrespective of the season. Interindividual distances tended to be greater in these groups; the Ruaha baboon troop was subjectively more dispersed while foraging and travelling than was the troop in the richer habitat at Gombe. Other social interactions were not compared between the baboon troops but among the vervets the mother–infant relationship and the time-course of weaning differed between the groups. In the stressed habitats, infants were weaned earlier than

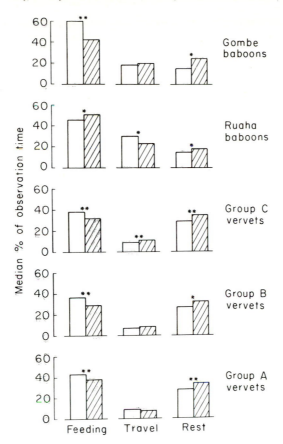

Fig. 10.6. Median percentage of observation time spent in feeding, resting and travelling through a 12-h day for each of five study groups for the dry (open columns) and wet (hatched columns) seasons. * = $P < 0.05$; ** = $P < 0.01$.

those in the better quality habitats but mothers continued to maintain strong affiliative relationships with infants and appeared to tolerate their infants suckling to an older age once the ecological and energetic constraints were reduced (Lee 1981). Since a mother's ability to suckle an infant for an extended period and to conceive with a short interbirth interval may be related to her nutritional condition (e.g. Berger 1979b), the nature and persistence of mother–offspring relationships can also be linked to the local environment. One might also predict that affiliative behaviour between close kin would be more evident in groups living under conditions of either seasonal or relative low food availability, resulting from the persisting closeness of mothers and offspring and from the benefits of concentrating relationships among kin when there are energetic costs attached to interactions. Among the vervets in this study, the most distinctive and

Table 10.7. Subjective relative assessment of habitat quality, food distribution and the frequency of social interactions for each of the studied groups. A mixed patch size is one where both small and large patches were available and exploited.

Food availability	Food distribution	Food quality	Frequency of social interactions	Time spent feeding
Baboons				
Gombe	Large patches	High	High	High
High (9.4 foods h^{-1})	contiguous			
Ruaha	Small patches	Low	Low	Low
Low (5.2 foods h^{-1})	widely dispersed			
Vervets*				
Group A	Mixed patches	Low	Low	High
Low (7.0 foods h^{-1})	widely dispersed	2898 kJ d^{-1}		
Group B	Mixed patches	High	High	Low
High (8.0 foods h^{-1})	contiguous	3486 kJ d^{-1}		
Group C	Mixed patches	Middle	Middle	Middle
Middle (7.5 foods h^{-1})	dispersed	3150 kJ d^{-1}		

*See Lee (1981) for caloric/quality estimates: medians for all seasons.

supportive relationships were maintained between close kin (Lee 1981).

Relationships and their structuring within the social group depend upon a framework of ecological constraints, some of which may be seasonal and which may affect different groups in different ways. Relationships, such as those between play partners, are directly constrained by the energy available from the diet (Lee 1981; Martin 1982). Relationships that are maintained through less energetically costly interactions, such as those between grooming partners, showed little seasonal variation and few differences between the different groups. However, among the immature vervets the duration of grooming bouts with peers tended to be longer during the dry season, when there was little peer–peer play, and shorter when there was more play. The immature Ruaha baboons showed a similar correlation with low rates of play, in that they shifted their partner preferences in grooming from mothers to peers when play rates declined (Oliver & Lee 1978). While grooming and play may serve different long-term functions, both interactions allow for relaxed and friendly contact between peers. With these seasonal shifts in partners and the types of grooming bouts, a high level of peer–peer interaction can be maintained despite habitat constraints.

Other aspects of social structure may also be affected by ecological factors. Wrangham (1981) has shown that the adult female vervets in the same three groups competed for different resources and that the abundance of these resources was a critical factor influencing female survival. Cheney (section 11.3) discusses the movement of breeding males between groups of vervets. Since the size and location of a potential new territory for a transferring male is related to the distribution and quality of the resources it contains, the number of males in a group, the length of their tenure in a group and the degree of competition for access to females may again be related to ecological factors.

It is not surprising that two different species of baboons in widely separated habitats showed differences in the structure of their relationships. However, it is more surprising that the three groups of vervets in small adjacent territories showed the same magnitude of intergroup differences which can be related to the same variables of food availability and distribution. What then, can be predicted about the effects of ecological variation on the social structure of primate groups? Among the five different groups studied, certain interactions, such as those between play partners, were affected, with other interactions taking their place, depending in part on their relative energetic costs. Relationships, such as those between competitors, may need very low frequencies of interaction to be maintained and seem to be less dependent on ecological conditions. Despite seasonal differences in rates of competitive interactions, both immatures and adults in all the groups had well-developed dominance hierarchies, which affected the distribution of other friendly and agonistic behaviour among individuals (Collins 1981; Lee 1981; Lee & Oliver 1979; Moore 1978; Seyfarth 1980; Wrangham 1981).

Thus, the interactions which form the basis for the stable relationships of immatures and the patterning of these relationships through time depend directly on ecological factors. Under conditions of low food availability, the frequency of interactions is reduced, their temporal patterning is restricted and their intensity and distribution among the different partners available is altered. The variety of interactions observed is reduced. Conditions of high food availability have the effect of increasing the frequency of interactions and affecting their patterning among individuals, resulting in an increased complexity of the observed social structure. While general 'rules' about primate social organisation (e.g. Clutton-Brock & Harvey 1976) can be related to general environmental quality, the details of social structures may ultimately be less stable, highly locally variable and change through time as a result of both the immediate environment and the composition of the group.

Chapter 11
Intergroup Relationships

11.1 Home Range, Territory and Intergroup Encounters
PHYLLIS C. LEE

The social units of many species of primates tend to move through a well-defined and familiar area—the home range (reviewed by Clutton-Brock 1977). Some species, particularly those dealt with in this book, forage together as a single unit through the home range, while others, such as the chimpanzee, forage on their own at some times and in groups at other times (e.g. Wrangham & Smuts 1980). The size of the home range differs between social groups of the same species and appears to be related to habitat variables—rainfall, food availability, food dispersion and renewal rate (Dunbar & Dunbar 1974; Gartlan & Brain 1969; Homewood 1978; Rasmussen 1979; Waser 1976; Waser & Homewood 1979). Population densities, the size of the social unit and the energetic requirements of the individuals in the group all affect the size of the area used as a home range (Clutton-Brock & Harvey 1979; Harvey & Clutton-Brock 1981).

However, the movements of a group through an area occur within a social environment composed of other such units. In such an environment, groups may contact each other and interactions between groups may take place. The nature of the interaction, whether that of friendly mixing of individuals, avoidance of other groups or aggressive defence, varies widely between species and between groups of the same species in different areas.

Hamadryas and gelada baboons both form large aggregations, composed of

231

many smaller units, at sleeping sites (Kummer 1968) or while foraging (Crook 1966; Dunbar & Dunbar 1975), and several hundred animals will interact peacefully. The Guinea baboon (*Papio papio*) also forms temporary aggregations of several hundred individuals from a number of troops sharing the same sleeping groves, grooming and playing and then fragmenting into smaller foraging units during the day (Sharman 1981). The other savannah baboons seldom form persisting aggregations and members of different troops will avoid each other (Altmann & Altmann 1970). Troops may sleep side by side but seldom forage as a larger unit. Adult males often herd females of their troop away from strangers (Packer 1979a) but occasional friendly interactions, especially on the part of young males, have been reported (Cheney & Seyfarth 1977; section 11.2). Territoriality, in the sense of combined group defense of an area, has been observed only rarely among the baboons and resulted from the defense of scarce water resources (Hamilton *et al.* 1976). In some species, there appears to be a hierarchy between groups, such that one group will leave a resource or an area at the approach of another without overt aggression. Hausfater (1972) observed aggressive encounters between groups of rhesus monkeys on Cayo Santiago which resulted either in high levels of aggression between the groups or in passive displacements. The frequency of contacts between groups was a function of the number of groups and the extent of their movements.

Encounters that define a territorial boundary may involve the expenditure of considerable energy (e.g. Robinson 1981) and the extent of territoriality varies with the profits and costs it entails (e.g. Lauer 1980; Mitani & Rodman 1979). Monogamous species, such as gibbons and titi monkeys, defend their territories by loud and regular vocalizations, which appear to signal other pairs that the territory is occupied and that mates will be defended (Robinson 1981; Tenaza 1976). Other species, such as vervets, use both vocalizations and aggressive contact to defend their territories (Struhsaker 1967d). An escalated aggressive encounter between groups may result when contact is made over a highly valued food resource, while the same groups may respond to other neighbouring groups with vocalizations alone if no resource is present (Lee 1981).

Territorial encounters may affect interactions within a group, leading to an increase in aggressive encounters within it (Marsden 1968). As Cheney (1981, section 11.2) points out, even individuals in groups such as vervets, who aggressively defend their home ranges, may interact in a friendly way when the costs and benefits of interacting with 'strangers', which vary through time or from individual to individual in the group, permit this.

Under conditions of population expansion, social units may grow in size leading to the fissioning of the group into new units. Fissioning may take place along lineage lines, as has been observed for rhesus and Japanese macaques (Chepko-Sade & Olivier 1979; Furuya F. 1968; Furuya Y. 1969; Sugiyama

1960). This can result in neighbouring groups who are related to each other; in mangabeys, interactions between groups who were more related as a result of fissioning tended to be less aggressive (Waser & Homewood 1979). Chacma baboon troops have recently been shown to fission on a temporary basis, forming one-male subtroops with consistent membership (Anderson 1981). The one-male subtroops interact peacefully when they coalesce into a larger troop. Cheney (section 11.3) has shown that the movement of males between groups is non-random and may lead to high degrees of relatedness between the members of different groups and that this again affects the nature of interactions between groups.

Thus, the relationships and social structure observed within a single social unit may also reflect the nature of that group's relations with other groups. The social structure of a group exists in a social context and this may affect interactions between groups in several ways. Aggression between groups depends on the frequency with which groups encounter each other, on the nature of the resources available in each group's range and on the characteristics of individuals in the groups. These in turn may depend on prior experience with strangers, on the degree of relatedness between individuals in different groups and other individual attributes, such as status, age or sex.

The two contributions that follow consider the relations between the individuals in neighbouring groups and the transfer of individuals from one to another. As stressed earlier, the issues of the immediate causation and the ultimate consequences of behaviour are closely linked and both are raised in these sections. Readers less familiar with evolutionary arguments may find it helpful to read section 12.1 before proceeding with 11.2 and 11.3.

11.2 Intergroup Encounters among Old World Monkeys
DOROTHY L. CHENEY

Introduction

In most sexually reproducing species, females contribute more energy and time to the production and rearing of offspring than do males. This is particularly true in the case of mammals, where pregnancy and lactation have largely eliminated the need for direct male parental care. As a result of the fundamental imbalance in parental investment between males and females, it has been suggested that the reproductive success of females is limited primarily by energetic and nutritional constraints, while that of males is limited by the distribution and availability of females (Bradbury & Vehrencamp 1977, Emlen & Oring 1977, Trivers 1972).

Applying these arguments to non-human primates, Wrangham (1980) has suggested that primate social structure is ultimately determined primarily by the distribution of food in space and time, because this determines the distribution of females. When food is evenly distributed, competition for food will be minimal and females are expected to forage singly or to form only temporary and unstable social groups. Wrangham has used this argument to explain the lack of permanent social bonds among female gorillas and chimpanzees. In contrast, when food is patchily distributed, such that one female by herself cannot easily defend the resources she requires, females are expected to form cooperative coalitions with their genetic relatives and collectively defend resources from other groups of females. Such female-bonded social groups are characteristic of most Old World monkeys (see section 12.2).

In such Old World monkey species as baboons, macaques and vervet monkeys, the stable core of each social group consists of a number of related adult females and their offspring. Females remain in the groups in which they were born throughout their lives, while males transfer to neighbouring groups at around sexual maturity. Within each group, females typically form linear dominance hierarchies, which determine access to food and water. Individuals generally acquire dominance ranks similar to those of their mothers and frequently retain these ranks throughout their lives (Chapais & Schulman 1980; Cheney 1977; Cheney *et al.* 1981; Hausfater *et al.* 1982; Kawai 1958; Koyama 1967; Sade 1967, 1972b). Wrangham has argued that this form of social structure reflects selective pressures acting on females to form cooperative coalitions in order to obtain and defend sufficient food resources for themselves and their offspring.

Social groups of baboons, macaques and vervet monkeys are typically multimale. Theoretical considerations of the factors that influence the reproductive success of each sex predict that the distribution of males should be determined primarily by the distribution of females. In species in which male parental investment is absent or only indirect, a male's reproductive success should be positively correlated with the number of females he is able to monopolize. However, although it may be in the reproductive interests of males to exclude all other competitors from the social group, it may not always be energetically feasible to do so, particularly when group size is large, as is the case with many Old World monkey species. In such cases, groups may become multimale and there should be intragroup competition for females. The consequence of such competition will be a male dominance hierarchy, in which high-ranking males have greater access to females than do low-ranking males.

The hypotheses described above lead to a number of predictions regarding the nature of intergroup encounters in Old World monkeys. First, because the defense of food resources seems critical to female reproductive success, females should actively participate in intergroup encounters and attempt to defend their

group's range against incursions by other groups. Second, because the reproductive success of males is primarily determined by access to females, male
aggression during intergroup encounters should be directed primarily toward
other males. Males should initially resist the immigration of many other
additional males into the group and attempt to prevent the members of other
groups from gaining access to their own group's females (Wrangham 1980).
Thus, while both sexes are expected to participate in intergroup encounters,
there should be differences in both the behaviour of each sex and the resources
that each sex defends. This section examines these predictions using data
derived from the intergroup encounters of baboons, macaques and vervet
monkeys. The generality of these predictions is then considered by examining
individual variation in the behaviour of males and females of different ages and
dominance ranks.

1. Behaviour of males during intergroup encounters

A number of studies of baboons, macaques and vervet monkeys have shown that
male reproductive success is strongly influenced by competition with other males
and that access to sexually receptive females is positively correlated with
dominance rank (Cheney this volume; Hausfater 1975; Kaufmann 1967; Packer
1979b; Seyfarth 1978a,b; Smith 1981; Struhsaker 1967b; Sugiyama 1976).
Because of the correlation between reproductive success and intrasexual competition, adult males might be expected to resist the entrance of additional males
into their groups and to behave aggressively toward members of their own sex
during intergroup encounters (see above). A number of studies have suggested
that this is indeed the case. Adult male baboons frequently chase males who
attempt to transfer into their groups (Packer 1979a; Slatkin & Hausfater 1976;
Stoltz & Saayman 1970). Similar aggression toward immigrant males has been
observed in rhesus macaques (Hausfater 1972; Vessey 1968), Japanese
macaques (Kurland 1977; Sugiyama 1976) and vervet monkeys (Cheney 1981;
Henzi & Lucas 1980; Struhsaker 1967a). In all cases, males appear to be more
aggressive toward males than toward females from other groups. Baboon males
have also been observed to chase or herd the female members of their own group
away from males in other groups (Buskirk *et al.* 1974; Cheney & Seyfarth 1977;
Hamilton *et al.* 1975; Packer 1979a; Saayman 1971). Such herding appears to
represent an attempt to prevent males from other groups from copulating with
females, an hypothesis that is supported by the observation that herding in one
baboon group was directed primarily toward sexually receptive females (Cheney
& Seyfarth 1977). Male herding of females appears to be less common among
macaques and vervet monkeys than it is among baboons. Moreover, while female
baboons rarely threaten adult males, female vervet monkeys frequently form

coalitions and chase males who attempt to herd them (Cheney 1981). Some possible causes for the difference between female baboons and female macaques and vervets are discussed below.

While resident adult males often resist attempts by additional males to transfer into their group, their behaviour contrasts sharply with that of natal males. In all Old World monkey species studied thus far, males and females appear to avoid mating with individuals with whom they have matured and this lack of sexual attraction seems to be ultimately related to the avoidance of close inbreeding (see section 11.3). Young baboon males who have not yet transferred from their natal group often provoke intergroup encounters by approaching other groups and attempting to initiate affinitive interactions with their members (Cheney & Seyfarth 1977; Hamilton *et al.* 1975; Packer 1979a). Similar behaviour has been observed in Japanese macaques (Itoigawa 1974; Norikoshi & Koyama 1975; Sugiyama 1976), rhesus macaques (Boelkins & Wilson 1972; Hausfater 1972) and vervet monkeys (Cheney 1981; Cheney & Seyfarth 1982; Struhsaker 1967a). In all cases, such interactions appear to precede and facilitate the natal males' subsequent transfer to other groups.

2. Behaviour of females during intergroup encounters

The behaviour of males during encounters with other groups appears ultimately to be related to gaining or maintaining access to females. In contrast, the hypotheses described above predict that while adult females should actively participate in intergroup encounters, their participation should be determined by competition over resources rather than sexual partners. This prediction is supported by observations of macaques and vervet monkeys. Rhesus monkey females frequently behave aggressively toward the members of other groups, even though they may sustain injuries during such interactions (Hausfater 1972). Japanese macaque females have been observed to chase and attack males who attempt to enter their groups and Packer and Pusey (1979) have argued that such aggression may represent an attempt to decrease the number of potential competitors for the group's food resources. Among vervet monkeys, females regularly chase and even physically attack the members of other groups and have been observed to injure severely young males who attempt to transfer into their groups (Cheney 1981; Cheney & Seyfarth 1982; Henzi & Lucas 1980; Struhsaker 1967a). Unlike vervet males, who are most aggressive toward members of their own sex and age, female vervets are aggressive toward males and females of all ages. Most intergroup encounters among vervet monkeys appear to occur over access to food or water (Wrangham 1981) and females actively defend their group's range against incursions by other groups. In contrast, while males often behave aggressively toward potential immigrants, they may ignore males who enter their group's range simply to feed or drink (Wrangham 1981).

According to Wrangham's (1980) arguments, there should be no significant differences between the behaviour of female baboons and that of female macaques or vervet monkeys during intergroup encounters. Surprisingly, however, there have been few reports of female baboon involvement in inter-group interactions. Indeed, at least two studies have commented upon the lack of female participation in intergroup encounters (Hamilton *et al.* 1975; Packer & Pusey 1979) and other studies have made no mention of female involvement (Cheney & Seyfarth 1977; Slatkin & Hausfater 1976). Thus far, only one field study of baboons has reported female aggression during intergroup encounters similar to that observed among macaques and vervet monkeys (Smuts pers. comm.).

There are at least two possible explanations for this difference in female behaviour. First, a female's ability to act aggressively toward adult males appears to depend upon the degree of sexual dimorphism within a given species. Female baboons are approximately half the body weight of males and this size differential may prevent them from successfully driving away potential migrants. In contrast, the body weights of female vervets and macaques more closely approximate those of males. Perhaps as a result, females in these species are frequently able to drive away males who threaten them or attempt to enter the group (Cheney 1981, this volume; Packer & Pusey 1979). As a consequence of this difference in sexual dimorphism, it is possible that female baboons exert less of an influence than do female macaques and vervet monkeys over the immigration of males into their groups.

A second factor potentially influencing the lack of baboon females' participation in intergroup encounters concerns the defendability of food resources. Vervet monkeys inhabit relatively small territories (averaging 0.4 km² in size) whose borders remain relatively stable over long periods of time (Cheney & Seyfarth 1982). Each group's range can therefore be easily patrolled and defended. In contrast, it is not unusual for baboons to range over areas greater than 16 km² (Altmann & Altmann 1970). Baboons do not defend fixed ranges or territories and a number of groups may forage over the same large geographical area. A group may encounter other groups in virtually any area of its range. The large size of a baboon group's range appears to prevent efficient defense and it may simply be too costly, in terms of energy and time, for females to exclude other groups from their group's resources. Perhaps as a result, females may avoid frequent participation in intergroup encounters, except on those occasions when a limited resource, such as a fruiting tree or a scarce water-hole, is being defended.

While groups of macaques do not defend fixed ranges, in the two areas in which macaque intergroup interactions have been studied (Cayo Santiago and Japan), groups range over relatively small areas that appear to be easily patrolled (Hausfater 1972; Packer & Pusey 1979). As is the case with vervet monkeys, therefore, food resources may be energetically less costly to defend for macaques than they are for baboons. It is possible that in areas where macaques range over larger geographical areas, female participation in intergroup encounters diminishes in frequency.

3. Individual variation in behaviour

The arguments described above offer plausible explanations for species-typical behaviour among male and female Old World monkeys. Social groups of monkeys, however, are composed of individuals of differing age and dominance rank. These differences may cause individuals to confront very different selective pressures, even when they inhabit the same social group and compete for the same resources. These differing selective pressures may be reflected in individual variations in behaviour.

(a) Behaviour of males

As mentioned above, the behaviour of males during intergroup encounters is highly variable and is strongly influenced by whether or not an individual has already transferred from his natal group. While adult males often chase other males from their own group's females, natal males frequently approach and attempt to interact with the members of other groups (see above).

Baboon males have been observed to behave most aggressively toward migrants who are roughly the same age and size as themselves (Packer 1979a). Since dominance rank is often correlated with size and age, males who are of approximately the same size may present the most competition for dominance and access to females (Packer 1979a; see also Dittus 1977; Noorikoshi & Koyama 1975).

The costs of migration, in terms of possible injury and loss of reproductive potential, appear to be higher for young males than for older, more-experienced males. Among vervet monkeys, natal males who are transferring for the first time are frequently not yet fully grown and therefore seem particularly vulnerable to attack and injury by adult females (Cheney pers. obs.). Perhaps as a result of the high rates of aggression they receive, males often make their initial transfer in the company of brothers or other natal group peers (vervets: Cheney this volume; baboons: Cheney & Seyfarth 1977; rhesus macaques: Boelkins & Wilson 1972; Drickamer & Vessey 1973; Meikle & Vessey 1981; Melnick 1981; Japanese macaques: Itoigawa 1974; Sugiyama 1976). The presence of such allies may cause young males to be less vulnerable to attack and may also increase the rate at which they rise in dominance rank (Meikle & Vessey 1981).

Lack of experience in the process of integration into a new group may also cause natal males to receive more aggression than multiply-transferring males. Among baboons, males who are transferring for the first time receive high rates of aggression from resident males and often remain peripheral to the group even days after transfer (Packer 1979a). In contrast, multiply-transferring males seem quickly able to recruit coalition partners and to establish bonds with females, an ability which may depend on experience. Such alliances appear to increase the rate at which older males are integrated into the group (Packer 1979a).

In contrast to natal males, adult male vervets who are transferring for a second or third time generally transfer alone and often migrate to distant or unfamiliar groups. It is possible that such solitary migration is facilitated by experience, as appears to be the case among baboons (see above). Multiply-transferring males are often subordinate individuals who have recently fallen in rank and whose access to females is restricted. The differential reproductive opportunities of subordinate and dominant males appear to influence their behaviour during intergroup encounters. For example, perhaps because they appear to derive the greatest benefits from defending the group's females from immigrant males, dominant males are often the most aggressive during intergroup encounters, even when such aggression may be of some potential cost to themselves (Cheney & Seyfarth 1977). In contrast, subordinate males are more likely to initiate affinitive interactions with the members of other groups (Cheney 1981). As is the case with natal males, such affinitive interactions may facilitate future transfer to other groups.

(b) Behaviour of females

Because the reproductive success of females is limited by food and nutritional constraints, it has been hypothesized that female Old World monkeys should actively defend their group's resources against the members of other groups (Wrangham 1980; see above). Groups of Old World monkeys, however, are also characterized by intragroup competition, with the result that females typically form linear dominance hierarchies in which high-ranking females are systematically able to exclude low-ranking females from scarce resources. A number of captive studies of macaques have documented a positive correlation between female rank and reproductive success (Drickamer 1974 a,b; Silk *et al.* 1981b), which may be associated with differential access to resources. No correlation between female rank and infant survival has been documented for vervet monkeys (Cheney *et al.* 1981). Nevertheless, high-ranking animals are able to exclude other individuals from food and water, with the result that the members of high-ranking families are less likely than the members of low-ranking families to die during periods of food scarcity (Cheney *et al.* 1981; Wrangham 1981).

Because of the prevalence of stable dominance hierarchies among female Old World monkeys, it seems likely that individuals benefit differentially from the resources contained within their group's range. This suggests that there may be individual differences in the frequency with which females of different dominance ranks engage in intergroup encounters as well as in the nature of their interactions with the members of other groups. High-ranking, vervet monkey females appear to be more aggressive during intergroup encounters than low-ranking females (Cheney *et al.* 1981). Moreover, low-ranking females are far more likely than high-ranking females to engage in affinitive interactions or to copulate with the members of other groups. Occasional affinitive gestures by females toward the members of other groups have also been reported in baboons and macaques, though no studies have commented upon the dominance ranks of the individuals involved (Boelkins & Wilson 1972; Cheney & Seyfarth 1977; Hausfater 1972).

It seems possible that high-ranking females, who benefit more than other females from the resources contained within their group's range, are more active than others in defending these resources from other groups. Low-ranking females, in contrast, seem to gain the least from the maintenance of the status quo within the group and may, therefore, be more likely than other females to engage in interactions which may have the effect of disrupting group cohesion. Studies of female participation in intergroup encounters among macaques and baboons have not commented upon any differences in the behaviour of females of different dominance ranks. It is worth noting, however, that when groups of macaques have been observed to fission, it is generally the lower-ranking genealogies which split off from the larger group (Chepko-Sade & Olivier 1979; Chepko-Sade & Sade 1979; Koyama 1970). Thus, among macaques, as among vervet monkeys, it appears that low-ranking females are more likely than high-ranking females to engage in behaviour that may disrupt group cohesion.

Conclusions

As is the case with many other aspects of their behaviour, the behaviour of male and female Old World monkeys during intergroup encounters becomes more comprehensible when considered in terms of the factors that contribute to the reproductive success of each sex. Data on parental investment and reproductive behaviour predict that although both males and females will participate in intergroup encounters, female involvement should be related to defense of food resources, while male involvement should be related to defense of females. Although these predictions are largely supported by data on Old World monkeys, it is clearly also important to consider individual differences in behaviour within each sex and the ways in which such differences may be related to individual variation in reproductive success. Group life imposes different constraints on individuals of different age and dominance rank, even when such individuals are of the same sex and live in the same social group. For both males and females, the costs and benefits of defending either females or food resources may differ according to age and dominance rank. Variation in behaviour during intergroup encounters appears to reflect behavioural adaptations designed to minimize the costs and maximize the benefits of each individual's different competitive abilities.

11.3 Proximate and Ultimate Factors Related to the Distribution of Male Migration
DOROTHY L. CHENEY

In many species of birds and mammals, dispersal is sex biased. Male dispersal predominates among mammals and numerous evolutionary hypotheses have been proposed to explain female philopatry as opposed to male migration (see e.g. review by Greenwood 1980). In most species of Old World monkeys, males migrate to neighbouring groups at sexual maturity, while females remain in their natal groups throughout their lives. It is generally assumed that the ultimate function of male migration is the reduction of inbreeding (Itani 1972; Melnick & Pearl 1982; Packer 1979a). The proximate cause of male migration in macaques and baboons appears to be sexual attraction to unfamiliar, presumably unrelated females (Enomoto 1974; Packer 1979a; Sugiyama 1976). The majority of male transfers among macaques and vervet monkeys occur during the breeding season and male baboons frequently transfer to those groups that include the largest number of sexually receptive females (Cheney & Seyfarth 1977; Drickamer & Vessey 1973; Henzi & Lucas 1980; Lindberg & Packer unpublished data).

While numerous studies of Old World monkeys have documented male dispersal, very little is known about the distribution of male movement between specific groups. Social groups of monkeys do not exist in isolation from one another and a given male usually has a number of different groups to which he might theoretically transfer. Do males always simply transfer to the group with the most sexually receptive females or do other factors influence the pattern of male movement? If so, what are the consequences of male transfer for inbreeding avoidance and future relations between groups? In this section these questions are addressed by examining male transfer among free-ranging vervet monkeys. The distribution of male transfer across a number of different groups is described and the effects of such transfer on inbreeding and intergroup relations are discussed.

Study groups

The vervet monkeys described here inhabit an area of semi-arid, acacia savanna in Amboseli National Park, Kenya. Three social groups have been studied since at least 1977 and an additional eight groups are censused regularly. Births, deaths, immigrations and emigrations have been monitored continuously. This section discusses male transfer between March 1977 and March 1982.

Between 1977 and 1982, the three study groups inhabited contiguous home ranges which averaged 0.4 km^2 in size. The areas incorporated by each group's range remained relatively stable over time and each group actively defended its range against incursions by other groups. Since male transfer usually occurred over a period of several days, it was possible to document the movements of almost all males into and from the study groups (see below). In Group A, the mothers and siblings of all individuals born later than 1974 were known. In Groups B and C, these kinship relations were known for all animals born after 1976. Males were considered to be adult at five years of age.

Recognition of individuals across groups

When considering the causes and consequences of the distribution of male transfer across groups, it seems important to determine the extent to which individuals are able to recognize animals in other groups. If males cannot monitor events in other groups and if they cannot recognize the members of other groups as individuals, male movement may simply be random and there may be no predictable pattern of male movement between groups. At least two factors, however, argue that vervets do recognize and distinguish among neighbouring groups.

First, vervet groups' ranges are small and it is physically possible for males to observe events in other groups without leaving their own group's range. Second, and more important, playback experiments have demonstrated that both male and female vervets recognize the vocalizations of individuals in other groups. Vervets associate the vocalizations of individuals in neighbouring groups with the ranges that those individuals' groups occupy, and respond very strongly to vocalizations when they are played from inappropriate ranges (Cheney & Seyfarth 1982). The ability to recognize the members of other groups may permit males to monitor neighbouring groups and perhaps to choose when and where to transfer. Similarly, such recognition may allow females to modify their behaviour during

intergroup encounters, depending perhaps on previous patterns of male movement and the history of their group's relations with each of its neighbours (see
below).

Timing of transfer

Between 1977 and 1982, it was possible to document the movements of 25 males
into and from the three study groups. Although it was not possible to determine
whether males who were adult by 1977 had previously transferred, no males
whose ages were known have ever remained in their natal group beyond seven
years of age. The youngest male known to have transferred was two-and-a-half
years of age. Dominance rank appeared to be important in determining the
probability of transfer for adult males migrating for a second time. In the ten cases
when the ranks of such males were known prior to transfer, all occupied one of
the two lowest ranks in their group prior to transfer. Whether or not males
transferred before reaching sexual maturity seemed to be determined by the
presence of close female kin (see also Itoigawa 1974). All males who transferred
before the age of five years were orphans with no known living maternal sisters.
All such immature males transferred in the company of their brothers or peers
from their natal group. Indeed, the presence of peers or siblings seemed to be the
most crucial determinant for transfer by immature males, since only males over
five years of age ever transferred singly.

Peer migration has been documented in baboons and macaques as well as
vervet monkeys (Boelkins & Wilson 1972; Cheney & Seyfarth 1977; Drickamer
& Vessey 1973; Meikle & Vessey 1981; Sugiyama 1976). Transfer in the
company of natal group peers appears to facilitate integration into the new group
by providing young males with allies during aggressive disputes. Adult female
vervets have been observed on a number of occasions to bite and severely wound
immature males who attempt to transfer into their group, although females never
injured fully grown males who were bigger than themselves (see also Packer &
Pusey 1979). The presence of natal group allies appears to reduce the aggression
that is directed against young migrants and may also increase the rate at which
such individuals assume high dominance rank. For example one study of rhesus
macaques has shown that males who transfer with their brothers support each
other in aggressive disputes and assume high dominance ranks more rapidly than
do males who migrate alone (Meikle & Vessey 1981). Thus, although the
presence of brothers or peers may represent increased competition for mates, the
advantage of potential allies probably outweighs the costs of such competition.
Moreover, since such males are often close genetic relatives, these costs may also
be partially offset by increases in the males' inclusive fitness.

Distribution of male transfer

Each of the three study groups had at least five contiguous neighbouring groups, in each of which the sex ratio and number of sexually mature females varied. In contrast to previous studies (Drickamer & Vessey 1973; Packer 1979a), there was no consistent tendency for males to transfer to groups which included the most females or in which the sex ratio was most skewed (Table 11.1). Thus, the distribution of male movement did not appear to be determined solely by the relative availability of females in different groups.

Table 11.1. Summary of male transfer to groups with differing sex ratios and numbers of females. Neighbouring groups were defined as groups with adjacent home ranges; each study group had at least five neighbouring groups. Significance values were calculated with two-tailed binomial tests.

	No. males		
	Yes	No	
Transfer to groups with more skewed female/male sex ratio than previous group	13	5	$P > 0.10$
Transfer to groups with more females than previous group	11	7	$P > 0.10$
Transfer to that neighbouring group with the most skewed sex ratio	9	8	(excluding one unknown) $P > 0.10$
Transfer to that neighbouring group with the most females	12	5	(excluding one unknown) $P > 0.10$

One possible factor influencing the probability of transfer to a particular group may be the likelihood of attaining high rank in that group. In many species of Old World monkeys, there is a positive correlation between male rank and access to sexually receptive females (Hausfater 1975; Packer 1979b; Seyfarth 1978a,b; Smith 1981). Among the vervet monkeys, between 1977 and 1980 (the period for which data on the frequency of copulations are available) there was an overall positive, but non-significant, correlation between male rank and copulation frequency (unpublished data). It therefore seems likely that there were reproductive benefits associated with high dominance rank. If males did monitor events in other groups, they might have been expected to transfer to those groups where the probability of rising rapidly in rank was high. Indirect evidence suggests that this might indeed have been the case. Of 12 cases in which the ranks of adult males were known, both prior to and following transfer, nine individuals assumed within the following three months higher ranks in their new group than they had held in their previous group. One such transfer occurred after a male in the new group was killed by a leopard. In at least three other cases, males transferred to a group following a period of instability in that group's male dominance hierarchy.

Although males in the study groups did not always transfer to the group with the most females, patterns of migration nevertheless appeared to be non-random, particularly among younger males. Of the 15 migrants who were either known to be natal males or not yet fully grown, 14 (93%) transferred to groups where others of their previous group had migrated. In contrast, this was true for only five of ten fully grown, adult males (50%), four of whom were known to be transferring for at least a second time (Figs 11.1 and 11.2). In the three study groups, natal or young males were significantly more likely to transfer to one specific neighbouring group than to any other (Binomial test, two-tailed, $n = 12$, $P < 0.05$). There was no such tendency among older males. In nine cases, peers or maternal brothers emigrated from the natal group together. In other

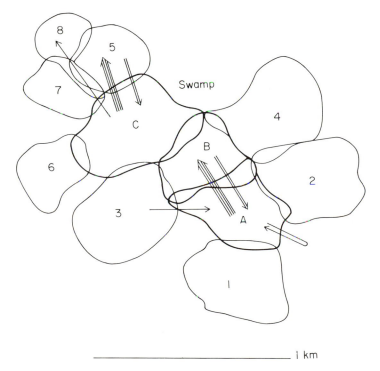

Fig. 11.1. Distribution of migration by natal and young adult males into and from the three study groups between March 1977 and March 1982. The arrows indicate the direction of movement and each line represents one male. The ranges of the study groups are lettered; the ranges of regularly censused groups are numbered. Only groups with ranges adjacent to the study groups are shown.

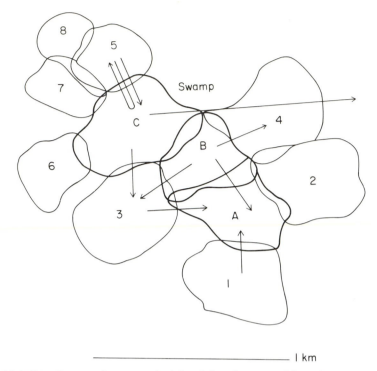

Fig. 11.2. Distribution of migration by fully adult males into and from the three study groups between March 1977 and March 1982.

cases, however, males transferred to a particular group as much as two years after other individuals from their previous group had transferred there.

Consequences of male transfer for relations between groups

Intriguingly, the non-random distribution of male migration between particular groups was correlated with qualitative differences in relations between groups. Each of the three study groups had regular encounters with each of its neighbours (see also Cheney this volume). The quality of intergroup encounters, however, varied, ranging from marked aggression to the exchange of friendly gestures between the members of opposing groups (Cheney 1981). Not all individuals in each group initiated friendly interactions with the members of other groups and adult females were generally less friendly than other age/sex classes (Cheney & Seyfarth 1982; see also Cheney this volume). Nevertheless, when vervets did initiate friendly interactions with the members of other groups, they were most likely to do so when interacting with the members of those groups to and from which males had previously transferred (Fig. 11.3).

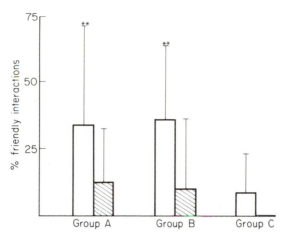

Fig. 11.3. Means and standard deviations for the frequency with which individuals in each study group initiated friendly gestures with the members of transfer (open columns) and non-transfer (hatched columns) groups. Transfer groups were groups from and to which males transferred between 1977 and 1980, while non-transfer groups were groups from and to which males were not observed to transfer. The data were gathered during 1980 over an eight-month period. ** = $p < 0.01$, using two-tailed Wilcoxon matched-pairs signed-ranks test. The frequency of friendly interactions was calculated as follows:

$$\frac{\text{No. friendly gestures given to members of other groups}}{\text{No. friendly and aggressive gestures given to members of other groups}}$$

The causal relation between reduced aggression during intergroup encounters and male transfer is not yet known. Males may have transferred to groups from which they received little aggression because reduced hostility facilitated integration into a new group (Cheney this volume). Males and females may also simply have been less hostile towards groups with which male transfer previously occurred. It seems probable that both males and females exerted an influence on whether or not transfer with a particular group would occur. As a result, male movement may have been constrained, both by the distribution of past transfers and by the history of each group's relations with its neighbours.

Relation between male transfer and inbreeding avoidance

Previous studies of male transfer in Old World monkeys have assumed that the ultimate cause of transfer is the reduction of inbreeding. Clearly, however, male vervet monkeys appear not to be avoiding inbreeding as much as they might if they simply transferred alone and at random. In other words, the tendency of two groups to exchange males may ultimately increase degrees of genetic relatedness between individuals of particular groups.

While non-random dispersal may eventually result in high levels of inbreeding within a local population, this dispersal pattern may nevertheless bring short-term benefits to individual migrants. By transferring non-randomly in the company of close relatives, males may minimize the costs of migration, both in terms of the risk of mortality due to predation and in terms of the loss of reproductive potential (see Bengtsson 1978). Migration with brothers or peers may be particularly beneficial to young males, who are vulnerable to attack by the members of other groups.

Because of its potential long-term genetic costs, however, persistent non-random transfer can only be evolutionarily stable if at least some males forego the benefits of this dispersal pattern and migrate to other groups. Population geneticists have shown that even small rates of immigration by outbred individuals from different groups are sufficient to reduce inbreeding coefficients (see reviews in Cavalli-Sforza & Bodmer 1971; Hartl 1980). Thus, as long as even a few males migrate to other groups, non-random transfer by other individuals should not result in inbreeding depression. Which males might be expected to adopt such an apparently altruistic strategy, especially if the costs of migrating to unfamiliar or distant groups are high? The data presented here suggest that males who deviate from the non-random dispersal pattern are older, fully grown males who are probably transferring for a second or third time.

Although it may be advantageous for a young male to transfer in the company of close kin, this benefit may diminish with age for a number of reasons. First, the risk of attack by resident males and females may decrease as a male becomes larger and more experienced, thus diminishing the need for allies (see Packer 1979a). Second, in any primate group in which paternity certainty is low, a long-term resident male eventually risks mating with his daughters. Although it might be advantageous for such a male to transfer for a second time, it could be genetically costly to return to the natal group. At this point, any costs of migrating to a distant or unfamiliar group might be outweighed by the potential reproductive benefits to be gained by such a movement. As a result, older males might be more likely than natal males to transfer outside the local population. Such dispersal might also be advantageous for those males who fail to attain high dominance ranks in their adopted groups and whose access to females is restricted. If the benefits of remaining in the group are low and if the presence of brothers or natal group peers fails to confer a competitive advantage, a male might be increasingly likely to migrate alone to an unfamiliar group. It may therefore be appropriate to consider male dispersal as an age-dependent polymorphism, in which older males are prepared to incur the costs of migrating outside the local population simply because there is little to be gained from remaining in more familiar groups.

In summary, therefore, there appear to be both social and genetic benefits associated with a dispersal pattern in which males transfer non-randomly with their brothers when young but alone and to more distant groups when older. These age-related differences in migratory behaviour should provide sufficient genetic variation to permit individuals to maximize the social benefits of non-random migration, while simultaneously minimizing the risk of inbreeding depression.

Chapter 12
Ultimate Factors Determining Individual Strategies, Relationships and Social Structure

12.1 General

ROBERT A. HINDE

While studies of the dynamics and development of social behaviour, relationships and social structure lead to consideration of causal factors earlier in time, it is necessary also to study their consequences: full understanding requires us to ask how observed behaviour has been selected for in evolution. Each behavioural act has many consequences, some beneficial, some neutral and some harmful in their effects on the survival and reproduction of the individual concerned and his close relatives. If the behavioural propensity is to be maintained by natural selection, it must have a heritable basis and the beneficial consequences of the acts in which it is manifested must outweigh the deleterious ones. This does not mean that every act must have beneficial consequences but only that on balance the behavioural propensities bring a profit.

Natural selection can act only when the profit is obtained to a different degree by different individuals. Thus, not all beneficial consequences provide material for natural selection: a given act may have many consequences, some of which are achieved equally by all individuals. The latter do not provide material for natural selection. Thus, the question, How is it useful? is not necessarily the same as, What are the consequences of this behaviour through which natural selection acts to keep it in the repertoire of the species? (Hinde 1982).

250

Two basic but as yet not wholly resolved theoretical issues must be mentioned here, both of which arise from the fact that animals sometimes seem to act against their own interests. Discussion of the first initially focused largely around the question of population control. Data suggest that in many species individuals disperse in such a way that competition for resources is less acute than it would be if all individuals concentrated in the areas of temporary abundance. Wynne-Edwards (1962) suggested that density was controlled by emigration or non-breeding of individuals such that local populations remained near the optimum. This view, i.e. that selection acts on groups of individuals as a unit, implies that some individuals (i.e. those who emigrate or refrain from breeding) act against their own interests for the benefit of the group. In general, group selection is unlikely to be important in most circumstances because it would be susceptible to invasion by individuals who cheated. If an extra individual refused to emigrate and stayed to breed, his genes might be perpetuated more effectively than those of his colleagues who moved elsewhere. It thus seems more profitable to attempt to explain the behaviour of those who emigrate in terms of selection for individuals who act to favour their own interests: perhaps those who emigrate would incur costs in attempting to stay not commensurate with their probable gains from attempting to breed, and would do better to move elsewhere (Lack 1954, 1968; Maynard-Smith 1964; Williams 1966). While there are some circumstances in which group selection may operate (Maynard-Smith 1976), it is best to seek for adaptive advantages at the individual level.

Thus, in examining the consequences of one type of group structure as compared with another, the advantages for the group as a whole must not be sought but an attempt to see the group's structure as emerging from the behaviour of individuals each trying to further his or her own ends. Thus, if achieving high dominance status involves costs, which it certainly does, it must also involve counterbalancing profits in the shape of enhanced individual (though see below) reproductive success. On the whole, the evidence suggests that individuals do in fact behave in a manner which optimizes their lifetime reproductive success. For example high dominance status is, in some species at least, associated with high reproductive success (e.g. Dunbar 1980b; Hausfater 1975; Packer 1979a; Sade *et al.* 1976; Saunders & Hausfater 1978; Silk *et al.* 1981a; Witt *et al.* 1981). However, the complexity of the issues is illustrated by the finding that in troops of free-ranging vervets, mortality from illness was concentrated among the low-ranking individuals but mortality from predation among high-ranking ones (Cheney *et al.* 1981).

As another example, infanticide by a new male when he takes over a troop can augment his own reproductive success relative to that of others by reducing that of one of his rivals, and augment his own directly by reducing the delay with which nursing mothers in the troop can again conceive: although lowering the troop's reproductive output, it augments that of the male (e.g. Angst & Thommen 1977; Chapman & Hausfater 1979; Hrdy 1977a, 1979; Struhsaker 1977). However, the generality of this phenomenon has been questioned and it may sometimes be a by-product of male take-overs rather than part of an evolved male reproductive strategy (Boggess 1979). Again, in every species, either males or females tend to emigrate from their natal group before reproducing, thus reducing the likelihood of mating with close relatives. The behaviour of individuals can be seen as enhancing their own chances of reproductive success, either because the costs of inbreeding depression incurred by staying would exceed those of transferring or because emigration provides access to more oestrus females (Packer 1979a; Rasmussen 1981; section 11.3).

Thus, in general, while group-living may carry beneficial consequences to its members, for instance in the transmission of information about predators or food sources (e.g. Azuma 1973; Cheney & Seyfarth 1980; Fairbanks 1975; Tsumori 1967), both the fact of group-living and the details of the interactions and social relationships which give rise to the group structure are to be interpreted in terms of costs and benefits to individuals and not to the group as a whole (see e.g. Crook *et al.* 1976; Wrangham 1979, 1980). There may sometimes be something to be said for living solitarily for a while (Slatkin & Hausfater 1976).

The second issue also concerns the behaviour of individuals who act to their own apparent disadvantage. Individuals sometimes behave altruistically, i.e. in biological terms they act so as to promote another individual's reproductive success at the cost of detriment to their own. Such behaviour poses a biological problem: given that the behavioural propensities involved are partly genetically determined, how could natural selection have promoted behaviour which decreased an individual's reproductive potential? Natural selection operates through the survival of genes, and in 1964 Hamilton pointed out that a genetic change leading to behaviour disadvantageous to the individual bearing the genes could be selected for, if it sufficiently benefitted others who were genetically closely related to it and therefore bore identical genes.

This has apparently happened in social insects whose colonies contain castes of sterile individuals. The workers propagate genes identical with their own more effectively by facilitating the reproduction of the queen than they would propagate their own genes by reproducing themselves. Thus, individuals are selected not to maximize their own fitness (measured in terms of the propagation of their own genes) but their inclusive fitness (measured in terms of the propagation of genes identical to their own whether carried in their own bodies or those of their relatives). In general, natural selection acts on individuals not just so as to maximize their own survival and reproduction but to maximize their own survival and reproduction *and* that of their relatives, devalued in the latter case by their degree of relatedness.

This principle may be of great importance for studies of social behaviour in monkey groups which contain numbers of closely related individuals. A high proportion of positive social behaviour is directed towards relatives and, even though involving some disadvantage to the actor, could be understood as the result of natural selection acting on individuals to maximize inclusive rather than individual fitness. Much of the evidence for this view involves data showing that individuals show positive behaviour to others in proportion to their degree of genetic relatedness (see section 10.1). There are, however, some difficulties with such evidence. For example while Kurland (1977) showed that individuals tended to groom others in proportion to their degree of genetic relatedness, Altmann and Walters (1978) pointed out that if benefits to the groomer accumulate in proportion to grooming time, all grooming should be directed to the closest relative and none to others. Furthermore, the extent to which inclusive fitness is enhanced by the grooming of a close relative will depend in part on the latter's reproductive potential: from this point of view it may not be profitable to groom a post-reproductive grandmother (though of course there may be other reasons why a relationship with her would be profitable). In addition, relatedness is often assessed only through the maternal line but age cohorts may share paternity and thus be more closely related than is often assumed (J. Altmann 1979). In any case, demographic and motivational factors, as well as kinship, may affect partner choice (Cheney 1978a). With such issues, kin selection theory has yet to come to terms.

In some cases, the costs incurred by altruistic behaviour may be outweighed by subsequent reciprocation (Trivers 1971). The benefits of grooming to a groomer may not lie in enhancing his inclusive fitness by the removal of ticks from his relatives but in increasing the probability that the groomee will aid him in future agonistic encounters (Dunbar 1980b; Seyfarth 1977). Giving aggressive aid may be reciprocated later (Packer 1977).

Finally, the profit derived from some actions may not lie so much in enhancing the reproductive potential of the actor or his relatives but in decreasing that of others. For instance bonnet macaque females sometimes kidnap infants from other females. The kidnapper seems to gain neither status nor maternal experience and the mothers certainly do not benefit. Silk (1980) thus suggests that the benefit may come through decreasing the fitness of the mothers.

A special case of inclusive fitness, but one of particular interest, concerns the relationship between parent and offspring. It is accepted as natural that parents should sometimes act for the good of their offspring, even though they may be risking themselves in doing so, for it should be expected that natural selection produces parental behaviour which promotes the survival of the young. Since each offspring obtains half his genes from each parent, it will be to the evolutionary advantage of a parent to care for his offspring so long as the costs to him (in terms of his or her inclusive fitness) are less than half the benefit to the offspring. Trivers (1974) has argued that immediately after birth parental care is essential, so that its benefits to the offspring are large, while its costs to the parent (e.g. in terms of milk production) are relatively small. As the infant develops, the costs to the parent increase and, as the infant becomes better able to care for itself, the benefits to the infant diminish. There thus comes a point when it is to the advantage of the infant to continue to obtain parental care and to the advantage of the parent to terminate it. Hence, there is a biological basis for a conflict of interests between parent and offspring.

Nothing in this discussion must be taken as implying that it is possible to specify 'ideal' relationships of maximal adaptive value, for what is best may vary with the circumstances. This has been documented in a field study of baboons by Altmann (1980). The infants of 'restrictive' mothers survive better in their early months but infants with '*laisser-faire*' mothers may achieve earlier independence and thus cope better if orphaned: the benefits of one or the other maternal strategy may vary with dominance rank. Here the different styles of mothering can be considered as conditional strategies (Maynard-Smith 1974) with the best type of mothering depending not only on aspects of the mother (such as her age or rank) but also on aspects of the infant (such as its sex and when it was born) and of the habitat (i.e. the probability of infant survival).

A related issue arises in situations of conflict. What are the ultimate factors that determine whether an individual should persist in an agonistic encounter or withdraw? Maynard Smith and Parker (1976) used a modelling approach derived from Games Theory to predict what an individual should do in a competitive or agonistic situation. The best course will depend in part on the probable behaviour of the opponent. If it is likely to fight viciously, it will probably be better to withdraw, but if it is likely only to threaten, an all-out attack may pay off. Given certain assumptions, a stable situation can arise where there are particular proportions of individuals who are vicious or conventional fighters or where each individual adopts each strategy in a particular proportion of contests (Maynard Smith & Parker 1976). Such a situation is said to involve 'evolutionary stable strategies' (ESS): an increase in the proportion of either type of individual (or strategy) would be disadvantageous to the individuals concerned and the original balance would tend to reassert itself. Although the argument involves a number of simplifying assumptions, the concept of the ESS has proved useful in, for example, assessing the conditions under which a display is likely to escalate into a fight between two individuals.

Parker (1974) has also shown that such stable strategies can be based on the assessment of another individual's ability to control a contested resource, such as food, females or a territory: this is called the 'resource holding potential' (RHP). Individuals in contests will be favoured when they can accurately assess another individual's RHP relative to their own and thus retreat or escalate the fight on the basis of this assessment. Some signals exchanged in contests are primarily concerned with the RHP (Maynard Smith 1979), though when acquaintances are involved, communication about immediate intentions is also important (Hinde 1981a, van Rhijn & Vodegel 1980).

Some aspects of these issues are exemplified in the following contributions. Section 12.2 addresses the basic question of why some species live in groups at all. The most probable hypothesis is that it is advantageous to females to ally with each other in order to compete successfully with other groups of females for food, the males attaching themselves to the female-bonded groups. As noted earlier, their presence adds considerably to the complexity of the social structure (section 10.4). There is, of course, still competition between individuals within groups and this helps to explain many aspects of relationships within the group. As has been shown, the characteristics of dyadic relationships require explanation in terms of the benefits and costs accruing to each of the participants: the relationships between male and female baboons are considered from this point of view in section 12.3. In sections 12.4 and 12.5 the ultimate consequences of high dominance rank for male and female rhesus adults are discussed.

The issue of alliances in agonistic disputes has come up in a number of contexts in previous sections (e.g. sections 6.7, 6.8, 9.3, 9.5, 10.2–10.5 and 11.3). The last three contributions in this chapter argue, from a functional perspective, that the benefits accruing from alliances or from interference in an agonistic episode will depend crucially on the characteristics of the individuals involved: the data indicate that the patterns of interferences that animals show do in fact reflect strategies which maximize benefits and minimize costs to the individual concerned.

12.2 Ultimate Factors Determining Social Structure
RICHARD W. WRANGHAM

The significance of some of the early studies of individual relationships among primates was clouded by their unnatural contexts—in captivity or in provisioned groups (Altmann 1962; Imanishi 1960; Mason 1963). However, now that it is known that primate social behaviour is fully as complex in the wild as it is under artificial conditions it can be treated as a natural characteristic and the question asked, Why did it evolve as it did? To do so we must think ourselves squarely into natural environments, where food supplies fluctuate erratically, predators may lurk behind every bush, and thorns, diseases and extremes of climate can bring untimely discomfort or death. The challenge is to understand the role of such crude natural forces in generating intricate patterns of social relationships.

Natural selection, of course, is the key. The door which it opens, unfortunately, reveals a maze. Some principles for finding the way are sketched here and the merits of four possible paths are discussed.

Asking appropriate questions

A natural approach is to select a given behaviour pattern (such as grooming) and ask what are its ultimate consequences, i.e. its effects on reproductive rate compared to an alternative behaviour pattern (such as resting). Unfortunately, social behaviours have such varied and inter-related consequences that it is difficult to assign reproductive results with any precision. In particular, if social acts occur within a social relationship, as they normally do, their significance depends on the type of relationship. Consequently, it is difficult to discuss the adaptive significance of social acts, except with reference to specific social relationships; it is similarly hard to discuss the significance of a social relationship without referring to other social relationships. Such reasoning leads rapidly to a loop: behaviour cannot be explained in isolation and it comes to be viewed as adaptive in the context of social structure. Useful though it is to understand the way in which different social interactions and relationships affect each other, such understanding stops short of the link to the environment. In order to determine the ultimate factors responsible for social complexity, therefore, it is necessary to begin with a simpler question than those concerned with behaviours such as grooming, aggression or play, or with the social relationships in which they occur. The question is, Why live in groups at all?

The occurrence of groups is not explained easily. There is no general agreement about the ultimate factors responsible for monkey groups or even about how such factors can be recognized. The following principles appear useful.

First, the advantages of grouping may be different for each sex. As described in section 12.1, it is important to ascribe fitness benefits to individuals rather than to the group as a whole and this raises a problem where groups contain adults of both sexes. Natural selection is a result of differential reproduction, and social behaviour is a mechanism to reduce the constraints on reproduction. However, the reproduction of females and males is constrained in different ways: for example females rarely lack mates, whereas males often do (Trivers 1972). Consequently, the benefits of group life are likely to be different for each sex. If so, the functions of group life must be analysed separately for females and males.

Second, different environmental pressures are expected to lead to different kinds of group. Equally, therefore, the adaptive significance of group life is indicated by the kind of group typical for the species. Rhesus monkeys, vervets and baboons, for example, share a number of special features which make their groups different not just from those formed by most insects and vertebrates but also from a number of other primates. These groups, which are 'female-bonded' (FB), are bisexual, with females as permanent members: only males routinely leave their natal groups around adolescence to breed elsewhere. Their groups are sedentary and xenophobic: from year to year they tend to occupy the same home range and they rarely tolerate the presence of other groups nearby. Regularities such as these need to be explained and they emphasize the importance of sex differences in group life. Since males are attracted to groups of females rather than vice versa, ultimate explanations of group-living cannot rely solely on the benefits to males. They must show either how females benefit, and then explain why males join female groups, or how both sexes benefit, and then why males leave their natal groups to breed elsewhere.

Third, the benefits of group life can be expected to be describable rather simply for each sex. The benefit of a strategy such as group-living, in evolutionary terms, is measured by how effectively it releases an individual from constraints on its reproduction: the more reproductive success it gives, the more important the strategy. Since a given strategy is unlikely to relieve two or more constraints with equal effectiveness, the benefit derived from one should be more important than those derived from others. The classic example of this principle is the effect of competition between breeding males for access to fertile females. Limited access to mates is argued to be the most important constraint on male reproduction in many species and male strategies are accordingly directed principally to obtaining matings (Trivers 1972). Thus, mating strategy is the 'key strategy' for males. An equivalent argument can be made for females. Female fitness, of course, is rarely limited by a shortage of mates. Instead, female reproductive success is constrained by resource availability, predation and other sources of ill-health or mortality. Hence, it is in these currencies that the benefits of group life are expected to accrue to females.

The following points arise from a summary of the above. First, it can be expected that the ultimate factors promoting group-living will be different for each sex. Second, females form the permanent membership of rhesus, vervet and baboon groups. Third, male strategies are likely to be adapted to finding fertile females. These points argue that in FB groups male grouping behaviour is adapted to the grouping patterns of females.

Why live in groups?

Four possible causes of group life are discussed in this section. These are the most promising or frequently advocated explanations relevant to female bonded groups of monkeys.

1. Increasing access to high-quality food patches

By this hypothesis, groups are viewed as alliances of individuals competing with other groups of the same species for access to the best food patches. Group life therefore has effects on reproductive rate and offspring survival by making more food available. It is supported as follows. First, food availability appears to limit reproduction in natural habitats, causing irregular birth rates within and between groups and between years (Altmann 1980; Cant 1980; Dittus 1977). Second, groups tend to orient their foraging paths around high-quality food patches, where individuals in larger or otherwise more dominant groups gain the most access to preferred foods, such as ripe fruit (Wrangham 1980). Third, only the primate species whose food is distributed at least sometimes in high-quality patches tend to form FB groups (Wrangham 1980). Several special features of FB groups are also explicable. Thus, because food availability affects the reproductive rate of females more directly than that of males, food-getting is a key strategy for females: hence it is females who form permanent alliances. Similarly, groups can be seen as xenophobic because they compete for access to food sources and as sedentary because they are embedded in a network of intergroup relationships which constrain long-distance movement. This hypothesis thus goes beyond accounting merely for the existence of groups.

It needs to be developed on several points. First, too little is known about primate nutrition to be sure that access to ripe fruits, which are the food items commonly found in preferred patches, is a critical variable affecting birth rate. Although several studies have shown that high-ranking females have a greater intake of preferred foods than lower-ranking females (Post *et al.* 1980; Wrangham & Waterman 1981) and that high-ranking females have higher birth rates than low-ranking females (Dittus 1979; Drickamer 1974a; Mori 1979), only one study has shown both relationships in the same population (Whitten 1982). Second, there are no quantitative data on variance in patch size and hence on the nutritional importance of access to preferred patches. Third, the correlation of food distribution with grouping patterns across the primate order is not perfect: at least two species appear to have FB groups which do not rely on high-quality patches (Wrangham 1980). Nevertheless, at present, this hypothesis appears to go further than others in explaining species differences in social structure and its significance is examined further below.

2. Reducing the costs of predation

Many models have shown that animals which form groups can thereby lower their susceptibility to predation, by mechanisms such as mutual aid in attacking predators and increased sensitivity to the presence of predators (reviewed by

Bertram 1978a). The risk of predation may therefore be a selective force favouring groups. This hypothesis is supported in macaques, baboons and vervets by three principal observations. First, predators do occur in the natural environments of these monkeys and several studies have recorded deaths of adults or young from predation (Cheney *et al.* 1981; DeVore 1963b). Second, groups are seen to respond effectively to predators, e.g. by concerted attacks (DeVore & Hall 1965). Third, for primates as a whole there are a number of correlations between apparent vulnerability to predators, such as being terrestial rather than arboreal, and characteristics helping to deter predation, such as having large groups, many males per group and males with big canines (Clutton-Brock & Harvey 1977; Crook 1972; Harvey *et al.* 1978). Despite these observations, however, the role of predation as an ultimate cause of group life is still highly uncertain. First, these species establish or retain their typical grouping pattern even in environments with little or no predation, such as on Cayo Santiago. This is odd, considering that in other animals grouping patterns are known to vary closely with environmental conditions (e.g. Macdonald 1979a; Stacey & Bock 1978). Second, groups would be expected to respond effectively to the approach of a predator regardless of the ultimate forces favouring group-living. Third, the correlations between apparent vulnerability and more effective ways of defending against predators can be explained without reference to predation. Thus large groups may occur because food patches are larger for terrestrial than arboreal primates and sexual selection may account for the number of males and the size of male canines (Wrangham 1980). The predation hypothesis also faces a difficulty in that it fails to explain why monkey groups are characterized by permanent females, migrating males and sedentary and xenophobic habits. Its value is therefore limited at present to accounting for the fact, rather than the form, of group-living and even here it is open to question. More sophisticated models of the role of predation need to be developed to clarify its effects.

3. Finding food more efficiently

The irregular distribution of primate foods means that individuals must work hard to find them. It has therefore been suggested that group-living aids individuals in locating new food patches as well as promotes efficient feeding (e.g. when individuals capture insects flushed by group foraging [see Clutton-Brock & Harvey 1977]). However, while this may be important for species such as flocking birds (Horn 1968), it is unlikely to apply to rhesus macaques, vervets or baboons. Groups of these species rarely seem to find foods by chance and insect-foraging is comparatively uncommon and tends to be an individual activity. The hypothesis ignores the effects of feeding competition (Dittus 1977) and fails completely to explain why groups are closed: indeed the opposite might be expected. No good evidence exists to make it an attractive idea at present, however reasonable its intuitive merits.

4. Travelling with care-givers

The hypotheses above have been concerned with the direct effects of ecological pressures. However, ecological pressures need not necessarily have positive effects on grouping. In certain environments they may merely be permissive, allowing, though not favouring, the evolution of groups. This appears to be the case for mountain gorillas, for example, in which groups function as permanent consortships of several females with a breeding male: females gain protection from the effects of male–male competition and the male benefits by predictable access to several mates (Wrangham 1979). This is allowed by a food distribution which makes feeding competition a trivial pressure (see section 13.3).

In principle, a similar argument could be applied to FB groups, but in practice it is difficult. Most importantly, group-living appears to be costly for monkeys because of feeding competition within the group (e.g. Dittus 1977). Ecological pressures do not, therefore, seem permissive. Second, the structure of FB groups is incompatible with any obvious source of grouping. For example arguments based on mate choice are inadequate for explaining why females stay in the same group throughout their lives: the strong tendency for close female kin to live and interact together requires an explanation in terms of female relationships. The best hypothesis is that females aid in the rearing of each other's offspring, providing an environment in which orphans can survive and infants develop more quickly and effectively than with isolated mothers.

Observations of the care of orphans (section 6.7 and 6.8) and other interactions between relatives and offspring make this possible. However, unlike the cases where male–male competition is supposed to generate groups (above), the relevant selective pressure (loss of mother) would often not apply. The hypothesis could be supported by data showing sufficiently high rates of maternal mortality and orphan adoption. High rates are known in some populations (Hamilton *et al.* 1982) but seem unlikely to be a widespread phenomenon. To solidify the argument, it would also have to be shown that the benefits of mutual care-giving outweigh the costs of intragroup competition.

Towards a synthesis

These hypotheses demonstrate not only the uncertainties in explaining why groups occur but also that it is often difficult to connect ecological pressures with relationships within groups: for example the view that predation is a critical pressure has been of little help so far in explaining internal group structure. In this respect, the most promising hypothesis to date is that groups are founded on alliances of females competing for access to food sources.

This is attractive because it reconciles the undoubted importance of intragroup feeding competition with the fact that close kin nevertheless forage in proximity. The proximity of kin is regarded as a result of the need not just for companions but for particular companions, namely individuals who can be relied on to provide support against conspecifics. Why should these allies tend to be kin? First, in triadic interactions where two individuals gain at the expense of a third, inclusive fitness tends to be maximized by supporting a closer against a more distant relative (Wrangham 1982). Second, individuals who are familiar from birth have repeated opportunities to test each other's reliability as partners, allowing the development of a reciprocally beneficial relationship (Axelrod & Hamilton 1981). Within groups, nested subgroups occur and compete together for access to group resources. Again, these subgroups tend to consist of close kin, who thus cooperate against more distant kin (Chepko-Sade & Sade 1979). A critical feature in each case is that females are bonded by the need for support against outside alliances. Relationships thus contain both competitive and cooperative elements, a condition conducive to complex social manoeuvers (Wrangham 1982).

With groups viewed as founded on female alliances, males enter the scene as secondary players, adapting to and not substantially changing the pattern of female relationships. Nevertheless, male strategies of course bring much complexity to social structure, principally through competition for mating rights. Males compete with and occasionally support other males (Hausfater 1975), develop long-term and short-term affiliative bonds with females (Smuts 1982) and threaten infanticide, which thereby favours male protection of infants they may have fathered (Busse & Hamilton 1981). How far male relationships can be explained in terms of sexual competition remains an open issue but it is surely the predominant factor. An important component of male strategies is that males leave their natal groups to breed, both because of the costs of inbreeding in their natal groups and because of the increased mating opportunities elsewhere (sections 9.4 and 11.3).

An understanding of the forces influencing female relationships is thus the key to explaining the ultimate sources of monkey social structure. If female bonds indeed evolved in the context of feeding competition, intergroup relationships deserve more attention than they have so far been given, for relationships within groups cannot be explained without them. In the distant past, a time can be envisaged when females whose mothers lived alone remained together to support each other when competing for food. The strategy paid and groups grew. With this scenario, the basis of monkey groups resembles a hypothesis applied to our own species, i.e. that the ultimate source of all social organization is the success of two against one (Caplow 1968). The problem of understanding the links between ecology and social structure in monkeys takes on new interest with so direct a link to human issues. The development of alternative hypotheses to the point where they can be confidently eliminated or appropriately combined is therefore an important next step.

12.3 Special Relationships between Adult Male and Female Olive Baboons: Selective Advantages

BARBARA B. SMUTS

Data on the frequency of grooming and spatial proximity revealed the existence of strong affiliative bonds or 'special relationships' between anoestrus adult female and adult male baboons (section 6.9; Smuts 1982). The females restricted nearly all of their non-sexual, friendly interactions to these 'special' males and they fed, rested and travelled near them. Different females had special relationships with different males and all but one of the 34 females had a special relationship with at least one of the 18 adult males. In this section, the selective advantages of these relationships to the participants are considered.

Benefits to the female of special relationships with males

In 49 of the 54 incidents (91%) in which an adult male was seen to intervene on behalf of a female or juvenile who was being threatened or attacked by another troop member, the male aided either a female with whom he had a special relationship or one of her offspring. Since only 12% of all possible adult male–female dyads were special dyads, it is clear that special males were disproportionately responsible for defence of females and their offspring. Possible benefits to females and juveniles of male defence include reduced interference with maintenance activities by higher-ranking females and other adult males (Altmann 1980) and decreased risk of injury by other adult males (Smuts 1982).

Special males were responsible for 97% of all friendly male–infant interactions that involved physical contact (holding, carrying and grooming; $n = 38$). Since special males spent a lot of time near the infant's mother, this result is not unexpected. However, when mothers were not near their infants, the infants were more, not less, likely to be found near affiliated males, and male–infant relationships persisted into the infants' second years of life after they had achieved independence from their mothers (Nicholson 1982; Smuts 1982). Possible benefits to the infant of a close relationship with a male include protection from predation, protection from injury by other troop members, especially other adult males, increased access to resources and a substitute care-giving relationship in case of the mother's death (Altmann 1980; Busse 1981; Busse & Hamilton 1981; DeVore 1963a; Nicolson 1982; Packer 1980; Ransom 1982; Ransom & Ransom 1971; Smuts 1982; Stein 1981).

These results suggest that when a female establishes a special relationship with a male, she and her immature offspring acquire an ally in the troop—an ally who, because of his larger size and superior fighting ability, may make a significant contribution to the fitness of the female and her offspring.

Benefits to the male of special relationships with females

If the offspring of a male's special females are his offspring as well, then special relationships with females may simply reflect a male's attempts to invest in offspring already sired. Results useful in evaluating the importance of this benefit are summarized in Table 12.1, which gives information on the special relationships, consort activity and dominance ranks of the 18 adult/subadult males in the Eburru Cliffs troop, grouped according to their length of residence in the troop. Groupings were based on residence rather than on dominance rank because in both the Eburru Cliffs troop and the adjacent Pumphouse Gang troop, residence is a better predictor of male consort activity than is dominance rank (Smuts 1982; Strum in press).

The findings in Table 12.1 and the quantitative data on which they are based (Smuts 1982) suggest several conclusions. First, although previous consort activity with the female is clearly one factor influencing the formation of special relationships, it is not the only important factor. Second, previous consort activity is rarely followed by the development of an affiliative bond between the male and infant unless the male also develops a special relationship with the mother. Third, special males who are *not* likely fathers (or siblings) are as likely to develop a bond with the infant as are special males who are likely fathers (or siblings). Thus, paternity is not a good predictor of male–infant interactions when the effects of the males' special relationships with the mothers are removed. This was precisely the result obtained in a study of male–infant relations in a group of captive rhesus monkeys in which paternity was determined by analysis of serum proteins and red-cell enzymes (Bernstein *et al.* 1981).

These conclusions need to be tested in studies of wild baboons in which paternity is determined by genetic analysis rather than estimated on the basis of consort activity. Assuming, for the moment, that the paternity conclusions are correct, they imply that investment in offspring is not the only benefit males derive from special relationships with females.

A second possible benefit is increased mating opportunities. If females and

Table 12.1 Special relationships, consort activity, interactions with infants and dominance ranks of Eburru Cliffs males.

Type of male ‖	No.	Dominance rank*	Mean no. of special relationships with females	Consort success†	Special relationships with females who were		Affiliative relations with infants	Characteristics of special females
					previous consort partners‡	not previous consort partners§		
Long-term residents (two or more years in troop); all were middle-aged or older	5	Low to moderate	5.2	High to moderate	Yes, except that the two most successful consort partners had special relationships with only some previous partners	No	Yes, but only if they also had a special relationship with the mother	Middle-aged and old; all dominance ranks
Short-term residents (six months to two years in troop) or recently matured natal males; all were young adults	6	Moderate to high	4.0	Moderate to low	Yes, always (nine cases, five males)	Yes, often (nine cases, five males)	Yes, but only if they also had a special relationship with the mother	Middle-aged and young; all dominance ranks

Newcomers (in troop less than three months at start of study); two were young adults, one was very old	3	High for the two young males; low for the old male	0	Low	No	No	No	No	None
Subadult, natal males (about 80% of adult weight)	4	Low	3.2	Low	No	Yes, often	No	Yes; some involved putative siblings, others unrelated infants of special females	Young females, usually lower ranking

*Dominance ranks were difficult to assign in this troop because of the rarity of dyadic agonistic encounters between males that had unambiguous outcomes. Only 51 such encounters were observed (excluding competition over meat and oestrus females) and ranks were based on the order of males that at minimized the number of outcomes to the left of the diagonal when the males were arranged in a standard dominance matrix.

†Consort success was based on the frequency with which the male formed consorts with females during the most fertile phase (one to seven days before detumescence of the sexual skin) of the oestrus cycles in which the females conceived ($n = 28$ conceptions).

‡Previous consort partners were defined as females with whom the male was observed in consort at least once during the fertile phase of her conception cycle.

§Females who were not previous consort partners were defined as females the male was not observed in consort with during the fertile phase of her conception cycle.

||Male age estimates were based on the general appearance and on the examination of tooth-wear patterns after males were trapped and sedated.

their immature offspring benefit from male investment and if a female tends to mate around the time of ovulation with males who have in the past demonstrated their willingness and ability to invest in her and her offspring, then a male who forms special relationships with females may achieve increased reproductive success, even when the females' current infants are not his own. Are special males more likely to mate with a female when she resumes cycling? To answer this question, each male's observed frequency of consort activity with 14 females who resumed oestrus cycles near the end of the study was compared with his expected frequency based on his consort activity with all oestrus females. The observed frequency was greater than expected in 13 of 21 special dyads but in only 30 of 105 non-special dyads (Chi^2 test for two independent samples, $n = 126$, $P < 0.01$). Unfortunately, few of these females conceived before the study ended, so paternity estimates for their next infants are not available.

Other possible benefits to males of special relationships with females include integration into a new troop (Packer 1979a; Smuts 1982), increased opportunities to use females and infants as buffers in agonistic encounters with other males (Packer 1980; Smuts 1982; Stein 1981; Strum in press a,b) and the acquisition of female allies (Chapais 1981, this volume). The first two may represent important benefits. The third, while important in macaques, seems unlikely to be important in baboons due to the much greater sexual dimorphism in size and the rarity with which females function as aggressive allies of males (Packer & Pusey 1979, pers. obs.).

Whichever benefits prove to be most important, there is increasing evidence that a male baboon's behaviour and, ultimately, his reproductive success are greatly influenced by his social (not just sexual) relationships with females. The same appears to be true in macaques (Chapais 1981, section 12.7), chimpanzees (Tutin 1979, 1980) and probably many other non-human primate species that live in multimale groups. As more attention is paid to the development, dynamics and consequences of these relationships, puzzles that have troubled primatologists for years, such as lack of a consistent relationship between male dominance rank and mating success, intra and interspecific variation in the intensity of male—infant relationships and in the number of males in a group, may be solved. Finally, the existence in our closest relatives of long-term affiliative relationships between males and females may provoke re-evaluation of some of the common assumptions underlying discussions of the evolution of the human pair bond (Smuts 1983).

12.4 Male Dominance and Reproductive Activity in Rhesus Monkeys

BERNARD CHAPAIS

Asymmetric contests reflecting the operation of sexual selection, such as the competitive interactions of mature males for receptive females in multimale primate groups, are expected to be settled on the basis of intermale differences in resource holding power (RHP) and/or asymmetries in pay-off from winning (Maynard Smith & Parker 1976). Considering that in the present case pay-off imbalances (other than those provided by female choice) are unlikely (Chapais 1983) and that dominance is probably the single most important component of RHP (factors such as age, size, physical condition and fighting ability being related to it) the present study assesses the explanatory power of male dominance for mate selection among rhesus monkeys when reproductive activity is analysed in relation to the likelihood of ovulation. The studies concerned with the relation between male dominance and reproductive activity in multimale primate groups vary in the accuracy of the measurement of reproductive success (i.e. the genetic contribution of males to future generations). Five categories can be recognized (Chapais 1981): (1) qualitative studies (e.g. Goodall 1968; Lindburg 1971); (2) raw quantitative studies (e.g. Kaufmann 1965; Stephenson 1975); (3) quantitative studies controlling for male differential observability (e.g. Drickamer 1974b; Taub 1980a); (4) quantitative studies further incorporating the likelihood of ovulation (e.g. Hausfater 1975; Packer 1979a); and (5) paternity exclusion analysis (e.g. Smith 1981; Witt *et al.* 1981). The evidence provided by the majority of the 29 studies reviewed points to a positive relation between male rank and reproductive activity, though this relation is often not linear, and the explanatory power of dominance is species-specific.

The present study belongs to the fourth category and is based on data on one group of rhesus monkeys (group F) observed on Cayo Santiago during the 1978 breeding season. Full details on subjects and methods are given elsewhere (Chapais 1983). Well-defined, linear dominance relations could be extracted from the distribution of submissive gestures among mature males. Group F comprised 30 sexually mature females (11 females aged three years and 19 older ones). A given proportion of these females were simultaneously receptive and interacting with males at any time. The observation procedure aimed at recording as precisely as possible for each oestrus female the sequence and duration of her associations with mature males. The most likely period of ovulation, termed the OV-1 period, was defined as covering the four days preceding the *attractiveness breakdown*. The latter is thought to coincide with the transition from the follicular to the luteal phase of the menstrual cycle and to be homologous with the onset of deturgescence of the sex skin in baboons; its diagnostic criterion was the cessation of consortships. The four days preceding the OV-1 period and the three days following it were included into the OV-2 period, presumed to correspond to lower probabilities of ovulation (e.g. Catchpole & Van Wagenen 1978). All other days of sexual activity are referred to as OV-3 days.

Table 12.2 permits comparison between the consorting activity of males during the fertile OV-1 period of the 18 females who became pregnant. The fertile OV-1 period was identified by backdating 168 days (the mean duration of pregnancy in rhesus monkeys) (Stolte 1978) from the known birth dates. The proportion of hours a male spent consorting a female, out of the total number of hours during which the observers were in contact with group F during the OV-1

Table 12.2. Number of hours spent by males consorting the 18 (to be pregnant) females during their fertile OV-1 period. The duration of consortships was estimated on the basis of data from all sampling sources (focal sampling on females, concurrent sampling during focal samples and *ad libitum* sampling). Total = the total number of observation hours during the female's OV-1 period. Not visible = the number of observation hours during which the interactions of males with the female could not be adequately observed. Males and females are arranged in descending rank order from top to bottom and from left to right.

Females	Total	Not visible	415	580	EE	TJ	IJ	Z2	339	433	481	285	491	557	282
FB	29	0			29										
510	30	8						6	10	6					
577	40	0						20	20						
511	28	7						4	17						
287	35	0		10			15	10							
503	23.5	0		18.5				3	2						
7L	28.5	11		17.5											
471	29	4		25											
692	27	14		6				7					2		
679	35	0		26		7									
N2	30	0		18	12										
411	25	0	3.5	14.5	1										
649	14	6		1.5		3		3.5							
K1	15	0		1		14									
697	20	0		14		6									
476	28	0		16		2						10			
543	38	0		33										5	
310	15	1							13	1					

period of this female (see Table 12.2), were assumed to correspond to the male's probability of fertilizing the female. Since males underwent changes in rank during the breeding season, rank-specific versus male-specific reproductive success was calculated. Rank-specific reproductive success was obtained by summing for all the males who had occupied the same rank the probabilities that they had fertilized females when occupying this rank (Fig. 12.1). A positive correlation was found between male rank and rank-specific reproductive success estimated by the time spent consorting ($r_s = 0.90$, $P < 0.01$).

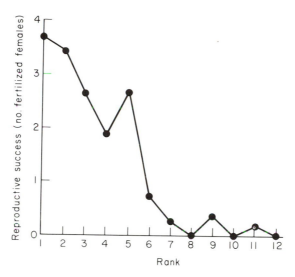

Fig. 12.1. Rank-specific reproductive success estimated on the basis of the time spent by males consorting the 18 (to be pregnant) females during their fertile OV-1 period.

The analysis of the 481 recorded copulations is presented in full detail elsewhere (Chapais 1983). Copulations include all instances of series mounts ended with ejaculation and all sightings of coagulated semen on a female's labiae which could be attributed with certainty to specific males. A positive correlation was found between the dominance rank of males and their copulatory frequencies irrespective of oestrus stage, cycle type, male differential observability and sampling source ($r_s = 0.60$, $P < 0.01$). However, when these raw estimates of mating success were corrected for male differential observability, the above correlation vanished, a result similar to observations made by Drickamer (1974b). In order to verify if males nevertheless mated with different frequencies depending on the stage of oestrus, the copulations of each male were partitioned according to their timing in relation to the attractiveness breakdown. A positive correlation was found between male rank and the proportions of copulations performed during OV-1 periods out of all copulations (OV-1+OV-2+OV-3 days) per rank ($r_s = 0.75$, $P < 0.01$). Thus, even if some low-ranking males copulated as frequently as middle-ranking males, they did so most often outside the presumed periods of ovulation. Similar observations were reported for baboons (Hall & DeVore 1965; Hausfater 1975; Packer 1979b; Saayman 1971). Finally, a positive correlation was found between the dominance rank of males and the frequencies with which they copulated during the fertile OV-1 periods of the 18 females who became pregnant ($r_s = 0.76$, $P < 0.01$). To summarize, male rank was found to correlate positively with raw mating success but not with mating success corrected for male differential observability. However, male rank correlated positively with mating success on fertile OV-1 days as a result of high-ranking males mating more often near ovulation time.

In order to check if male rank explained all the variance in male reproductive activity, a version of Altmann's (1962) 'priority of access model' was tested. Translated into time (Hausfater 1975), this model states that if dominance consistently confers priority of access to females, the proportion of total oestrus time when at least r females had been simultaneously receptive would equal the proportion of oestrus time when a male of rank r had access to receptive females.

In the present study, Altmann's model was tested on the basis of time spent by males consorting on the OV-1 days of all cycles of the first half of the breeding season. The number of hours during which at least r females were simultaneously going through their OV-1 period and were consorted was compared to the number of hours males who occupied rank r had consorted these females. Figure 12.2 compares graphically the expected and observed rank-specific amounts of time spent by males in consortship with females on OV-1 days. Although there is a positive correlation between male rank and time spent consorting on OV-1 days ($r_s = 0.79$, $P < 0.01$), the alpha-males consorted OV-1 females less often than expected, whereas males ranking lower than third consorted more often than expected (chi^2 on the first four ranks $= 156.8$, $P < 0.001$).

The behaviour of males departed from the two main predictions of the model. First, males did not always consort the female presumed to be the most likely to be ovulating. Second, male rank did not consistently confer priority of access to receptive females. In some cases, males abstained from consorting seemingly available females. This could not be entirely explained in terms of male fatigue, sperm depletion, female rejection, incest avoidance, short periods spent out of the group or rank reversals resulting from asymmetries in pay-off from consorting. In some other cases, females actively preferred lower-ranking male suitors. Part of the explanation may lie in the existence of a conflict of interest between the sexes over mate choice. Females were hypothesized to have an advantage in mating with higher-ranking males assumed to carry genes conferring the ability to achieve high ranks. However, females might do better if they

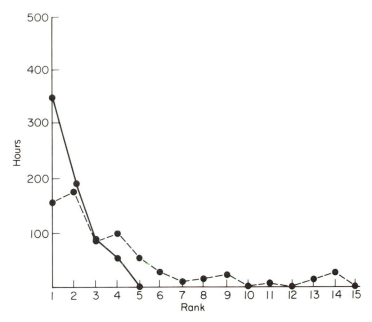

Fig. 12.2. Expected (solid line) and observed (broken line) rank-specific numbers of hours during which males consorted females on the OV-1 days of all cycles of the first half of the breeding season.

were able to evaluate the males' lifetime reproductive success and mate preferentially with males likely to achieve the highest (e.g. young males occupying *relatively* high ranks).

To conclude, though there was a positive correlation between male rank and two estimates of reproductive success among group F rhesus monkeys, male dominance could not explain all the variance in reproductive activity. Male–male competition for oestrus females appeared less stringent than expected on theoretical grounds. Furthermore, the possibility of a conflict of interest between males and females over mate choice calls for more refined hypotheses incorporating criteria of female choice.

12.5 Fitness and Female Dominance Relationships

BERNARD CHAPAIS and STEVEN R. SCHULMAN

In this section, the effects of dominance rank and age on fitness in female primates is briefly reviewed. Several functional relationships linking social dominance and inclusive fitness are described. Some proximate mechanisms involved in dominance relations among female macaques are then examined and an attempt made to interpret their consequences from a functional perspective.

Effects of rank on fitness

Rank has been shown to affect the fitness of females in primates through effects on: (1) survivorship; (2) reproduction; (3) offspring survival. Concerning survivorship, Wrangham (1981) reported that during a period of water scarcity in Amboseli, Kenya, mortality among low-ranking female vervet monkeys was significantly higher than among high-ranking females (but see Cheney *et al.* 1981). In reviewing the literature on social behaviour and demography in Cercopithecines, Dittus (1980) concluded that the main factor accounting for greater mortality among low-ranking individuals was their reduced competitive ability.

Rank-related effects on fecundity were inferred by Dunbar (1980b), who found a positive correlation between female rank and the number of living descendants in gelada baboons. Low-ranking females were thought to have experienced more infertile cycles than high-ranking females as a consequence of harrassment by high-ranking females. Indirect evidence supporting such an interpretation is provided by Sackett *et al.*'s (1975) observations of behaviourally induced abortions, stillbirths and neonatal deaths in pigtail macaques. High rank has also been shown to enhance reproductive success by various other means, e.g. high-ranking female rhesus monkeys matured earlier and exhibited shorter interbirth intervals (Drickamer 1974a).

Finally female rank has been found to affect offspring survival in captive groups of rhesus monkeys (Drickamer 1974a) and bonnet macaques (Silk *et al.* 1981a). Daughters of high-ranking females had a higher probability of surviving to the age of one year in rhesus and six months in bonnet macaques.

The above studies suggest that high rank can influence a female's fitness in many ways (see also section 10.5). Sade *et al.* (1976) reported that high-ranking matrilines were significantly more productive overall than lower-ranking matrilines in the Cayo Santiago rhesus monkeys.

Effects of age on fitness

At least two explanations could account for the observation that mortality is higher in younger age-classes. First, it could be that there is an association between age and rank such that younger individuals tend to occupy lower ranks and thus have limited resource access. Dittus (1980) suggested that the high mortality rates suffered by juveniles in toque macaques and other Cerco-pithecines is a consequence of resource competition. Second, it could be that younger individuals are more vulnerable than adults independently of their rank. Low rank could affect a young individual disproportionately more than an adult if, for example, younger individuals have different energetic requirements, suffer more from stressful situations or are generally less experienced and able to compensate for lower rank through alternative behavioural responses. This possibility follows from the more general principle that the costs and benefits of competitive interactions are partly determined by differences between the values that a given resource has for different individuals (Clutton-Brock & Harvey 1976; Parker 1974).

Age–specific fecundity varies considerably among age-classes in the population of rhesus on Cayo Santiago (Sade *et al.* 1976). Primiparous mothers are less successful than multiparous mothers in rhesus and bonnet macaques (Drickamer 1974a; Silk *et al.* 1981a). Such differences may result from differences in maternal experience. There are also likely to be age-dependent differences in the energetic cost of pregnancy and susceptibility to socially induced disruption of menstrual cycles. Offspring survivorship was found to vary with maternal age in bonnet macaques (Silk *et al.* 1981a). For offspring of high-ranking females, it increased with age, whereas for low-ranking females, offspring survivorship was highest for mothers between six and ten years old. In captive langurs, infant mortality rates were higher for young primiparous females (Dolhinow *et al.* 1979).

Factors involved in dominance interactions

Outcomes of dyadic dominance interactions occurring outside any influence from third parties are affected by the interplay of at least four factors. The first factor is the relative individual competitive ability of the contestants. Other things being equal, the stronger female should win. Second, there may be age-dependent asymmetries in benefits derived from dominance. Other things being equal, the female with the larger expected pay-off should be victorious (Maynard Smith & Parker 1976). Third, for related females, competitive interactions affect the inclusive fitness of each opponent. Factors affecting interactions among related individuals that need to be considered are: (1) the degree of relatedness between contestants; (2) the exact nature of the effect of rank on fitness; and (3) possible age-dependent asymmetries in benefits derived from dominance. Furthermore, related females competing for dominance may incur a dual fitness cost: a direct cost for the interaction itself and an indirect cost attributed to the subordinance of the genetically related loser of the interaction. Both costs are more pronounced the greater the degree of relatedness. Fourth, there may be asymmetries in the value of dominance as a consequence of rank inheritance, e.g. in rhesus and Japanese macaques, females rank below their mother and above females ranking lower than their mother (see below). If rank is transmitted socially, a reproductive female would derive larger benefits from dominance than a post-reproductive female, not only because rank affects survivorship and fecundity but also because dominance may enable a female to transmit a higher rank to her descendants. Hence, a small difference in rank may be amplified through generations. Furthermore, if being outranked implies that a female's living descendants are also outranked, the female with the larger number of living descendants (the older female) stands more to gain from dominance. On the other hand differences in genealogy size produce a counter-asymmetry in the value of dominance. The matriarch with a smaller genealogy (the younger female) has a relatively greater interest in her genealogy outranking a larger one. Assume, for example, that dominance is correlated with priority of access to various resources and that a_i and b_i are descendants of A and B respectively. Now consider the alternative rank orders $A:a_1:a_2:a_3:B:b_1$ versus $B:b_1:A:a_1:a_2:a_3$. Matriarch A ranks first or third while B ranks first or fifth. Clearly, the smaller genealogy incurs a disproportionately larger cost by subordinance when priority of access to depletable resources is competitively determined.

Interest of individuals in rank relations among relatives

It can be expected that individuals will attempt to manipulate where possible rank relations involving kin. Irrespective of the precise nature of the effect of rank on fitness, and other things being equal (e.g. age), an individual benefits by favouring the dominance of a close relative over a more distant relative. If potential recipients of aid are equally related to female A (e.g. the female's two daughters) but differ in age, female A's favouritism will have different effects on her inclusive fitness, depending on whether dominance affects survivorship, fecundity, offspring survival or a combination of these factors (Charlesworth & Charnov 1981). Thus if the only effect of high rank is to increase survivorship, female A should favour the relative with the highest reproductive value (defined as the age-specific expectation of future reproduction) unless the recipient with the lower reproductive value can benefit disproportionately from aid (e.g. due to her age [see above]). If, instead, dominance primarily affects fecundity (e.g. via reducing the frequency of infertile cycles), female A's interest depends on how age-specific fecundity changes with age as well as the shape of the function relating fitness to aid received. Other things being equal (i.e. relatedness, fitness accrual rates, etc.), female A has an interest in a relative in an age-class of higher age-specific fecundity ranking above a relative of lower fecundity, e.g. if fecundity increases with age, female A would do better by favouring an older relative over a younger relative, unless the younger one stands to benefit disproportionately from dominance. Finally, if dominance affects fitness primarily through enhancing infant survivorship (e.g. through the better nutritional status of higher-ranking mothers), female A's interest will depend on whether infant survivorship changes with maternal age. If it does, she has an interest in the dominance of the relative possessing the highest probability of caring for an offspring over a prolonged period of time and which would thereby benefit most from dominance: for instance it might be advantageous to favour a multiparous and more experienced female relative over a primiparous relative unless the latter can benefit disproportionately from being dominant.

The above evidence on the effects of rank and age on fitness, and the functional relationships deduced, suggest that the structure of a competitive hierarchy depends on multiple factors. Hence, interspecific differences in life-history parameters and the nature of the influence of rank on fitness might lie at the origin of interspecific differences in ranking systems. Using this perspective, rank relations among female rhesus monkeys are next examined.

Dominance relations among female rhesus monkeys

Rank relations among female rhesus follow a highly predictable pattern. Females acquire a rank immediately below that of their mother. Among sisters, younger sisters rise in rank above the older sisters as they mature (Datta 1981; Missakian 1972; Sade 1972b). Age and kin-dependent asymmetries in the way females intervene in each other's fights seem to reinforce this structural pattern (see also section 6.8).

Young females engaged in fights with immature or adult females are frequently aided by their mothers and other relatives; consequently they can win fights against females ranking below their mother (Berman 1980b; Datta 1981; sections 6.8, 6.9 and 12.8). Presumably, through such a long-term conditioning process, females come to rank above all females ranking below their mother (younger sisters excepted). If the mother is dead, support may still be provided by other relatives, a common observation on Cayo Santiago: thus two orphaned, three-year-old females from Group F on Cayo Santiago were often supported by a higher-ranking sister or a higher-ranking aunt against lower-ranking females. Other factors, such as observational learning (Berman 1980b), are likely to be involved in rank inheritance as well. However, fight intervention may be primary, in the sense that if it were non-existent, other processes might be less efficient (see sections 6.8 and 6.9).

Females will generally not aid a daughter against a mother. In contrast, mothers have been observed to be aided in fights with daughters by higher-ranking and intermediate-ranking (a younger daughter) females. In fights between sisters, younger sisters are usually, but not always, favoured by their mother or by other relatives, e.g. other sisters, an aunt (pers. obs.; Datta 1981; Kurland 1977).

The observation that a female inherits her mother's rank, and the underlying patterning of fight interventions, can be interpreted in terms of mothers increasing their inclusive fitness by facilitating their daughters' acquisition of the highest possible rank within the female dominance hierarchy. Without the assistance provided by a mother, orphaned females can nevertheless attain the same rank with the aid of other relatives.

In dyadic encounters, young immature females rank below older sisters. However, since they are supported by their mother and other relatives, younger sisters may secure priority of access to resources, provided they remain in proximity to their supporters. In dyadic encounters between adult sisters, the younger one is dominant. Observations of interventions by mothers and other relatives on behalf of younger adult sisters suggests that this reliable network of supporters assures ascendance of younger sisters and the resulting permanence of their adult dominance status among their sisters.

In previous papers (Chapais & Schulman 1980; Schulman & Chapais 1980) the authors presented a simple mathematical model predicting the direction of support and favouritism that a female might be expected to demonstrate in the dominance relations of female kin. Different types of dominance systems were predicted to follow from different life-table-parameter values. The model was based on several of the principles stated above. When applying the model to demographic data from Cayo Santiago, it was found that the observed female dominance hierarchy followed from the dynamics predicted by the model. Other feasible dominance systems resulted in suboptimal inclusive fitnesses. The model demonstrated that a mother, as well as all collateral and ancestral relatives, have an interest in a daughter with high reproductive value ranking above a daughter with low reproductive value. Among adults, the younger of two females always has the higher reproductive value. This might explain why among adult sisters the younger of two sisters is generally favoured by supporters.

If benefits derived from dominance are independent of age, females should always support the sister with the higher reproductive value. This might mean supporting a young adult female against her immature sister. However, if immature individuals can benefit disproportionately from aid, the immature female might become the favoured sister. Suppose, for example, that a population of rhesus monkeys goes through a period of food scarcity during which time mortality is rank related, and that immatures are more vulnerable to this stress than adults. Consider the following rank orders: A,d,D versus A,D,d where d and D are immature and adult daughters of A, respectively. If dominance is associated with priority of access to food and the third-ranking female's survivorship is strongly age-dependent during food scarcity, then the immature's greater ability to gain from dominance may offset her lower reproductive value. Hence, A will support d against D. To summarize, consistently supporting a younger female against her older sister may be adaptive because initial age-related differences in benefits derived from dominance offset differences in reproductive value. With the maturation of the younger daughter, this initial asymmetry is expected to disappear and differences in reproductive value may become the primary factor.

It was also demonstrated (Chapais & Schulman 1980) that an old, but not post-reproductive, female has an interest in remaining dominant to her daughters in order to remain in a position to manipulate the rank relations of daughters yet to be born. Most relatives will share this interest. Consequently, this might explain why an old mother remains dominant to her daughters who hold high reproductive values.

Investigation of correlations of dynamic life history and demographic variables with social behaviour is currently the subject of intensive research by a large number of theoreticians and field-workers (e.g. Charlesworth & Charnov 1981; Cheney *et al.* 1981; Gadgil 1982; May 1982; Rubenstein 1982). A promising collaboration between empirical and theoretical researchers is being fueled in part by the growing understanding of the ecology and behaviour of the primates discussed in this volume. Field-workers studying macaques, baboons and vervets have developed greater appreciation for the importance and feasibility of collecting data on the demographic and genetic structure of their study populations. Melnick's (1981) recent study of a Himalayan rhesus population provides an example of the value of coupling population genetic studies with behavioural studies. Conjointly, theoretical population biologists are showing renewed interest in the evolution of social behaviour.

It has been suggested in this section that some of the more salient attributes of rhesus social structure may be an epiphenomenal consequence of a particular aid-giving strategy within a network of genetically related individuals. The hypothesized strategy is not fixed, rather it is a function of important demographic parameters, notably reproductive value. There has been an attempt to emphasize the importance of carefully distinguishing between the effects of aid on survivorship and on fecundity, an important distinction recently the subject of formal treatment by Charlesworth and Charnov (1981).

As discussed in this section, the differential ability of individuals to benefit from aid is a very important consideration. The models of Chapais and Schulman (1980) and Weigel (1981) are based on the simplifying assumption that individuals have similar abilities to benefit from aid. In reality, the extent to which individuals benefit from aid, which is in essence a fitness accrual rate, probably shows great phenotypic variability among individuals. Schulman and Rubenstein (1983) have investigated the consequences of allowing accrual rates to vary and have provided an analytical solution to the problem of the distribution of aid among kin of various relatedness and possessing varying fitness accrual rates. A variety of factors contribute to differences in accrual rates. Age-dependent vulnerability is one factor suggested in this chapter. Disease and injury will also have an influence. There are also likely to be genotypic differences in accrual rates. Schulman and Rubenstein (1983) found that small changes in accrual rates may result in major alterations of the optimal allocation of aid. Thus, field sociobiologists will need to develop clever techniques for assessing this component of variability.

12.6 Extrafamilial Alliances among Vervet Monkeys
DOROTHY L. CHENEY

Macaques, baboons and vervet monkeys often form alliances during aggressive interactions. Most such alliances occur among related individuals but they also occur among unrelated individuals who reciprocate one another. As a result, both kin selection and reciprocal altruism have been invoked to explain the evolution of alliances in Old World monkeys (e.g. Kurland 1977; Massey 1977; Packer 1977). Not all alliances, however, can be explained in terms of these two functional hypotheses: for example in a study of free-ranging chacma baboons, both adult females and juveniles were found to form a high proportion of alliances with the members of high-ranking families, even though such individuals seldom reciprocated the alliances of low-ranking animals (Cheney 1977). Moreover, two allied, low-ranking animals were seldom able to defeat a higher-ranking female or juvenile, apparently because high-ranking animals were able to recruit more alliance partners than low-ranking animals. In attempting to explain these observations, the author suggested that the distribution of alliances may have been determined by the assymetrical benefits to be derived from bonds with animals of high and low rank. Since bonds with high-ranking animals appeared to be potentially more beneficial than bonds with low-ranking animals, individuals were valuable as alliance partners in direct relation to their rank. As a result, the rate at which individuals were chosen as alliance partners was positively correlated with rank (Cheney 1977; see also below).

In this section, an attempt is made to determine whether similar patterns of alliances are evident among free-ranging female and juvenile vervet monkeys. Like baboons and macaques, vervet monkeys form a large proportion of alliances with close relatives. Moreover, the majority of alliances among adult females occur among individuals of adjacent rank (unpublished data). Since similarly ranked females are often closely related, hypotheses based on kin selection appear to explain the distribution of most vervet alliances. Nevertheless, vervets of all ages and either sex also form alliances with unrelated animals and, in general, high-ranking females and juveniles tend to be chosen as alliance partners most often. Kin selection, therefore, cannot adequately account for all aspects of alliance formation in vervet monkeys. It is with the apparent exceptions to the kin selection hypothesis—alliances formed by adult females and juveniles with unrelated animals—that this section is chiefly concerned.

Background

The vervet monkeys described here live in groups with adjacent home ranges in Amboseli National Park, Kenya (see Chapter 2). Both adult males and adult females can be ranked in linear dominance hierarchies, based on the direction of approach–retreat interactions. High-ranking animals have priority of access to food and water and are less likely than low-ranking animals to die during periods of food and water scarcity (Cheney *et al.* 1981). While the ranks of adult males may change over a period of several months, adult female ranks remain relatively stable over time. Over a three-year period in Amboseli, each adult male changed ranks at an average annual rate of 0.75, while each adult female switched ranks at an average annual rate of only 0.11 (see Cheney *et al.* 1981). Offspring assume ranks similar to their mothers' during aggressive interactions and daughters acquire their mothers' ranks even from infancy (unpublished data; section 4.4). The ranks of adult males become less dependent upon maternal rank as they approach sexual maturity. For the purposes of this section, however, all juveniles have been assigned their mothers' ranks.

Unless otherwise stated, the alliances described involve 'unrelated' partners, where related animals are defined as adult females and their immature offspring. Animals against whom an alliance is formed have been termed the 'targets' of the alliance.

Rank-related attributes of alliance formation

When two vervets were involved in an aggressive dispute, a third animal could 'choose' to form an alliance with one of them by intervening in the dispute and threatening that individual's antagonist. When intervening in disputes, adult females and juveniles formed more alliances with high-ranking than low-ranking animals. The tendency to choose alliance partners of high rank was most apparent when the target of the alliance was another female or juvenile. When aggressive interactions involved only adult females or juveniles, there were consistent positive correlations in the three study groups between rank and the rate at which individuals were chosen as alliance partners by unrelated animals (Table 12.3). These positive correlations were less evident, however, when disputes involving adult males were considered. When the targets of their alliances were adult males, females and juveniles did not consistently choose

Table 12.3. Spearman correlations (r_s) between dominance rank and the rate at which individuals were chosen as alliance partners by unrelated animals. There were no juveniles in Group B during 1977–78. Rates of alliances received were calculated as follows:

No. alliances received from unrelated females and juveniles

No. aggressive interactions in which females and juveniles (or adult males) involved

Actor	Recipient of alliance	Study group	No.	Adult female or juvenile	Adult male	Is correlation less positive when target is adult male?
					Target of alliance	
Adult females	Adult females	A, 1977–78	8	0.96†	0.81*	+
		B, 1977–78	7	0.84*	0.02	+
		C, 1977–78	8	0.52	0.19	+
		A, 1980	2	1.00	1.00	0
		B, 1980	7	0.67	0.29	+
		C, 1980	4	0.11	0	+
Adult females	Juveniles	A, 1977–78	7	0.81*	0.47	+
		C, 1977–78	11	0.55	0.66*	−
		A, 1980	8	0.47	0.42	+
		B, 1980	5	0.40	−0.87	+
		C, 1980	11	0.70*	0	+
Juveniles	Adult females	A, 1977–78	8	0.85*	0.04	+
		C, 1977–78	8	0.79*	−0.05	+
		A, 1980	2	1.00	1.00	0
		B, 1980	7	0.81*	0.37	+
		C, 1980	4	1.00	−0.26	+
Juveniles	Juveniles	A, 1977–78	7	0.91*	0.81	+
		C, 1977–78	11	0.64*	0.03	+
		A, 1980	8	0.23	0.58	−
		B, 1980	5	−0.22	0.35	−
		C, 1980	11	0.81*	−0.63*	+

$P = 19, P < 0.05$
(two-tailed binomial test)

*$P < 0.05$.
†$P < 0.01$.

partners of high rank (Table 12.3).

Why did adult females and juveniles choose high-ranking animals as partners when forming alliances against other females and juveniles but not when they formed alliances against adult males? As with baboons, alliances among female and juvenile vervets rarely involved partners who both ranked lower than the target of the alliance. In 89% of the 666 cases when vervets formed alliances against other females and juveniles, they chose partners who already ranked higher than the target of the alliance. As a result, alliances seldom affected the outcome of a dispute. Instead, it seems likely that vervets, like baboons, may have chosen alliance partners on the basis of the potential benefits to be derived from the partnership.

There are a number of possible ways by which low-ranking animals might benefit from alliances with high-ranking individuals: for example they may increase the probability of future reciprocal aid. Such aid need not occur at equal rates. Since the efficacy of high- and low-ranking animals in disputes is asymmetrical, the rate at which high- and low-ranking animals form alliances with each other may also be expected to be asymmetrical. Moreover, reciprocity need not always occur in the same behavioural currency. If alliances increase the probability of feeding near or establishing a bond with a high-ranking animal, it may be advantageous to choose alliance partners of high rank. By forming alliances with high-ranking animals, a low-ranking individual may also reduce the probability of future harrassment by those animals (Silk 1981) or ensure that they do not support an even lower-ranking animal against itself (Chapais & Schulman 1980).

Effects of a rank-based system of alliances

1. Alliances formed against adult females and juveniles

A number of implications of a rank-based system of alliance formation are worth emphasizing. First, alliances with high-ranking animals usually have the effect of strengthening and perpetuating the existing dominance hierarchy. The hypothesis that low-ranking animals form alliances with high-ranking animals for their own potential benefit therefore appears initially to be counter-intuitive. Why do not low-ranking females and juveniles simply collaborate with each other against high-ranking animals and rise in rank through fighting, as adult males occasionally do? While rank reversals among females are rare, they do occasionally occur (see above) and alliances among low-ranking animals appear to be one mechanism for such reversals in rank (Gouzoules 1980; Koyama 1970). Why, however, are they not more common?

In considering this question, it may be useful to regard alliances among females and juveniles in terms of frequency-dependent behaviour or evolutionarily stable strategies (ESS), in which each individual's behaviour depends on the behaviour of others (Maynard Smith & Parker 1976). This, in turn, depends in large part on each individual's resource holding power (in this case, familial rank), which is typically asymmetrical (see Chapais & Schulman

1980). As Maynard Smith and Parker (1976) have demonstrated, the ESS will usually be to permit the asymetric cue (i.e. rank) to settle the contest without escalation.

A hypothetical low-ranking female can pursue two different alliance strategies. She can ally herself with other low-ranking animals against high-ranking females and rise in rank through fighting (strategy 1). While the benefits (B_1) of such a strategy are clearly high, the costs (C_1) are also high, since the alliance has a high probability of failing (see below). Alternatively, the low-ranking female can ally herself with high-ranking females against low-ranking animals who are certain to be defeated (strategy 2). Although such alliances may not lead to increments in rank, benefits (B_2) may be derived indirectly through future bonds or support from the high-ranking females, as described above. Moreover, since such alliances will almost certainly be successful, the costs (C_2) of this second strategy will be minimal. In order for strategy 1 to be an ESS, $B_1 - C_1$ (the net benefit of strategy 1) must be greater than $B_2 - C_2$ (the net benefit of strategy 2) or, since C_2 approaches 0, $B_1 - C_1$ must be greater than B_2. The costs of strategy 1 may include injury, loss of rank or loss of future support from the high-ranking female against whom the alliance is formed. At the very least, therefore, the costs of pursuing strategy 1 will be equal to or greater than B_2. Thus, in order for strategy 1 to be an ESS, B_1 must be at *least* twice B_2. ($B_1 - C_1$ must be greater than B_2. If $C_1 = B_2$, $B_1 - B_2$ must be greater than B_2 or B_1 must be greater than twice B_2; see also Table 12.4.)

Since female ranks are relatively stable over time, subordinate females can potentially derive predictable, long-term benefits from forming alliances with high-ranking females. In other words, alliances with high-ranking females (strategy 2) are 'safe bets'. B_2 values may therefore normally be expected to be high and in most cases it may simply not be economical for females to pursue strategy 1. Because of the predictable nature of B_2 values, high-ranking females

Table 12.4. Pay-off matrix of types of benefits from alliances with dominant and subordinate partners. Alliances with dominants are formed against animals who are subordinate to the alliance partner, while alliances with subordinates are formed against animals who are dominant to the alliance partner. The pay-off matrix assumes that alliances with dominant partners do not cause increments in rank and that dominant animals do not benefit from the support of subordinate partners.

	Alliance partner	
Actor	Dominant	Subordinate
Dominant	B_2	0
Subordinate	B_2	$B_1 - C_1$

will typically be able to recruit more alliance partners than will low-ranking females and should therefore be able to defeat most alliances that are formed against themselves. It should be emphasized that such recruiting abilities do not depend on high-ranking females having more relatives than low-ranking females: thus in Amboseli there is no positive correlation between female rank and reproductive success (Cheney *et al.* 1981). Instead, the ability of high-ranking females to recruit even unrelated partners may depend on the alliance's minimal cost and the high probability of deriving at least some benefit from it.

In contrast, successful alliances by low-ranking animals against high-ranking females may be expected to be rare. Although the potential benefits of rising in rank may be high, the costs of such alliances are also high, since other animals will tend to support the higher-ranking female. This does not mean that low-ranking animals will never form successful alliances against high-ranking females but rather that such alliances will occur at low rates and that propensities to form such alliances will not usually be evolutionarily stable. Rank reversals that occur as a result of such alliances will be infrequent.

2. Alliances formed against adult males

Why do alliances that are formed against adult males differ from those that are formed against adult females and juveniles? Among baboons, macaques and vervet monkeys, adult male rank seems to be less dependent on maternal rank than upon such factors as age, fighting ability and the presence of alliance partners (Boelkins & Wilson 1972; Kawanaka 1973; Packer 1977; Sugiyama 1976; unpublished data). Adult male rank is also less stable than adult female rank. Unlike bonds with high-ranking females, bonds with high-ranking males may not bring predictable long-term benefits to adult females and juveniles. Not only is the probability of a male retaining his rank less than it is for females but there is also some probability that a male will transfer from his group. When forming alliances against adult males, therefore, B_2 values may typically be expected to be low. As a result, B_1 values may easily exceed B_2 values and it may be economical to pursue strategy 1 and to ignore the relative ranks of alliance partners. Indeed, for many species of Old World monkeys, the outcome of disputes involving male–male alliances seems to be determined by the number of alliance partners rather than by their relative ranks (Hall & DeVore 1965; Packer 1977; Struhsaker 1967b; unpublished data). In other words, when males form alliances against each other, the ESS seems to be the number of individuals recruited by each side rather than the males' relative ranks.

While baboon females and juveniles rarely form alliances against adult males, such alliances are common among macaques and vervets. This difference may be due to the relative lack of sexual dimorphism in the latter species as opposed to the former (Packer & Pusey 1979). Among vervets, any two allied females or juveniles can successfully drive away an adult male, apparently because additional animals are easily recruited against males. Since dominance among females and juveniles is determined matrilineally (and possibly also through bonds with other females), alliances with males should usually have little effect on female rank. Indeed, rather than presenting females with an opportunity to increase their ranks, adult males may simply represent increased competition for scarce resources (Cheney 1981; Packer & Pusey 1979).

Males are usually dominant to females and juveniles in dyadic interactions and frequently exclude them from food and water (Dittus 1977; Hausfater 1975; Packer 1979b; Seyfarth 1978a; Sugiyama 1976). Alliances therefore provide a means by which females and juveniles may increase the ambiguity of their dominance relationships with adult males and improve their competitive abilities. Thus, even high-ranking females and juveniles benefit from forming alliances against males and all individuals may be expected to cooperate with each other against males. Since the outcome of a dispute involving males is determined by the number of alliance partners rather than by their relative ranks, the rate at which individuals are recruited as alliance partners should be unaffected by rank. Indeed, when forming alliances against males, the rate at which female and juvenile vervets formed alliances with each other was relatively independent of rank (Table 12.3).

It might also be argued that females cooperate with each other against males because the females in a given group are, on average, more closely related to each other than to the resident adult males. Thus, kin selection may have operated in such a way that females and juveniles of all ranks support each other in alliances against males. At least two factors, however, argue against this hypothesis in the case of the Amboseli vervets. First, rates of mortality in Amboseli are high and most adult females probaly have few living, close female relatives (Cheney *et al.* 1981). Average degrees of relatedness among females may therefore be quite low. Second, and more important, patterns of transfer among vervet males are remarkably non-random, with the result that at least some males may be as closely related to the females in their adopted groups as the females are to each other (section 11.3). It seems more likely that alliances against males benefit all females and juveniles because they provide them with a competitive advantage over males. Since animals of all ranks are equally effective alliance partners, the benefits each individual has to offer are symmetrical. As a result, patterns of alliances are also symmetrical and relatively independent of rank.

3. Effects of rank on family cohesion

When forming alliances with unrelated animals against adult males, adult females and juveniles chose partners irrespective of their ranks. In contrast, when the targets of alliances were other females and juveniles, females and juveniles chose alliance partners of high rank over those of low rank. Females and juveniles of all ranks, however, formed more alliances with their close relatives than with unrelated animals (unpublished data). What were the implications of these three distributions of alliances for relations within families?

For high-ranking females and juveniles, the tendency to form alliances with close relatives and with high-ranking animals seem to complement and reinforce one another. For low-ranking females and juveniles, however, the two tendencies appear to contradict one another and may cause low-ranking families to be less 'cohesive' than high-ranking families. Indeed, when forming alliances against other females and juveniles, high-ranking vervets formed alliances with their close relatives at higher rates than did low-ranking families (Table 12.5). The difference between high- and low-ranking families disappeared, however, when alliances against adult males were considered (Table 12.5). Thus, the relative

Table 12.5. Spearman correlations (r_s) between the familial dominance rank and the rate at which related individuals formed alliances with each other. In families with more than one immature offspring, rates of alliance formation have been averaged. Rates of alliances were calculated as in Table 12.3.

Actor	Alliance	Group	No.	Adult female or juvenile	Adult male	Is correlation less positive when target is adult male?
Mothers	Offspring	A, 1977–78	5	1.00*	0.41*	+
		C, 1977–78	6	0.62	0.09	+
		A, 1980	2	1.00	0	+
		B, 1980	5	0.87	−0.90	+
		C, 1980	2	1.00	0	+
Offspring	Mothers	A, 1977–78	6	0.63	0.54	+
		C, 1977–78	10	0.62*	0.09	+
		A, 1980	3	0	0	0
		B, 1980	5	0.64	−0.21	+
		C, 1980	4	0.78	−1.00	+
Juveniles	Siblings	C, 1977–78	8	0.63	0.35	+
		A, 1980	5	0	0.55	−
		C, 1980	8	0.47	0.52	−

$n = 12, P < 0.05$
(two-tailed binomial test)

*$P < 0.05$.

'cohesiveness' of families of different ranks varied, depending upon the nature of their aggressive interactions.

Bonds among related individuals appear to be manifested in different ways, depending as much upon individual costs and benefits as on benefits to close relatives. When a low-ranking vervet is threatened by a high-ranking female, it could potentially benefit from its relatives' aid. The cost of such aid to its relatives, however, is potentially high, since there is only a small probability that an alliance against a high-ranking animal will be successful. In contrast, almost all alliances formed by females and juveniles against adult males are successful. As a result, when low-ranking animals are involved in aggressive interactions with adult males, the cost of aid to their relatives appears to be low and rates of aid are correspondingly high. The potential costs incurred by aiding kin therefore vary, depending upon both familial rank and the context in which the aid is given. Patterns of alliances appear to reflect the constraints imposed on individuals by their relative ranks.

12.7 Adaptive Aspects of Social Relationships among Adult Rhesus Monkeys

BERNARD CHAPAIS

Observational studies of primate behaviour often produce large amounts of interactions of the type: A does X to B or A does X to B in relation to C. Such data are then partitioned according to certain characteristics of the participants and the conclusions derived are specific to the categories used, e.g. it is possible to look for a relation between the dominance rank of adult females and the total frequency of grooming directed to other adult females. However, since one cannot assume that the acts of grooming a higher-ranking female versus a lower-ranking one or a related female versus an unrelated one have identical consequences on the inclusive fitness of the actor, the above relation may not be easily interpretable in functional terms.

Perhaps the best illustrations of this principle are found in the area of fight interventions. Consider for example the possible functional consequences of aiding among females. First, females might increase their inclusive fitness by favouring related females against unrelated ones (e.g. see sections 12.5 and 12.6). Second, the same females might further increase their inclusive fitness by defending female relatives against unrelated males. Third, they might benefit from supporting unrelated higher-ranking females against lower-ranking ones if the aid given is reciprocated (e.g. see below). Finally, they might also benefit from forming coalitions with unrelated females against unrelated males if this enables them to drive non-central males away from food sources (e.g. see below).

Although these four possibilities refer to aid given by adult females to other adult females, the ultimate causes of each type of aid differ. If one or more of the above hypotheses are true, gross data partitioning (e.g. lumping the targets independently of their sex or lumping the beneficiaries independently of their degree of relatedness to interferers, etc.) can be misleading.

Generally speaking, the functional interpretation of social behaviour can be considerably hampered by the categorization of interactions into functionally *heterogenous* classes of recipients, i.e. classes including individuals differing by at least one characteristic (e.g. sex, age, rank, relatedness, reproductive state, etc.) which affects their value as social partners. The various ways by which the characteristics of individuals determine their usefulness as partners in affiliative interactions are just beginning to be understood. Briefly, the actor may benefit by enhancing either the kin component of its inclusive fitness (if related to the recipient) or its personal fitness (whether related or not to the recipient). In the latter case, the interaction can increase the likelihood that the recipient will cooperate with the actor or that the recipient will reciprocate the actor or the interaction can simply increase their familiarity, a condition which may benefit the actor in a number of ways, for instance by facilitating social learning or access to scarce resources or the use of the recipient as a buffer in agonistic interactions (e.g. Strum in press b).

The foregoing principles of data partitioning were applied to the analysis of social interactions between adult females and between adult males and females in one group of rhesus monkeys on Cayo Santiago (data presented in sections 10.4 and 10.5). The distribution of affiliative acts and fight interventions were strongly rank related, a finding which supports a model proposed by Wrangham (1980) for the evolution of female-bonded primate groups. The model suggests that these groups have evolved as a result of females competing for high-quality food patches containing a limited number of feeding sites (e.g. fruiting trees). Females are hypothesized to have acted together to supplant other females at food patches and to have maintained such cooperative relationships on a long-term basis. Furthermore, females would compete not only with other groups but also within their own group. The existence of well-defined female dominance relations where rank correlates with fitness (see section 12.5) supports this view.

It is suggested that a further step in the evolution of competitive interactions *within* female-bonded primate groups may have been the use by females of other group members (e.g. higher-ranking females and high-ranking males) to increase and maintain their position in the order of access to resources. Rank-related patterns of affiliation and aiding will now be discussed in relation to this hypothesis.

Female–female relationships

At least three scenarios could be at the origin of the principle that females seek to affiliate with higher-ranking females as a means of avoiding being outranked by lower-ranking females. First, if one assumes that there is a positive correlation between the degree of relatedness and the degree of familiarity among females so that differential familiarity is the criterion on the basis of which females decide how to apportion their aid (Chapais 1981), an initially low-ranking female needs only to increase disproportionately her familiarity with a high-ranking female in order to be favoured by the latter over an intermediate-ranking female. Therefore, cheating on true relatedness might be at the origin of the attractiveness of high rank. A second possibility refers to the transactional nature of social relationships (see section 6.1). An initially low-ranking female C behaving in a way profitable to a high-ranking female A (e.g. by directing grooming to A) would increase her value as a social partner to A, who might in turn tolerate C's access to resources otherwise monopolized by A and intermediate-ranking females. A third possibility concerns a ranking system where coalitions of low-ranking relatives can outrank higher-ranking females, provided they outnumber them. In such a system, a female forming alliances with unrelated higher-ranking females would become independent of variations in the number of related supporters and be at an advantage over females restricting their affiliation to relatives.

   Although these three scenarios differ in the selective mechanisms involved, they have in common that when facing a situation where lower-ranking competitors threatened to outrank them through alliances with higher-ranking females, intermediate-ranking females have no choice other than affiliating with the high-ranking females more than do lower-ranking females. The observation that females support unrelated females down the dominance hierarchy (see section 10.5) and in doing so contribute to stabilize rank relations supports the present hypothesis. Thus, whatever the precise origin of the need to affiliate with high-ranking females, the function of the attractiveness of high rank appears to be the same—the maintenance of rank.

Male–female relationships

The nature of the relationships between males and females during the birth season (see section 10.4) could not be explained either in terms of the continuation of sexual cycling or in terms of a long-term process of mate selection (very little concordance was found between the structure of the birth-season male–female relationships and the patterning of sexual activity [Chapais 1981; contrast section 12.3]). A third hypothesis finds support in the data. If priority of access to resources is crucial for females (see above), it is likely that females have an advantage in feeding before males as well as before other females. Consequently, while on the one hand, females may benefit from the presence of males as allies against neighbouring groups, on the other, they also have an advantage in minimizing the number of group members, *including males*, eating before them. Since differences in size and strength may prevent females from dominating all the adult males, a compromise might be for the females to ally with a fraction only of the adult males present in their group against the other fraction. Allying with the highest-ranking males would allow females to minimize the number of males comprising this fraction and, as a result, to maximize the number of males over which they have priority of access to food (i.e. mid-central and peripheral males). In this context, high-ranking males have an advantage in defending and supporting adult females against lower-ranking males (as observed), since which males the females ally with ultimately depends on the willingness of males to defend them. Males might also benefit from the grooming received from females.

In conclusion, much of the social structure of rhesus monkeys seems interpretable in terms of the polyadic nature of resource competition and resulting dominance relations.

12.8 Patterns of Agonistic Interference

SAROJ B. DATTA

Introduction

The phenomenon of interference in disputes among primates is by now well documented. The interferer typically threatens one antagonist, thereby allying with the other. (More than three individuals may often be involved but the situation remains fundamentally triadic: the two parties to the dispute and the third, interfering, element).

There have been a variety of conjectures about the 'function' of interference. One common until recently (Bernstein & Gordon 1974; Kaplan 1977) suggested that 'breaking up' fights maintained group order, reducing the wounding, stress and social disruption which might result from frequent fighting. This approach (which often implies group selection) has now been abandoned, partly due to the cumulative evidence of data showing that interference is often in aid of *close kin* being beaten in fights (e.g. Kaplan 1978; Massey 1977; de Waal 1977; see also below) and may therefore be 'altruistic', in the sense of Hamilton's (1964) kin-selection hypothesis (see also section 12.1).

The idea that interference generally serves to defend the victims of aggression still prevails, albeit such aids have also been seen as helping subordinate immatures in the acquisition of rank (e.g. Cheney 1977; Datta 1981; Kawai 1958). However, there is increasing evidence that interference often favours *aggressors*. Suggestions (when made at all) for the function of such supports are various: (1) the regulation of relations between the alliance partners (de Waal 1977); (2) attempts by low-ranking animals to form profitable relationships with high-ranking animals (Cheney 1977); (3) the acquisition of higher rank by immatures (Datta 1981; Walters 1980); and (4) the maintenance of the rank of the animal aided (Datta 1981; section 6.8).

Clearly, interference is not a unitary phenomenon: a variety of patterns and functions may coexist. Previous attempts to distinguish these are subject to two related criticisms:

1   By creating large and probably heterogeneous units they lump functionally distinct categories. For instance interferences may be classed as aids (supports to victims) or alliances (supports to aggressors) (Cheney 1977; de Waal 1977). This ignores the possibility that support to aggressors, for example, may differ, depending on whether the interferer is subordinate or dominant to the target: in the first case the interferer may be trying to challenge the dominance of the target under relatively safe conditions, in the second it may be helping the beneficiary maintain dominance over the target (Datta 1981).

2   They may result in rather frail generalizations. Studying Group F of the rhesus macaques on Cayo Santiago, Kaplan (1977) concluded that females were more likely to support the victims of aggression, while non-natal males were more likely to support aggressors. The author's study of Group J of the same species on Cayo Santiago yielded very different data (Table 12.6). Yet these differences need not imply that different 'principles' of intervention are at work, simply that a variety of strategies have been sampled in different proportions in the two sets of data. Other broad-based differences between groups of the same or different species may be similarly explicable.

Table 12.6. Number of times the aggressor or victim was supported by adult female and non-natal male interferers intervening in intragroup fights in groups F (Kaplan 1977) and J (Datta 1981) on Cayo Santiago.

Sex of interferer	Interferer supports aggressor	Interferer supports victim	Study
Female	339	390	Datta (1981)
	140	546	Kaplan (1977)
Non-natal male	42	108	Datta (1981)
	34	51	Kaplan (1977)

There seems strong justification, therefore, for distinguishing patterns and strategies of interference at a more detailed and sensitive level. This is a complex task which cannot be dealt with here at any length or depth, but a possible approach to the more detailed study of interference is outlined below.

Consequences of interference

If the immediate and ultimate functions of interference are to be clarified, accurate records of its consequences, both long- and short-term, and for each of the three individuals concerned (interferer, target and beneficiary) are needed. Why the consequences are what they are depends, as will become apparent, on features of the relationships of the three individuals with one another and with others. Perhaps the most important concern:

1 Whether the original dispute continues or not. If it is halted, the victim can justifiably be said to have been defended from further aggression. Not surprisingly, disputes are more likely to end when the *victim* is supported (Datta 1981; Kaplan 1977; de Waal 1977). Interferers subordinate as well as dominant to targets successfully halt disputes—but for different reasons. The former typically succeed by diverting the direction of attack (the target attacks them instead) (Datta 1981; Kaplan 1977), while the latter intimidate targets. Kaplan's (1977) conclusion that interferers *subordinate* to targets more successfully halt disputes is not justified by his data, in which there is a strong association between the *dominance status* of the interferer (relative to the target) and *whom* (victim or aggressor) it supports. The fact that subordinate interferers intervene at considerable risk to themselves appears to have a marked effect on patterns of interference (see below).

2 Who is defeated or made to submit as a result of the interference. If consistent in direction, this effect is likely to be extremely potent for the long-term consequences of interference. If A consistently supports subordinate-victim B against dominant-aggressor C, causing C to submit, then C might be expected to learn that disputes with B are likely to be costly and B to learn that it has powerful support against C. Such support may eventually encourage B to rebel against C: indeed it is associated with the raising of rank by subordinate immature monkeys (Cheney 1977; Datta 1981; Kawai 1958). 'Effective' support (see also section 6.8) is usually by an interferer dominant to its target but, as Datta (1981) shows, more complex characteristics may be involved: effective interferers typically not only dominate targets but are high born to targets. Interferers dominant to but lower born than targets are likely to be threatened and defeated (in a secondary wave of interference) by supporters of the latter.

If it is interferer A who is defeated (by C and/or C's supporters) as a result of the interference, it might be expected that A would become more circumspect about supporting B against C. There is some evidence to suggest this (Cheney 1977; Datta 1981; see below). Moreover, B apparently learns it has no powerful support against C; this appears to make B reluctant to challenge C (Datta 1981).

Finally, the long-term consequences of powerful and effective interference are expected to be quite different where it favours aggressors: it is then expected to reinforce the status quo (Datta 1981).

Support may also have other, more subtle consequences:

1   It informs members of a social group who is likely to ally against whom, how consistently and with what immediate effect(s).

2   It may 'confirm' and indeed strengthen alliances or associations. Thus, sexual partners often seek, and receive, 'token' support from one another while threatening harmless bystanders (like the observer) (Datta unpublished).

3   It may ensure continuation of a relationship or activity beneficial to the interferer. Thus, adult male rhesus defend their sexual consorts against threats (which may disrupt the relationship) by females dominant to the latter (Datta 1981).

4   It may benefit the interferer more than the beneficiary, e.g. high-born subordinates (see section 6.8) in the process of challenging low-born dominants sometimes support individuals dominant to and threatening the latter, perhaps because low-born dominants are less likely to retaliate under these circumstances (Datta 1981).

Finally, it is worth pointing out that a given episode of interference may have several consequences, both short- and long-term: thus effective support of subordinate B against dominant C may not only prevent injury to B (short-term effect) but promote B's confidence in challenging C (long-term effect).

Determinants of interference and interferer choice

As suggested above, the precise consequences of interference depend on who supports whom against whom in what circumstances. The distribution or patterning of such consequences both reflects and affects social relationships and social structure. Thus, the eventual dominance status of an individual may depend on the power of its allies, which may in turn depend on prevailing power relationships between rival subgroups (Datta 1981; section 6.8).

By what criteria do interferers 'choose' one antagonist over another or indeed decide to interfere at all? Several have been suggested and are considered briefly below.

1. Relatedness within the interference triad

The *close* relatives of an individual provide a disproportionately large share of the total support it receives (Datta 1981; Massey 1977) and are (allowing for spatial and social availability) significantly more likely to intervene than more distantly related individuals (Datta 1981). There is also a strong tendency for interferers to support closer relatives against more distant (including non-) relatives (Datta 1981; Kaplan 1978; de Waal 1977).

2. Role in dispute

Interferers do not appear to be impelled to help the victims of aggression as was once thought but evidence is still somewhat conflicting: some studies find support to aggressors much less common than support to victims (Kaplan 1977; Massey 1977), others find it about as common (Datta 1981) or even more common (de Waal 1977). These differences may be due partly to differences in sampling methods and partly to demographic differences, resulting in different contributions to the data of different interferer strategies. (See p. 297 for a possible example of demographic differences affecting the *kinds* of interference seen.)

There is evidence that the more vulnerable to injury the victim of aggression, the more likely it is to receive support (Kaplan 1977; section 6.9).

3. Age

The age (of the interferer or antagonists) is reputed to influence patterns of interference. This is difficult to judge since age may be correlated with factors such as dominance status (section 6.8) and hence the likelihood of being a victim or aggressor. Such factors are not usually controlled for (but see below).

Both Massey (1977) and Kaplan (1978) found mothers significantly more likely to aid offspring than the reverse. Massey ascribed this to the fact that the reproductive potential (RP) of an individual declines with age, hence benefits to it from the receipt of altruistic acts must also decline. In fact, RP rises to a peak in late adolescence and only then declines. However, aid to young individuals of comparatively low RP may be favoured if it protects them from injury or in any other way prevents or mitigates setbacks to normal development. A general tendency to aid the younger of two contestants is therefore conceivable.

4. Relative family ranks of antagonists

Cheney (1977) found immatures from high-ranking families more likely to receive support than those from low-ranking families; immatures were also more likely to ally with high-ranked adult females. Cheney postulated a tendency to form alliances with high-ranked individuals, possibly enabling individuals to share in the perquisites of high rank.

5. Cost of interference

Any interference incurs cost to the interferer in expenditure of energy but potential costs (retaliation by the target or its supporters) may also need to be taken into account. Interferers subordinate (Kaplan 1977) or subordinate and low-born (Datta 1981) to targets are more likely to suffer such retaliation, so the risk they take in interfering is greater. There is evidence (Cheney 1977; Datta 1981) that interferers at considerable risk of attack are less likely to interfere than those at little or no risk.

6. Sexual and other benefits

Interferers may sometimes act to prevent disruption of an activity, such as grooming, or a relationship that benefits them in some way. For instance in supporting their sexual partners against aggressors (e.g. Bernstein 1963; Carpenter 1942) adult males may ensure maintenance of the sexual relationship.

Unlike previous studies (see p. 289) most of those mentioned above invoke kin selection (Hamilton 1964) and reciprocal altruism (Trivers 1971) as ultimate explanations of interference. The implication is that interferers maximize benefit by, for example, supporting the closer relative (1 above), the more endangered of the two relatives (2 above) or the relative with higher reproductive potential (3 above). Yet in a complex social group several, often undoubtedly conflicting, 'tendencies' may be expected to operate: for instance what happens when the closer relative is also the aggressor? Possible interactions between benefits have usually been ignored and data treated as if they were independent of them. Yet it is probably by the resolution of different benefits that one antagonist is supported over another or that interference occurs at all. A general 'rule' for the resolution of such conflicts may be the maximization of inclusive benefit (see below).

Data on interactions between the parameters given above are few, perhaps partly because of the daunting complexity of possible interactions and partly because relevant data (e.g. on (matrilineal) relatedness) are not always available. Kaplan (1977) used multidimensional analysis to explore interactions between some parameters. However, the results were only somewhat revealing, perhaps because different (and conflicting) patterns were lumped, obscuring many meaningful associations.

An alternative (more pedestrian but possibly more fertile) approach is to use a framework based on some parameter which splits the data into a series of logically related and biologically meaningful subsets and to explore interference strategies within these. (If sufficient data are available, multidimensional analysis may be carried out on subsets.)

Kinship is one suitable parameter: in natural groups of many primate species it is a strong predictor of dominance–subordinance relations (e.g. Cheney 1977; Kawamura 1965; Lee & Oliver 1979; Sade 1967; Walters 1980), proximity relations (e.g. Berman 1978a; Kurland 1977) and affiliative behaviours (Sade 1965), as well as patterns of interference. Table 12.7 provides a possible frame of reference for discussing the interaction of relatedness with other variables. With its aid some evidence that observed patterns of interference are compatible with expectations based on the assumption that interferers are maximizing inclusive benefit are now summarized. Unless otherwise specified, detailed data are available in Datta (1981), as are further instances of such interaction.

Table 12.7. Frame of reference for considering the interaction of relatedness between an interferer and contestants A and B with other variables in the study of interference.

Relatedness of interferer to contestant A (the closer relative)	Relatedness of contestant A to contestant B			
	Close relative (sib, mother)	Other relative	Natal non-relative (other matriline)	Non-natal non-relative (adult)
Close relative	1	2	3	4
Other relative	5	6	7	8
Natal non-relative (other matriline)	9	10	11	12
Non-natal non-relative (adult)	13	14	15	16

1. Interaction between relative relatedness, family rank and vulnerability in dispute

Interferers supported the contestant more closely related to themselves significantly more often than the other in cells 2, 3, 4, 5, 6, 7 and 8 of Table 12.7.

Choices against the trend, though relatively few (see below), appeared to result from conflicts between relatedness and other parameters. Consider cells 2 and 6. Disputes in these cells are between members of different families; opponents are high-born (H) or low-born (L) relative to one another (see section 6.8), dominant-aggressors (D) or subordinate-victims (S) and younger (Y), or older (O) or the same age (E) as their opponents. When all these factors were taken into account the following pattern emerged.

When the interferer was the mother of contestant A (cell 2), she almost invariably supported it (in 149 [98.1%] out of 152 triads), irrespective of its other characteristics. Neither the siblings (cell 2) nor other relatives (cell 6) of contestant A were as consistent as mothers. They supported it in 74 (84.1%) out of 88 and 36 (78.3%) out of 46 triads respectively.

If individuals are concerned to maximize inclusive fitness, support of the closer relative may not always be the best strategy: for instance if the less-close relative is also the younger and the victim of aggression, it may in some cases be more profitable to aid it. This appears to be happening when interferers, more closely related to a low-born dominant-aggressor older (LDO) than its high-born subordinate-victim (HSY), support the latter: sibs of the former supported the latter in 11 out of 12 triads ($P = 0.05$, Sign test, two-tailed) and less closely related interferers who were closer relatives of the former supported the latter in seven out of seven cases ($P = 0.05$, Sign test, two-tailed).

However, this cannot be the whole story, because interferers more closely related to a *high-born*, older dominant-aggressor (HDO) than to its low-born, younger subordinate-victim (LSY) are in fact more likely to support the HDO: siblings supported it in 15 out of 18 triads ($P = 0.05$, Sign test, two-tailed) and less closely related interferers who were closer relatives of the HDO supported it in 16 out of 18 triads ($P = 0.02$, Sign test, two-tailed).

In fact, these interferers treat HSYs differently from LSYs—the latter are less likely to be supported. Consider three families, A, B and C, in descending order of rank. In all the triads above where a high-born from A (or B) was supported against a low-born from C, the interferer, from B (or A) was itself high-born to the individual from C. In other words, individuals from A and B aided one another against members of C: such aids disclose the mutual alliance of high-ranking families against lower-ranking families—an alliance that may enable both A and B to maintain their higher ranks.

To conclude, therefore, interferers may experience a conflict between supporting the more closely related (or more vulnerable) antagonist and the one from a more powerful family. The resolution of the conflict suggests that inclusive benefits may be maximized: the more closely related the interferer to the low-born antagonist, the less likely it is to betray it for a higher-born individual.

2. Interaction between the cost of interference and relatedness

Interferers subordinate and/or low-born to their targets are likely to be attacked by the latter (or supporters of the latter) (see above). There were two particularly clear examples of the effect of this on the readiness of individuals to interfere. First, in cell 1 of Table 12.7 it was found that siblings subordinate to the dominant sibling of a contesting pair were significantly less likely to interfere than siblings dominant to it. (This effect was independent of age, since sibs subordinate to the dominant sib were older as well as younger than it.)

Second, in interfamily disputes, only close relatives (mother and siblings) were likely to intervene as low-born and subordinate to the target (cell 2). Less close relatives were significantly less likely than close relatives to interfere in this way. Furthermore, the less close relatives of an individual were significantly more likely to support it when it was dominant and high born to its opponent than when it was subordinate and low born: in the first case interferers were safe(r) from retaliation by targets. Potential interferers appear, therefore, to balance the benefits of interference against its costs. When the benefits are likely to be high (as when a close relative being beaten by a less close relative is defended), the willingness to incur risk appears to increase.

Conclusion

Individuals intervene in the disputes of others with a variety of consequences for themselves, the individuals they support and those they intervene against. Interference appears, therefore, to have a variety of 'functions', ranging from the protection of relatives being attacked by dominants to the maintenance of alliances and perpetuation of the status quo.

While all these may benefit the interferer, interference usually also incurs some cost. It might be expected that interferers behave in such a way that profit (benefit less cost) is maximized. Indeed, there is evidence to support this. Thus interferers more readily risk attack by targets when intervening on behalf of threatened *close* relatives than on behalf of *less close* relatives, and individuals who are not closely related to contestants often support, not the more closely related contestant, but the one from the higher-ranked family. Alliance with dominants, with its potential advantages for the acquisition and maintenance of rank, may well be more profitable than aiding a low-ranked if closer relative.

If, as seems likely, such trade-offs often occur in social groups, then considerable variation between groups (of the same or different species) in the patterns of interference observed must be expected, since these will depend on the unique configuration of a social group and the possibilities for, as well as obstacles to, maximization of inclusive fitness inherent in that configuration. The principles and functions of interference may remain the same but their manifestation may differ. For instance it is possible that in extended matrilines, such as those on Cayo Santiago, immatures can depend on close relatives to help them attain and maintain rank, but where immatures have few relatives (close or otherwise) in the social group they may need to acquire and maintain rank by *actively seeking* alliances with high-ranking (and often unrelated) individuals (e.g. Cheney 1977; Walters 1980).

Chapter 13
Generality of the Approach to Other Species

13.1 Applicability of the Approach to Other Species
ROBERT A. HINDE

This chapter is concerned with the extent to which the general approach to the study of social behaviour used in this book, and the explanatory principles that have been found useful, are applicable to other species. Before proceeding, three characteristics of the approach should perhaps be re-emphasised. First, it is recognized that social behaviour must be studied at several different levels, including those of individual behaviour, interactions, relationships and social structure. Each of these levels may show emergent properties not relevant at lower levels.

Second, practically every contribution has explicitly or implicitly stressed the importance of studying dyadic relationships. Fifteen years ago, most field studies concentrated on describing the behaviour shown by age/sex classes of individuals. There were of course pioneers, like Hans Kummer, who stressed the importance of individual relationships, but mostly the emphasis was on categories of individuals. Now there is a major focus on the relationships they form. Those relationships influence and are influenced by the individual characteristics of the participants: they also influence and are influenced by the social nexus in which they are embedded. Thus, as has been shown, the study of dyadic relationships isolated from the social situation still involves simplification and must be regarded as only a stage on the road to full understanding.

Third, the four questions of causation, development, function and evolution are in principle of equal interest, independent, yet interrelated and interfertile.

The preceding chapters have been concerned with three species of Old World monkey, all of which live in multimale, female-bonded groups. Such a narrow data base could be misleading (Dolhinow 1982; Zucker 1982). How far is the approach more widely applicable? How widely can the specific principles elucidated for those species be applied? In section 13.2, the relationships and social structure of two other species of baboons are contrasted with those of the savanna baboons discussed in previous sections: superficial similarities in surface structure conceal marked differences, which are revealed only after examination of interindividual relationships but could depend on principles similar to those found useful for the other species.

The great apes show obvious differences from the other primates discussed but again similar conceptual tools can be applied: studies of the great apes both confirm and extend the principles found useful for other species (section 13.3). In section 13.3 Harcourt and Stewart properly emphasize the need to relate studies of social structure to such characteristics as individual- and group-ranging patterns: it is of course a matter of definition whether social structure is to be thought of as comprising those patterns. This contribution also raises the interesting issue of 'negative' relationships: if orang-utan females seem not to interact and have individual ranges scattered at random through the forest, they could not be said to have relationships with each other, but if their ranges are non-randomly not overlapping, they presumably do. Relationships involving near-total avoidance may be important in other species, even though difficult to study.

Social behaviour with a complexity comparable with that found in primates is found in at least some other mammals. This is exemplified by data on the African elephant (section 13.4), a species in which the need to distinguish between different levels of social functioning is even more apparent than in primates. Nevertheless, in the broader perspective of the whole animal kingdom, primates are special (section 13.5) and it remains to be seen how far the principles found useful here will be applicable. In the final section, the value and limitations of this approach for understanding human social behaviour are discussed.

13.2 Relationships and Social Structure in Gelada and Hamadryas Baboons

ROBIN I.M. DUNBAR

In this section, two particular aspects of the structure of relationships in gelada and hamadryas baboons are explored in order to illustrate how some of the ideas so far developed in this book can aid the understanding of social relationships in other, rather different species. A good part of the discussion has focused on the savannah baboons (see Chapter 2), a species group which behaves in much the same way, both socially and ecologically. They are usually contrasted with the socially very different hamadryas baboon, *Papio hamadryas*, and the gelada, *Theropithecus gelada*, a species that is taxonomically intermediate between the African baboons and the Asian macaques.

The main features of the gelada and hamadryas social systems will first be compared to show that, though superficially very similar in demographic terms, the 'surface' structure of relationships is quite different, being based on completely contrasting 'deep' structures. Some of the ways in which demographic and other factors can affect the structure of these relationships will then be discussed in more detail, drawing in the main on the author's work on the gelada. One key point that will emerge from this discussion is the immense fertility of a combined functional and causal analysis in guiding us to a deeper understanding of a species' social system.

1. Structure of relationships

Both gelada and hamadryas baboons live in one-male reproductive groups (i.e. permanent social units that consist of a single breeding male and a number of reproductive females and their offspring) (see Dunbar & Dunbar 1975; Kummer 1968). Both species also live in higher-level groupings termed *bands* that consist of a number of one-male groups together with those mature males that do not have their own harems of breeding females. In the gelada at least, the latter form stable all-male groups. One might be tempted to conclude from these similarities (as many early writers did [e.g. Crook & Gartlan 1966; Jolly 1972]) not only that the underlying dynamics were probably similar but that the two societies were functionally similar and owed their origins to similar evolutionary pressures.

In actual fact, the first supposition is certainly wrong and the second remains in considerable doubt.

Figure 13.1 shows in stylized form the typical pattern of social relationships among the adult members of gelada and hamadryas bands. The lines joining individuals indicate the most frequent interactions. It is clear that the general pattern of relationships is quite different in the two species. In the hamadryas (Fig. 13.1a), male harem-holders interact quite extensively with each other, whereas gelada males very seldom do so (Fig. 13.1b). Conversely, hamadryas females interact mainly with their harem males and rather rarely with each other, whereas the reverse is true in the gelada.

If the nature of the relationships themselves are examined in more detail, comparable contrasts between the two species are found. The hamadryas one-male unit is held together mainly by the aggressive herding of the harem-holder (Kummer 1968). Kummer found that older males tended to lose their females to

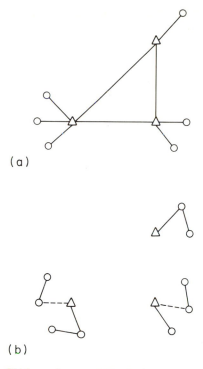

(a)

(b)

Fig. 13.1. Sociograms of (a) hamadryas and (b) gelada bands, showing the most frequent patterns of interaction between adult animals. In each case, three one-male units are shown. Triangles = males; circles = females. Note that while the hamadryas male interacts more or less evenly with all his females, the gelada male interacts rarely or even not at all with many of his and invariably has one clearly predominant social partner. (a after Kummer 1968; b from data in Dunbar & Dunbar 1975.)

younger males, partly as a result of kidnapping and partly because the females just seemed to drift away. When a unit leader disappeared or was experimentally removed, his group of females was broken up and the females dispersed among other one-male units. The females have only weak social bonds with each other, their 'loyalty' being strictly to their unit leaders. In contrast, gelada reproductive units are held together over time by the strength of the social bonds among the females, these bonds themselves being largely based on kinship (Dunbar 1979b; Dunbar & Dunbar 1975). In a very real sense, the male is a social supernumerary whose role in the group is mainly that of breeder. He is virtually excluded from interactions with his females because the latter prefer to interact with their immediate female relatives. When a unit leader dies or is removed, the group of females stays intact and continues to behave as a discrete, self-contained social unit until another male takes it over (Dunbar & Dunbar 1975; Mori 1979).

Similarly, at the band level, the strong male–male relationships that characterize the hamadryas are quite unique. These have since been shown to depend on male kinship (Abegglen 1976). They serve important social and ecological functions within hamadryas society (Kummer 1968; Stolba 1979): through these relationships, the males coordinate band movement during the day's foraging, communicating decisions about direction and the timing of resting and watering stops and generally preventing the individual one-male units from becoming too dispersed and getting completely separated (Stolba 1979). The relationships are subserved by a unique set of behaviour patterns known as 'notifying' (Kummer 1968): males wanting to move in a particular direction approach a neighbouring male, peer closely into his face and then abruptly turn round and present their rears. This behaviour serves both to attract the other male's attention and to act as a channel through which the coordination of movement can be maintained. No such complex decision-making process occurs in the gelada. Gelada herds are much more fluid in their structure, with individual one-male units constantly leaving and joining. The movements of herds are determined more by a gravitational effect than by collective decision making: the more animals that seem to be going in a given direction, the more likely it is that units in the rear of the herd will go that way too. As a result, gelada herd movements often appear to be confused and aimless and lack the conspicuous decisiveness of the hamadryas (Dunbar & Dunbar 1975). This is perhaps not too surprising from a functional point of view: hamadryas baboons live in an arid environment where water and good feeding places are scarce, whereas the gelada live in a habitat that is (at least from their point of view) relatively rich in terms of food availability, with a superabundance of sleeping and watering sites. One might expect a far higher premium on fast, reliable decision making among the hamadryas, since errors of judgement would be more likely to result in disaster than they would for the gelada.

These comparisons suggest that gelada and hamadryas societies, although superficially similar, owe their appearance to quite different behavioural mechanisms. Indeed, it is not implausible to suggest that the two social systems have converged on a similar superficial structure from different evolutionary starting points for quite different functional reasons: the hamadryas by the break-up of the classic *Papio* multimale troop into smaller units to facilitate foraging under conditions of scarce food resources, the gelada by the congregation of isolated one-male units into larger groupings under conditions of better food availability in response to predation pressure.

2. Relationships as dynamic processes

(a) Structure in relation to demography

Permanent social groups are not static entities: they are subject to constant change as animals are born and die, mature, emigrate and immigrate. As a result of these demographic processes, the composition of a group changes with time and the availability of social partners of a given type will inevitably fluctuate (Altmann & Altmann 1979; Dunbar 1979a). One can hardly expect the pattern of relationships to remain stable under such conditions.

The structure of social relationships in gelada reproductive units undergoes quite marked changes as the unit increases (or decreases) in size. Figure 13.2 shows how a measure of the social cohesiveness of the unit is affected as the

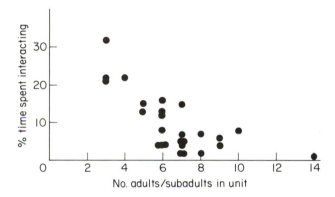

Fig. 13.2. Effect of changing unit size on a measure of the social cohesiveness of the unit. This measure is the minimum frequency of interaction required to split the unit into at least two quite discrete subgroups whose members do not interact at that frequency of interaction. This can be done by using the matrix of interaction frequencies for a unit to construct a tree dendrogram (see Morgan *et al.* 1976) and finding the lowest frequency at which the dendrogram can be split into two discrete clusters.

number of animals in it changes. The index is the minimum frequency of interaction between individuals required to split the unit into two, discrete, non-interacting subgroups. As the unit size increases, the subgrouping discriminant declines, indicating that the individuals are interacting less frequently outside their immediate circle of interactees, with the result that the unit as a whole becomes increasingly fragmented socially.

Other important changes consequent upon this can also be observed. As the number of females in the unit increases, a point is reached at which the harem holder no longer has the time available to groom all his females with any regularity

(Dunbar 1983). In effect, he is caught by a problem in time-sharing that is compounded by his females' preferences for interacting with their own immediate female relatives rather than with him. This might be of only academic interest were it not for the fact that a female's 'loyalty' to her male (i.e. her reluctance to desert him in favour of another male) is dependent on the frequency with which she grooms with him (Dunbar 1983; Kummer 1975). The male is thus caught in an invidious bind, for as his unit size increases he cannot groom regularly with all his females, yet if he does not, he risks losing them all as a result of a take-over by another male. (Take-overs are all-or-none events in which the females desert or remain loyal *en masse*, depending on the number of females that 'want' to desert on any particular occasion [for further details see Dunbar 1983; Dunbar & Dunbar 1975].) In fact, the proportion of males who acquire units by take-overs (as opposed to any other way) is a linear function of the mean unit size in the population (Dunbar 1979a, 1983).

Thus, changes in the demographic structure of the group can here be seen to have important consequences, not only for the structure of relationships within the unit but also on the dynamic aspects of social structure (including, in this case, the processes whereby males acquire reproductive units). Thus, the ramifications of a single factor can be seen to permeate throughout the species' social system. As a result, the general impression that an observer gets of the species' style of social life will depend on the demographic structure of the population at the time he happens to study it and this, in turn, can be traced back to an underlying complex of environmental, social and demographic factors that determine birth and death rates in preceding years (see Dunbar 1979a, 1980a,b).

It may also be noted that the same demographic factor (the number of females in the unit) can have important reproductive consequences for the females, this time acting via a complex physiological route (Dunbar 1980b). In this case, the accumulated harassment by more dominant females appears to result in the disruption of a low-ranking female's reproductive physiology, such that she fails to ovulate although undergoing normal behavioural (and morphological) oestrus; as a result, she takes longer to become pregnant and thus has a lower birth rate than higher-ranking females. This effect becomes progressively more noticeable as harem size increases, since the lowest-ranking females have an increasing number of females dominant to them.

The story is brought full circle at this juncture, since, from a functional point of view, it is probably this reduction in reproductive rate that makes it more likely that females in large units will desert their males in the 'hope' of improving their chances elsewhere. Thus, we find that the whole structure of society, indeed the very tenor of the animals' social life, is the outcome of an extremely complex web of demographic, social and physiological factors acting on social relationships.

(b) Effects of other factors on social structure

Aside from any demographic changes that occur within a group over time, other changes may take place which are likely to have a more direct impact on the social relationships of individual animals. In exploring the effects of some of these events, comparisons with the savannah baboons will be used to draw attention to the fact that these effects depend on the general biological context within which the animals are embedded and cannot be assumed to be of universal occurrence.

It has been shown (sections 8.5 and 10.5 above) that new babies are a source of considerable interest to the juvenile and adult females of many species of primates. Seyfarth (1978b) found that chacma baboon mothers were a source of attraction to other females and that the pattern of their social relationships changed as individuals with whom they did not normally associate began to interact more frequently with them. Although other females do become interested in newborn infants and their mothers, such dramatic changes in the social relationships of the mothers and in the overall pattern of relationships within the unit as a whole are not observed among gelada baboons (Dunbar 1979b). This is partly a consequence of the different social and ecological conditions prevailing in the two species: as a result, the constraints and rules that govern the animals' behaviour by determining their preferences for social partners are very different. Strong social bonds with female relatives are of overriding importance to a gelada female, since these help to maintain her dominance rank in the long-term (though not necessarily in the short-term), thereby influencing her lifetime reproductive output (Dunbar 1980, 1983); nothing can be allowed to interfere significantly with the process of maintaining these relationships. Consequently, only related females show a marked and persistent increase in the time spent interacting with new mothers among the gelada: this is insufficient to cause a significant change in either a female's pattern of relationships or in the overall distribution of relationships within the unit as a whole.

A rather similar contrast between the *Papio* baboons and the gelada emerges from a consideration of the effects of oestrus on social relationships. In the *Papio* species, oestrus tends to result in a marked change in the female's relationships, since she usually spends a great deal of time in consort with one or more males: time spent interacting with the male(s) thus increases while time spent interacting with other social partners decreases (see e.g. Hausfater 1975; Kummer 1968; Rasmussen 1980; Seyfarth 1978a). In some cases, the presence of an oestrus female in the troop may cause fighting among the males and can be very disruptive for the troop as a whole (Hausfater 1975; Packer 1979b). Among the gelada, in contrast, oestrus has a minimally disruptive effect, both on the relationship between the male and the female and on the structure of relationships within the unit as a whole (Dunbar 1978). These contrasts between the species are again a consequence of differences in the fundamental ecological and social constraints under which the animals live. Living in one-male groups obviates any risk of immediate competition from other males for access to an oestrus female: instead, gelada males compete only for long-term hegemony over units as a whole and this makes it unnecessary for a male to devote time and energy to guarding oestrus females as closely as *Papio* males do. This difference can be shown to have wide-reaching effects even on the minutiae of behaviour (including for instance which sex is most responsible for initiating sexual interactions and the rate at which copulations occur [for details see Dunbar 1978]).

More generally, perhaps, it is found that the overt (i.e. surface) structure of relationships may change with time (and in some cases the deep structure may also be influenced), simply as a result of maturational changes in the animals themselves, though these may often be compounded by simultaneous demographic change. Thus, older male gelada do seem to be genuinely less diligent in their attempts to interact with their females, over and above the fact that older males tend to have larger units than younger males. The same appears to be true of hamadryas baboons: older males are apparently less vigorous in herding their females and, in consequence, are more prone to lose females as a result of kidnapping (Kummer 1968, pers. comm.).

One last example is of particular interest in this context since it suggests that surface structure can change without any changes occurring in the deep structure rules that govern behaviour. Among gelada females, the frequencies of interaction and coalitionary support during agonistic encounters are related as a step-function rather than as a linear correlation (Dunbar 1980b). I interpret this to imply that close female relatives are still willing to support each other, even though they have little time to spare from grooming their own offspring in which to interact with each other. In other words, it seems that the *fact* of interaction is more important than the actual *amount* of interaction in eliciting coalitionary support. If this interpretation is correct, it implies that these animals are able to maintain relationships in the abstract with minimal social interaction required to service them.

3. Structure and function

I have attempted to show, through examples from my own work on the gelada, how some of the ideas developed in earlier sections can help us understand the social biology of other, very different species. There has also been an attempt to show that in the complexly interwoven biology of the gelada, knowledge of one biological level has been required in order to understand observations at another level. In particular, to achieve an adequate functional explanation of many aspects of gelada society, it has been necessary to delve very deeply into the underlying complex of relationships and, in some cases, even down to the level of specific behavioural interactions and the physiological mechanisms that underlie them.

Had this not been done, the explanations of observed phenomena at the individual and supraindividual levels of society would not merely have been the poorer but would in many cases have been incorrect. One example will serve to underline the point.

During our first, more superficial field study of the gelada in 1971–2, we found that rates of female–female aggression were very low. It was concluded that dominance hierarchies among the females were weak, if not altogether absent, and were unlikely to play an important part in determining female behaviour and relationships (see Dunbar & Dunbar 1975). More detailed investigation of female relationships during the second study in 1974–5 not only revealed that dominance hierarchies were present but also showed tht they were very strictly linear, though maintained on a relatively subtle level requiring little overt aggression. Moreover, these hierarchies turned out to be the key to understanding the functional significance of many of the behavioural phenomena observed in gelada society, including not only female–female relationships but also many aspects of the structure of gelada reproductive units and even the processes whereby males acquire females with whom to breed.

The short lesson is that a proper understanding of observed phenomena can often only be achieved after a very detailed analysis at all relevant levels of the animals' biology has been carried out. It will often be surprising just how many levels turn out to be relevant.

13.3 Interactions, Relationships and Social Structure: the Great Apes

ALEXANDER H. HARCOURT AND KELLY J. STEWART

While it is now obvious that social relationships must be described in terms of social interactions and, in turn, social structure in terms of social relationships, the influence of higher levels on lower, social structure on relationships and relationships on interactions, is not so clear. It is here that interspecific comparison is perhaps particularly useful, because species with one type of social structure can be used to test the precision, validity and especially the generality of rules abstracted from study of a species with another type.

The three species whose description forms the main basis of this book are typical group-living mammals in that they are all 'female resident', i.e. females tend to stay and reproduce in the group of their birth, while males leave and transfer to other groups. In this narrow sense, they thus all share one aspect of social structure. It is, therefore, possible that the approach, and in particular the specific principle abstracted, lack generality. The apes are a useful group with which to test the validity and pervasiveness of some of the rules drawn from study of these other species because their own varied social structures are different from those of female-resident species, both in the narrow sense above and more widely. They thus provide a range of social contexts in which to test the explanatory principles found useful in understanding the social structure and relationships of female-resident species.

The main aim of this section is, therefore, to show how examination of diverse social structures can be important and useful in the testing, refinement and exploration of the generality of principles of social behaviour abstracted largely from study of a few female-resident species. The account concentrates on the interface between social structure and social relationships and examines the causes of different female-residence patterns.

The apes considered are the orang-utan, *Pongo pygmaeus*, the common chimpanzee, *Pan troglodytes*, the pigmy chimpanzee, *Pan paniscus*, and the gorilla, *Gorilla gorilla*. Hylobatids (gibbon and siamang) are not discussed because their monogamous social system is of less heuristic value in this context.

One basic argument behind the conceptual framework of much of this book is that social structure needs to be understood in terms of social relationships. As described by Hinde (1976) and emphasized by Crook (1975) and Altmann and Altmann (1979), social structure is also more than this. It comprises and is influenced by population composition and individual- and group-ranging patterns. A species' social structure cannot, therefore, be fully described or explained without reference to these parameters or to the overall ecological and species' biological variables (such as body size) that control the distribution and movement of individuals within a population.

The orang-utan is the second largest of the primates and is primarily frugivorous and arboreal (Mackinnon 1974; Rijksen 1978; Rodman 1973, 1977). Its food is therefore effectively scarce and widespread. To utilize such a distribution of resources, the orang-utan has become an unusual primate, being essentially solitary (Galdikas 1979; MacKinnon 1974; Rijksen 1978; Rodman 1973, 1977). Except for mothers and their most recent offspring, individuals meet rarely and, when they do, interactions are often aggressive, if the animals bother to interact at all. In fact, so infrequent are most interactions that analysis of orang-utan social relationships lags far behind other aspects of its biology.

This species demonstrates the first level of the framework: relationships consist of interactions and where these are very infrequent, it is doubtful whether anything that could be usefully called a social relationship exists. Nevertheless, even without a description of the network of social relationships in an orang-utan population, it is valid and useful to speak of orang-utan social structure and to compare it with other non-group-living mammals as consisting of small, overlapping female ranges overlaid by larger male ranges in which resident males tend to be dominant to others and through whose ranges younger or subordinate males are transitory.

The orang-utan has shown that with few interactions it is difficult to speak of relationships. The next example concerns the influence of relationships on interactions. In the human being, the nature of a relationship may influence the course of future interactions, for individuals evaluate their relationships and this affects how they behave subsequently (Kelley 1979). A difference in inter-individual relationships between common and pigmy chimpanzees, connected with the species' different grouping patterns, is associated with, and presumably affects, a difference in their interactions. Pigmy chimpanzees have been observed for only a few hundred hours, compared to thousands of hours for the common chimpanzee, but the differences in grouping patterns appear to be great enough to be reliable.

Grouping patterns of the common chimpanzee vary from area to area (Itani 1980; McGrew *et al.* 1981) but in the most-studied populations in Tanzania, males and anoestrus females tend to range apart. Parties that form are unstable but those of males are more common and more stable than parties of males and females or of females (Goodall 1968, 1975; Nishida 1979; Wrangham & Smuts 1980). In the pigmy chimpanzee, by contrast, mixed-sex parties are the most frequent and these seem to be far more stable even than the all-male parties of the common chimpanzee (Kuroda 1979). A consequent difference between the species is that individuals in parties of the pigmy chimpanzee have been in one another's presence for longer and are more familiar with each other than is the case for individuals in parties of the common chimpanzee. Associated with this greater familiarity and stability of party composition and hence presumably greater stability of relationships, is a much reduced level of excitability and tension during interindividual interactions in the pigmy chimpanzee compared to the common chimpanzee (Kuroda 1980). Overt greeting gestures are frequent among common chimpanzees (Goodall 1968) and agonistic behaviour between meeting animals is common and often extreme, as if the animals were testing for changes in their partner after an absence of some time and also attempting to assert themselves (Bygott 1979; Goodall 1968). In the pigmy chimpanzee, however, agonistic interactions seem to be comparatively infrequent and mild and overt greeting gestures are subdued and uncommon (Kuroda 1980). It seems that having more stable groups and knowing one another better, pigmy chimpanzees have less need for overt testing of relationships, self-assertion, and dissolution of tension within the relationship (Kuroda 1980). The transition from unstable relationships and overt interindividual interactions to stable relationships and more subtle interactions is of course a well-known and common one in Western human society as couples progress from courtship to marriage.

Moving from interactions and relationships we come to the interface between relationships and social structure. The following two examples are concerned with showing how the unusual social structures of the common chimpanzee and the gorilla are associated with unusual patterns of social relationships.

The stable, long-lasting and complex relationships of females rather than males in female-resident species are the starting point. Are these due to inherent propensities to form such relationships being present in primate females but not males or are they a consequence of the social context preventing expression of propensities in males that could be shown under other circumstances? (cf. Vaitl 1978).

The common chimpanzee is described as living in communities rather than in groups because males and anoestrus females range separately yet within prescribed areas (e.g. Goodall 1975). Males, however, are the resident sex: they stay in the area of their birth while females normally emigrate to another area and community (Kawanaka & Nishida 1975; Pusey 1980), in contrast to most other group-living mammals where females are the resident sex (Greenwood 1980). Chimpanzee communities are therefore recognized by their male membership in the same way that female membership defines groups of female-resident species. While males in a chimpanzee community often travel alone, they are also often found in parties and it is in these that the similarities to female relationships in female-resident species are most evident.

Frequent affiliative and agonistic interactions, a linear dominance hierarchy and correlation of rank with grooming relationships are characteristic of females in female-resident species. Just these characteristics are also found among chimpanzee males, in contrast to chimpanzee females among whom relationships are, as for the orang-utan, so difficult to recognize that they are almost undescribed (Bygott 1979; Goodall 1975; Simpson 1973). With males in chimpanzee communities and females in female-resident species showing such similarities, it seems likely that social context rather than, or as well as, sex-linked propensities is an important factor in the expression of the behaviours described.

One pervasive characteristic of female-resident species is the great stability of the females' hierarchy, even across a number of generations (e.g. Kawai 1958; Sade 1972b). The dominance hierarchy among male chimpanzees, however, is unstable. As in baboons (Hausfater 1975) and macaques (e.g. Sugiyama 1976), male rank is continually changing (Bygott 1979; Riss & Goodall 1977). Aspects of ecology and functional consequences of high rank among males could explain why the difference between the sexes in the stability of their hierarchy persists across different social structures. Thus, the unstable male hierarchy could be explained by the fact that male chimpanzees often range alone in search of food (Wrangham 1977) and so cannot count on the proximity of supportive relatives, while the instability of any male hierarchy in comparison to female ones could be explained by the fact that males of high status in all polygynous species can achieve comparatively greater reproductive success than can females and therefore face greater and eventually irresistible competition for that high status.

If one looks merely at group and population composition, the gorilla is the ape that most clearly resembles female-resident species: it exists in heterosexual groups that contain a number of immature animals and more adult females than males (Harcourt *et al.* 1981; Schaller 1963), a description that would fit vervets, baboons and macaques. However, only in this superficial view is it similar. In particular, the gorilla, like the chimpanzee, is not a female-resident species (Harcourt 1978b; Harcourt *et al.* 1976). It appears that at least in the most studied population of gorillas, the majority of females leave the group of their birth. One consequence is that adult females in a group are largely unrelated and have not known one another from immaturity.

If female-residence and hence close consanguinity and familiarity from immaturity among females are associated with stable, complex relationships among females, then lack of the above factors (if it is indeed they and not inherent propensities of females that are the cause) should be associated with lack of close relationships among females in species that are female-resident. It has already been shown that this is the case for chimpanzees and the same is true for gorillas: females in gorilla groups have been described as merely tolerating one another (Harcourt 1979a). The strongest bonds among adult gorillas appear not to be between females, as in female-resident species, but between females and the leading male (Harcourt 1979b), the adult with whom the females are probably the most familiar. That this is due to an aspect of the gorilla's peculiar social structure (female emigration) rather than that gorillas operate according to different rules from other primates is indicated by two study groups where daughters stayed with their mothers into adulthood and after parturition. Affinitive interactions were as or more frequent between these related females as between females and the leading male. In other words, the gorilla's normal social system has constrained the expression of certain behaviours among individuals. The rule, be friendly to a familiar adult female, operates in gorillas but is usually irrelevant because emigration of females means that such females are normally not present (cf. Vaitl 1978).

In female-resident species, consanguinity and rank are often inextricably linked as explanations of observed patterns of social relationships because of the inheritance of the mother's rank by daughters. Partly as a consequence and partly as a result of present fashions in biology (Wilson 1975), consanguinity often comes to assume overriding importance as an explanatory principle. However, comparative study of the apes, with their different combination of factors controlling social relationships, can allow a separation of these influences and can therefore provide some insight into their differential importance in the control of social relationships in female-resident species.

Among females in female-resident species, it is found that high-ranking partners give less and receive more grooming than low-ranking ones and also that most grooming is between animals adjacent in rank. Since adjacently ranked animals are often sisters or mother and daughter, it can be difficult to know whether this pattern of grooming is due to competition to groom dominant animals or is simply a function of the frequent grooming of close relatives. However, the fact that the same pattern is found even in captive groups, where females are unlikely to be related, suggests that competition to groom the dominant animal is at least as important a factor as consanguinity (Seyfarth 1977; section 10.2). Grooming in a different sex of a different species of different social structure is used here to emphasize this point.

Simpson (1973) described exactly the same pattern of grooming among male common chimpanzees as reported above for female Cercopithecines. In the chimpanzee, however, closely ranked animals are not necessarily closely related. Thus, among male chimpanzees the frequent grooming of closely ranked animals, along with the predominant direction of grooming up the hierarchy, strongly supports Seyfarth's (1977) suggestion that competition to groom dominant animals is an important factor in the observed pattern of grooming animals among female Cercopithecines.

The above examples have demonstrated how the unusual social systems of the gorilla and common chimpanzee are correlated with unusual patterns of social relationships and in doing so have confirmed and extended findings from female-resident species. The examples have shown how similar principles acting in different social structures produce patterns of social relationships different in themselves or different in the sex in which they are expressed. The final example on the subject of the interaction between social structure and social relationships is chosen to demonstrate the fact that similar propensities, even in different social structures, can nevertheless produce similar patterns of relationships. It concerns grooming by immature females.

In many female-resident primate species, immature females, especially subadults, are characterized by the high number and variety of their grooming partners, in contrast to immature males who by comparison rarely groom anyone except immediate family members (e.g. Cheney 1978). This pattern has been explained functionally by the suggestion that grooming is for future aid, which the resident immature females need and can benefit from but which the immature males who are going to emigrate obviously do not need (Cheney 1978). Nevertheless, exactly the same pattern is found among gorillas (Stewart 1981). If it is correct that female emigration is normal in gorilla populations, then the above functional explanation does not suffice for gorillas. The implications are, therefore, that either such a pattern is due to an inherent difference between the sexes that is of fortuitous advantage to female-resident species but irrelevant to ones in which females emigrate, or that the explanation offered in terms of female residence and male emigration is inadequate. In either case, study of a species with a different basic aspect of social structure has raised questions that might not have been raised if only the more normal female-resident species had been investigated.

It has been suggested here that the emigration of females from the group of their birth is a major factor contributing to the importance of the study of the apes for understanding of the general principles that control the interplay between social structure, relationships and interactions. In the same way that study is useful at these levels of analysis, so investigation of the correlates of female emigration in the apes is of value in testing the explanations offered for female residence in other species (Wrangham 1980). That is, study of the causes of a basal condition (such as female residence patterns) of one type of social structure in one set of species can be used to test hypotheses about the evolution and maintenance of the same condition in another set of species with a different social structure.

While female residence has been explained by factors such as predation and defence of females by males, it is now suggested that it is better explained by benefits accruing from cooperation between relatives in competition for clumped, economically defendable food sources (Wrangham 1980; section 12.2). Support for this hypothesis is provided by the fact that dispersal of females in the common chimpanzee and orang-utan is associated with scattered, economically undefendable resources (Wrangham 1980).

However, by this hypothesis, gorilla females too ought to be dispersed (Wrangham 1979, 1980). A study of why they are not provides support for the contention that factors other than the distribution of food can be important in the determination of the evolution and maintenance of female residence patterns. In addition, the analysis emphasizes the importance of knowledge of the demography and basic biology of species for full explanation of its social structure (cf. section 12.2).

One aspect of gorilla behaviour commented on by all observers is the importance of the adult male in defence of the group against potential or actual predators (e.g. Schaller 1963). Given that the distribution of food should predispose females to disperse, it seems likely that protection by males is an important reason for the association of females with them, even if it is of secondary importance in female resident species (cf. Wrangham 1979; section 12.2).

If, however, it is advantageous for females to associate with a male, and thus necessarily with one another, it seems likely that it would be yet more advantageous to associate with relatives than with non-relatives. The reason why gorillas appear not to could lie in the species' demographic and reproductive biology. Groups are small (with a median of ten) (Harcourt *et al.* 1981) and adult females come into oestrus during three or four months only once every three or four years (Harcourt *et al.* 1980). It is therefore comparatively easy for adult gorilla males to prevent other males entering their group and mating with their females, in contrast to the difficulties faced by baboon, vervet and especially macaque males. As a consequence, gorilla males, once they are associated with a group of females, appear to be able to stay with them for life. Thus, if a daughter is not to mate with her father she is forced to leave. Based largely on a study of female-resident species, Wrangham (1980) suggested that the distribution of primate males is mainly controlled by female distribution, since females (as opposed to food) are the males' most cost-effective resource. In the gorilla by contrast, the opposite seems to be the case: males determine the distribution of females.

In conclusion, it is hoped that it has been shown that, because of the unusual structure of the apes' social systems, their study has enabled confirmation, extension and some modification of the principles found useful in understanding the social behaviour of female-resident species. These principles acting in the non-female-resident apes, especially the gorilla and common chimpanzee, are associated with different patterns of social relationships (although sometimes simply because the appropriate partners are not present), similar patterns but in the opposite sex and similar patterns in the same sex despite the different social structure. Study of the apes has also indicated limitations to generalizations, which while applicable to female-resident species are not so to the apes. Comparison of people with other animals is vitally important to the understanding of our own species. We are, like the gorilla and common chimpanzee, not a female-resident species (Murdock 1957). The implication of this comparison of the apes with the more normal female-resident primates is that only by study of a wide range of species can the limits of general biological rules be found and therefore the validity of the extrapolation to our own species be known.

13.4 Relationships and Social Structure of African Elephants

CYNTHIA J. MOSS and JOYCE H. POOLE

Introduction

In most primate species an individual's relationships rarely extend beyond its natal group or at most to neighbouring groups. Males transfer to other groups and in doing so interact with a greater number of individuls than do females, but in the course of a lifetime even a male primate interacts with a restricted and relatively small number of individuals. While elephant social structure is similar to that of some primate species in that elephants are born into stable groups of related females, an elephant's relationships radiate well beyond the family group through a multitiered network of relationships encompassing a whole population.

The basic unit of elephant social organization is the family unit, which consists of one or more related adult females and their offspring (Buss 1961; Douglas-Hamilton 1972; Laws & Parker 1968; Moss 1977). Females born into a family unit remain there upon reaching sexual maturity, while males leave their families or are forcibly ejected shortly after reaching puberty (Douglas-Hamilton 1972; Laws & Parker 1968; Moss in prep.). Family units are matriarchal in structure and long-lasting bonds are formed among female members. Long-term studies of female–female relations and association patterns in Lake Manyara National Park, Tanzania, and Amboseli National Park, Kenya, reveal levels of organization above that of the family unit consisting of 'kin' or bond groups, clans, subpopulations and populations (Douglas-Hamilton 1972; Moss 1981). Studies of adult male social organization in Serengeti National Park in Tanzania and in Amboseli show that adults males have entirely different patterns of distribution, association and social interactions from those of the females and calves (Croze 1974; Hendrichs 1971; Poole 1982).

In this section, some findings will be described resulting from the Amboseli Elephant Research Project which was started in 1972 and is still in progress. Amboseli National Park covers an area of 390 km² and consists of semi-arid, wooded, bushed and open grassland interspersed with a series of swamps (see Chapter 2). The Park and the surrounding area comprise an ecosystem of 3500 km² of semi-arid savannah in which water availability is highly seasonal (Western 1975). This ecosystem is inhabited by a free-ranging population of elephants presently numbering 615 individuals. All adults and juveniles in the population are identifiable individually with the aid of a photographic recognition file, and young calves can be recognized within the context of their families. In the Amboseli population, family units average 9.4 individuals (with a range of two to

29) and are typically composed of two to three adult females (with a range of one to nine) and their offspring (Moss in prep.). There are 48 such families in Amboseli, accounting for approximately 451 animals. Another 164 animals are independent adult males who have left their natal family units. Males are found singly, in all-male groups and with females. Groups consisting of two or more adult males (with a mean group size of 3.8 and a range of two to 25) are short lived and relatively unstable in composition (Poole 1982). Whether a bull associates with males or females depends both on the bull's age and its sexual state (Poole 1982; Poole & Moss in prep.).

In many areas in East Africa there is a tendency for elephants to aggregate on a seasonal basis (Buss 1971; Douglas-Hamilton 1972; Hendrichs 1971; Leuthold 1976; Lindsay 1982; Moss 1981). In Tsavo, Leuthold (1977a) found that during drier periods family groups utilized small, dry-season home ranges. During the wet season, elephants moved from these ranges into habitats with abundant vegetation where they formed aggregations (Leuthold 1976) and elephants from different dry-season home ranges come into contact with one another (Leuthold 1977a). Radio-collared elephants in Tsavo were seen to travel long distances in response to localized rainfall and subsequent new growth of vegetation (Leuthold & Sale 1973). A similar situation pertains in Amboseli where family groups and bulls concentrate in and around the permanent swamps in the dry season, moving in small restricted ranges (Moss 1981; Poole 1982). Soon after the rains begin, family groups tend to aggregate, forming larger groups with males in attendance and these aggregations move throughout the eco-system (Lindsay 1982; Moss 1981).

Thus, the frequency with which members of the population come into contact and interact with those outside their family unit depends upon rainfall and subsequent food availability. Within this ecologically dependent framework of possibilities exists an intricate network of social relationships. In Amboseli, at some point during the course of a year, each individual associates with most other members of the population but the quality and quantity of their interactions vary with the age, sex, relatedness and sexual state of the individuals. For the purposes of this section, the way in which resource availability affects the patterns of distribution and association will be briefly reviewed and then, within this ecological framework, an attempt will be made to explain the existing social structure in terms of social interactions and relationships.

Distribution and association patterns

Females

The females may be divided into two subpopulations based on the way they utilize the available resources during the dry season (Moss 1977). Members of the 'central subpopulation' concentrate around the swamp system during the dry season, while members of the 'peripheral subpopulation' utilize the surrounding bushland and use the swamps primarily for drinking and bathing. In the wet season, the groups making up these two subpopulations abandon their dry-season ranging patterns and mix freely together, often in large aggregations of over 100 individuals, moving to areas where food is plentiful (Moss 1981). These movements sometimes take the animals as far as 40 km outside the Park boundaries (Croze pers. comm.).

Within each subpopulation a further division can be discerned, based on the home ranges of individual family groups during the dry season (Moss 1977, 1981). Each family returns to the same home range during the drier months (Fig. 13.3). Those groups that use the same dry-season home range have been called 'clans'. The central subpopulation of 27 family units consists of four clans: Longinye, Il-Mberisheri, Ol-odo Are and Southern, consisting of nine, eight, five and five family units respectively. The peripheral subpopulation of 16 family units is still unresolved on the clan level but appears to consist of two to three clans. One other clan which is now considered part of the central sub-population—the Namalog clan—consists of five family units who in the first

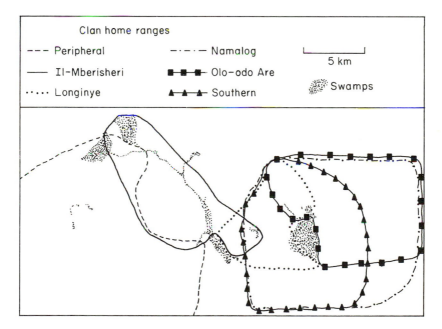

Fig. 13.3. Dry-season home ranges of the Central (Longinye, Il-Mberisheri, Ol-odo Are, Southern and Namalog clans) and the Peripheral subpopulations in the Amboseli Basin.

years of the study were peripheral and then gradually became central, changing their ranging patterns and behaviour. They have moved into the Ol-odo Are and Southern clan ranges. Agonistic behaviour between adult females from different families or different clans is rarely observed but long-term records may show that high-ranking families may be utilizing the optimal habitats during the dry season and excluding the lower-ranking families (Moss in prep.).

Finally, within the clan itself yet another level of social organization is revealed by the association patterns of the family groups. Individual family units spend significantly more time with certain other family units and when together they show particular spatial and behavioural patterns which indicate that they have close social relations (Moss 1981). The family units which are found in frequent association with each other make up what has been called a 'bond group' (Moss 1981). For example the Longinye clan is made up of three bond groups and a single family unit that does not form a bond group with any other family unit. Most family units form a bond group with at least one other family unit and some bond groups are made up of as many as five family units.

The distribution and degree of aggregation of female groups are patterned in response to temporal and spatial fluctuations in food availability. The frequency of social interactions between members of the population are limited during the dry season when groups are smaller, while during the wetter months feeding constraints are lifted and families aggregate. As the dry season progresses, groups of females appear to try to maintain social groupings by moving to areas where the absolute quantity of food is high and evenly distributed (Lindsay 1982).

Males

Among sexually mature males, older individuals spend less time in association with females than do younger males (Poole 1982). The older males (those over the age of 25) show very definite periods of sexual activity and inactivity and for the purposes of this section we concentrate upon the social relationships of these individuals.

An analysis of the ranging patterns of individual males in the older age classes reveal four bull or 'retirement' areas: Ol-keluniet, Ol-tukai Orok, Ol-engia and O-lolarashi (Fig. 13.4). These are used when bulls are sexually inactive, either on their own or in association with other bulls (Poole 1982). During sexually active periods, males leave their retirement areas and search out and associate with females throughout the female range. Each male always returns to the same retirement area after his period of association with females (Poole 1982).

Sexually inactive males in their bull areas may distribute themselves so as to use efficiently seasonally changing food resources, and so that they can choose either to move and feed alone or in association with other bulls. When they are

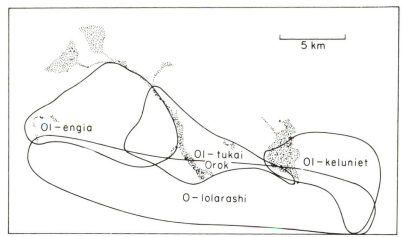

Fig. 13.4. The four bull retirement ares: Ol-engia, O-lolarashi, Ol-keluniet and Ol-tukai Orok. These areas overlap with each other and with the female clan ranges.

sexually active, their movement patterns depend on the degree of female aggregation and mobility and they will travel from group to group in search of oestrus females. In doing so, they will meet with males from other bull areas. The quantity and quality of social interactions between a pair of males and the relationship formed between them will depend upon the social context of these interactions.

Social interactions and relationships

Females

Among primates, a variety of interactions, such as grooming and play, function to maintain and reinforce social bonds between individuals living in relatively permanent groups. In a social system where units aggregate and break down with varying degrees of frequency and duration, one would expect to see the evolution of specific interactions that reflect the importance of reunion to individuals and which maintain and reinforce the types of bonds that these social units share. Among female elephants, reunions are marked by specific greeting behaviour and the form that a greeting takes can be used to indicate the strength of the social bond between individuals (Moss 1981).

All elephants may greet one another but the nature, frequency and intensity of the greeting will vary depending upon the age, sex and relatedness of the elephants involved (Moss 1981). Greetings between some individuals may occur rarely and will involve at most a reaching of trunks into each other's mouths, while other members of the population will frequently greet one another in an excited

performance even when they have been separated for less than an hour
(Moss 1981). Individuals performing an intense greeting ceremony run together
and upon meeting raise their heads and ears, turn and back into one another,
entwine trunks, while often urinating and defecating (see illustration p. 298).
This behaviour is accompanied by deep rumbling, trumpeting, screaming and
loud ear flapping (Moss 1981). With rare exceptions, only those animals
belonging to the same family unit or bond group greet in this manner.

An examination of the association patterns of one family unit in the
population reveals the structure of its relationships (Fig. 13.5). The EB family
unit, a Longinye clan member with ten individuals including three mother–
offspring units, has been sighted by Moss on 254 occasions. The EB family has
been seen in association with every other family unit in the population but
sightings with the majority of the 48 family units are relatively infrequent.
However, the EB family has been seen with each of 19 other units for more than
15% of its total sightings. All of these groups are central subpopulation families
and include at least one family from each central clan. Eighteen of these families
were seen in close association with the EB family on 15.4–29.1% of the
occasions that the EB family was sighted, while one family, EA, was seen with the

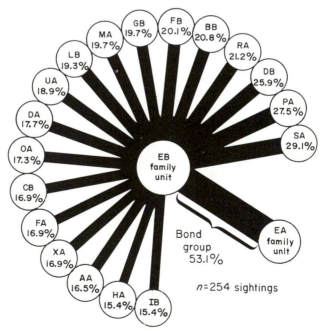

Fig. 13.5. Association patterns of the EB family unit. The family units with whom the EB
family was seen on more than 15% of their total sightings are presented. The EB family
associated with the EA family unit significantly more often than with any other family.
These two families make up a bond group.

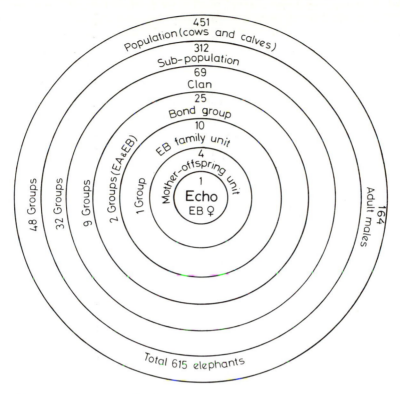

Fig. 13.6. Multitiered network of relationships for the adult female Echo. These relationships change in quality and content as they radiate out through each different level.

EB family on 53.1% of sightings. Not only are the EA and EB families spatially and temporally close but when they are together the strong bonds between the members are manifested by their social interactions. Members of the two units greet each other after separation with the intense greeting ceremony and when together animals from both families mix freely, moving, feeding and resting close together with a coordination of activities. Young animals from the two family units play together and, during periods of rest or comfort activity, individuals from the two families lean and rub against one another. The EB family members were never seen to greet any other family's members in the intense greeting ceremony. Thus, the quality of the EB family members' interactions with the EA family members is substantially different from that with other units. Physiognomic similarities in both ear and tusk shape suggest that the animals in these two family units are related.

Taking a single adult female, Echo, the matriarch of the EB family, as the centre of a network of relationships, it can be seen that she lives in a multitiered social system (Fig. 13.6). These relationships change in content and quality as they radiate out through each level. First are Echo's immediate offspring, second the other adult females in her family and their offspring and third the members of her bond group. These first three levels include the animals with whom she has strong amicable relations, which are renewed through frequent contact and greeting ceremonies. The fourth level includes members of her clan, the fifth her subpopulation and finally the sixth covers the whole population including the other subpopulation and the adult males. Her interactions with the cows and calves on these final three levels will for the most part be brief and perfunctory, with some amicable and some aggressive interactions. The quality and quantity of her interactions with the adult males depend on the male's age and sexual state and her own reproductive state.

Males

Interactions between individuals are context specific and the quality and quantity of interactions between male elephants change dramatically depending upon their sexual state. Sexually inactive males associating with other males exhibit low rates of aggressive interactions, while sexually active males interact aggressively more frequently (Poole 1982).

The older, high-ranking males have a period of heightened sexual activity, known as musth (Poole & Moss 1981), which has been likened to rutting behaviour among ungulates (for *Elephus maximus* see Eisenberg *et al.* 1971). Musth is characterized by copius secretion and swelling of the temporal glands and a continuous discharge of urine (Poole 1982; Poole & Moss 1981). During this period, males have higher levels of urinary testosterone (Poole *et al.* in prep.) and higher rates of aggression than either sexually active or inactive non-musth males (Poole 1982). Musth periods of males are not synchronized but occur throughout the year (Poole & Moss 1981). The relative rank of males in the population changes depending upon the temporal and spatial occurrence of individual males' musth periods (Poole 1982). Consequently, the types of relationships formed between male elephants are highly dependent upon the temporal and spatial occurrence of the musth periods of individual males. Males who only associate with one another when they are with females and are sexually active will form different relationships from males who only associate with one another when they are in all-male groups and are sexually inactive (Fig. 13.7).

Males who share a similar bull area associate with one another frequently in small all-male groups. They greet one another by reaching their trunks into one another's mouths. This greeting is often extended to smelling each other's temporal glands and genitals, presumably to test each other's sexual state. Greetings between males may lead to gentle sparring. Males may rumble to one another but they do not greet in the excited manner exhibited by female bond-group members. Interactions between sexually active males are more aggressive and greetings are rarely observed (Poole 1982).

Most authors (e.g. Croze 1974; Laws & Parker 1968) have stated that males in bull groups associate with other males in a random fashion and that they do not form any long-term bonds with other individuals. It is certainly true that males do not form temporally and spatially stable groups but some individuals do appear to have strong associations with one or more other males. Several pairs of males were observed in association greater than 30% of the total time that each was observed in the company of other males.

Males who form close associations when in bull areas are rarely seen together when in the company of females, while those from different bull areas are unlikely to come into contact until they are sexually active and in association with females (Fig. 13.7). Bulls who come from the same bull area tend to look very similar and these males may be more closely related than they are to bulls from outside their own area. It may be that related males avoid competing with one another by becoming sexually active at different times. Alternatively, high-ranking males may be able to suppress the onset of musth among subordinate males utilizing the same bull area. However, it will be many years before it is possible to say whether relatedness plays a role in determining the types of relationships established between male elephants.

Male elephants live in a complex social world where the interactions and relationships between individuals are dependent upon their use of a large and variable habitat, their age, their sexual state and possibly their degree of relatedness.

Conclusion

Elephant social organization is dynamic. The spatial occurrence and size of social groups is highly flexible and patterned in response to ecological conditions. Relationships among both male and female elephants are based on a hierarchy of social and spatial interactions. Among the females complex greeting ceremonies are found, the nature and frequency of which distinguish between closely associated or usually closely related individuals and those less so (Moss 1981). Family-unit and bond-group members have strong social ties which are maintained throughout the year through frequent contact and greeting ceremonies. Between other clan members social bonds are less strong; although they share the same dry-season home range and come into contact regularly, they do not greet one another in the excited manner exhibited by bond groups. Relationships between individuals who utilize different dry-season home ranges are relatively weak: they come into contact only when ecological conditions permit and when in association they interact infrequently.

Male associations

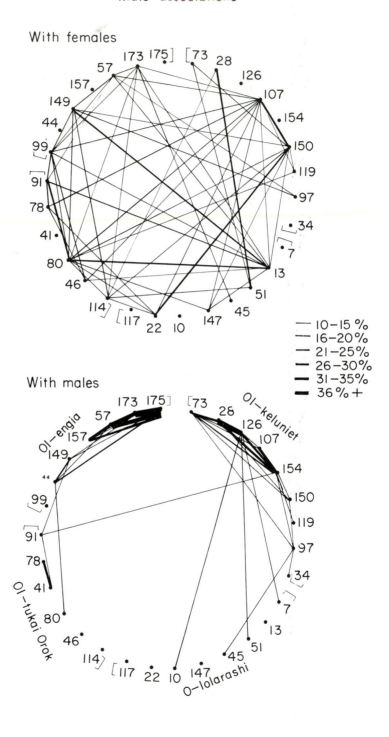

With females

With males

10−15 %
16−20%
21−25%
26−30%
31−35%
36 % +

The spatial separation of bull areas and the temporal separation of sexually active periods means that an individual male will interact with other bulls on two very different levels. Interactions between males in bull groups are relaxed and amiable, while those between sexually active males are competitive and aggressive. The temporal pattern of musth periods and the spatial distribution of musth bulls changes with the season and through time and the dynamic nature of male interactions, relationships and social stucture is a reflection of these fluctuating patterns.

13.5 Social Relationships in Comparative Perspective
RICHARD W. WRANGHAM

Social relationships differ dramatically between species and because they have widespread effects on individual fitness they demand explanation in terms of their adaptive significance. Yet there have been few attempts to incorporate the nature of relationships into a comparative sociobiology. The traditional objects of comparative analysis have been features such as group size, sex ratio, territoriality and the repertoire of displays and social interactions (Wilson 1975). Those features are relatively easily observed and compared and they tend to demon-strate similarities between the social behaviour of primates and other animals. Empirical data on relationships, by contrast, are difficult to obtain but suggest that primates differ from other animals in having more highly differentiated and individualistic social networks (Eisenberg 1973). To clarify such comparisons, primate social relationships are put into a wider context here according to five characteristics: species constancy, individuality, the openness of social net-works, the role of kinship and the complexity of social structure. The focus is on relationships among breeding adults, both because these are comparatively well known and because they have the most influence on the evolution of mating systems and hence on social structure.

Species constancy

Though social relationships vary within and between groups of a given primate species, there are clear differences between species (e.g. sections 13.2 and 13.3).

Fig. 13.7. Association patterns of individual males when they are sexually active (with females) and inactive (with other males). Interactions between males in all male groups are relaxed and amiable, while those between males who are associating with females are competitive and aggressive. The width of the line indicates the proportion of sightings (> 10%) that two males were in association out of the total number of times that both were observed either in all-male or female groups. Males who show strong associations in bull groups are unlikely to be observed together when sexually active.

In other animals, relationships also commonly follow a consistent pattern. At all the sites where African elephants, for example, have been studied, their social structure is broadly similar (section 13.4). Likewise, in four different habitats, acorn woodpeckers, *Melanerpes formicivorus*, live in groups containing up to 12 adults who cooperate in defending year-round territories (Emlen 1982).

Intraspecific variation does occur in species other than primates however, and some cases leave primate relationships looking comparatively conservative. Acorn woodpeckers provide an example. On one occasion, more than half the adults at one site switched to breeding in simple male–female pairs rather than in communal groups and did not maintain the stable long-term bonds of their species. A temporary shortage of food appeared responsible (Stacey & Bock 1978). The most extreme examples of natural variation in primate social relationships, by contrast, concern the number of individuals per group rather than whether groups are temporary or permanent (e.g. Yoshiba 1968). A second kind of variation is shown by species with two or more radically different types of reproductive strategy within a given age/sex class in the same social network, such as occurs among male ruffs, *Philomachus pugnax*, (Rubenstein 1980). Provided such variation is taken into account, species can usefully be compared, with the variation itself as a dimension of comparative analysis.

Individuality

In species where individual recognition is not apparent, personal relationships are confined to a set of interactions between temporary associates. Such brief relationships may be reproductively significant. For instance a male dragon-fly, *Hetaerina vulnerata*, remains with his mate after copulation and defends her from other males even when she flies to adjacent male territories (Alcock 1982), and in frogs male–female pairs may remain in amplexus for as long as 12 days (Wittenberger & Tilson 1980). Even if they are important, however, temporary relationships do not allow the development of complex social networks. The mechanisms by which animals recognize each other are in general poorly known but learning is a major component (Holmes & Sherman 1983). This suggests that species differences in the degree of differentiation of relationships are associated with differences in learning abilities.

In mammals and birds there is much observational and some experimental evidence of individual recognition between parents and offspring, siblings and unrelated peers (Beer 1970; Harrington 1976; Holmes & Sherman 1983; McArthur 1982; Porter & Wyrick 1979; Radesäter 1976). Though individual relationships can therefore occur, the development of complex social networks may be subject to at least two kinds of constraint. First, there could be limits to the number of other individuals that can be recognized. This might contribute to explaining, for example, why birds tend to have smaller closed groups than primates (though other explanations are certainly important). Second, within groups of a given size, learning abilities may influence complexity. Cheney and Seyfarth (1980) found evidence that vervet monkeys recognized not just other individuals but also relationships between others. This ability is almost certain to differ between species, with important consequences.

Evidence of relationships is limited for other animals and comes from three main sources. First, monogamous pairs are found among a few fish, crustaceans and insects (Wittenberger & Tilson 1980). This implies recognition of the partner but it is possible that mates stay together because they are attracted to a common environmental stimulus rather than to each other. Second, dominance hierarchies are known in various lizards, amphibians, fish and social insects (Wilson 1975). Again, their significance is uncertain because in theory they could be established by mechanisms other than individual recognition: for instance a defeated combatant might become subordinate to all individuals of the same size as his or her superior. Third, recognition has been established in some species but as yet only at class level: for instance tadpoles of *Rana cascadae* associate more with siblings than with non-siblings, even when reared with both (O'Hara & Blaustein 1981). Similarly, many social insects recognize members of their own colonies (Hölldobler & Michener 1980; Ross & Gamboa 1981). Differentiated social relationships can therefore occur but they are known to be personal only when a class is represented by a single individual, such as a monogynous queen, and even here the relationship is impersonal from her point of view.

How open is the social network?

Whether or not relationships are personal, social networks can be either closed or open. Most primates live in closed groups. Within these, members are linked by a continuous series of affiliative relationships. Relationships between individuals in different groups, by contrast, exist primarily at the group level and only to a lesser extent from individual to individual (Chapter 11). Closed groups are less common in other animals, though many cases probably remain to be described.

Monogamous pairs occur in about 5% of mammals (Kleiman 1977) and represent the simplest form of the closed group. In a few species, a proportion of offspring remain temporarily with their parents as non-breeding adults and assist in the rearing of young (e.g. Macdonald 1979b). Larger closed groups with several breeding females are known mainly among carnivores, rodents, ungulates and bats (Bradbury & Vehrencamp 1976; Kleiman & Eisenberg 1973; Leuthold 1977b; McCracken & Bradbury 1977; Michener 1982; Rood 1978). Many mammals, however, live in open groups between which individuals move either with complete independence, as in female elephant seals, *Mirounga leonina,*

(McCann 1982) or tend to travel with particular companions, as in female red deer, *Cervus elaphus*, (Clutton-Brock *et al.* 1982). Males commonly defend individual territories in such species but females and young need not be confined to a single territory (e.g. Leuthold 1977b). Primates appear to be unique in the degree to which social boundaries are emphasized, with the carnivores coming closest in this respect.

About 90% of birds breed in monogamous pairs (Lack 1968). Many of these spend the non-breeding season in open groups where there is no evidence of long-term personal relationships, though in some species individuals form stable winter flocks within which there are linear dominance hierarchies (e.g. black-capped chickadees, *Parus atricapillus:* Smith 1976). More permanent closed groups occur in at least nine orders (Brown 1978). Typical groups have fewer than 20 members and live on a cooperatively defended territory. Most species differ from primates in having only one pair of breeding individuals but in at least nine species two or more females breed in the same group. In several of these, as in other communally breeding birds, there is a second important difference from primates: a large proportion of adults not only forego breeding attempts but also provide direct assistance to others by feeding and guarding their young. These differences aside, communally breeding birds show clear parallels to primates in having closed groups with differentiated relationships within them (see references in Brown 1978).

Such closed groups are rare in other animals. The sporadic cases of monogamy in fish and invertebrates (above) provide the clearest examples, especially where helpers aid the breeding pair, as in the anemone fish, *Amphiprion akallopsis*, (Fricke 1980). The only large closed groups are in the social insects, where they are widespread but lack personal relationships (Wilson 1971). All social insects aid colony members and in a proportion of species outsiders are attacked, killed or even 'enslaved' (Hölldobler 1976). Like birds but unlike mammals, species with hostile intergroup relationships tend to be those with only one breeding female (Hölldobler & Wilson 1977).

Role of kinship

Two questions are relevant to understanding the role of kinship in social networks. First, Do kin associate? and second, What kinds of relationship do associating kin have?

So far as breeding adults are concerned there are no cases of systematic association of close kin between sexes, though isolated instances have been described (section 9.5). The tendency for kin to associate within sexes varies widely, however. Some closed groups of mammals and birds are composed primarily of unrelated adults, with young of both sexes dispersing from their natal groups (e.g. mountain gorillas: Harcourt & Stewart 1981; plains zebra, *Equus burchelli*: Klingel 1972; Trinidad fruit bat, *Phyllostomus hastatus*: McCracken & Bradbury 1977; Mexican jay, *Aphelocoma ultramarina*: Brown & Brown 1981). Others consistently show natal residence by one sex and dispersal by the other, e.g. female kin live together within clans of spotted hyaenas, *Crocuta crocuta*, and coteries of prairie dogs, *Cynomys ludovicianus*, (Kruuk 1972; Michener 1982), whereas male kin associate, while females disperse, in groups of African wild dogs, *Lycaon pictus*, and babblers, *Turdoides* spp. (Frame *et al.* 1979; Greenwood 1980). Lions, *Panthera leo*, show association by adult kin of both sexes, though in different groups: related females remain in their natal ranges, while groups of related males sometimes disperse together and jointly enter another pride (Bygott *et al.* 1979).

There are no easy explanations of such differences (Greenwood 1980) but one consequence is clear. Markedly different patterns of kin association can occur in superficially similar groups. This is also true of primates, where species differences in kin association patterns are reflected in their social relationships. For instance in species with female kin groups and male dispersal, females groom each other more than males do and play an active role in group leadership, whereas males groom more and tend to lead groups in species with male kin groups and female dispersal (Wrangham 1980). This suggests that in other animals social relationships may again reflect patterns of kin association.

Unfortunately, this is difficult to test, partly for the lack of observations, partly because some species have fewer types of interactions than primates and partly because rates of interactions are often low: for instance among spotted hyaenas, where a primate model would predict more grooming among females than among males, no allo-grooming occurs (Kruuk 1972). Nevertheless, associating individuals often appear to have more cooperative relationships when they are kin than non-kin, e.g. female participation in intergroup interactions is greater in species with female kin groups (e.g. spotted hyaenas, lions, prairie dogs) than in those with females who are unrelated to each other (e.g. plains zebra, Trinidad fruit bat) (Kruuk 1972). Exploration of such patterns is an important task and already it is clear that the rules are not simple. Female wild dogs, for example, occasionally associate and breed together (Frame *et al.* 1979). Cohabiting bitches tend to be sisters but their relationships are nevertheless tense and hostile, unlike those between female lions or hyaenas.

Care-giving relationships also illustrate that social and kinship patterns are not always well correlated. Sometimes they are, as in certain cases of suckling another's offspring (elephants: Douglas-Hamilton & Douglas-Hamilton 1975; lions: Bertram 1976) and guarding another's young (foxes, *Vulpes vulpes*: Macdonald 1979b; black-backed jackals, *Canis mesomelas*: Moehlman 1979). Even such clear examples of altruism, however, need not be based on kinship.

Thus, in several bats there is evidence of indiscriminate communal suckling (Vehrencamp 1979) and in the dwarf mongoose, *Helogale parvula*, offspring are often guarded by young females, including unrelated immigrants (Rood 1978).

Discrimination between individuals of differing degrees of relatedness has been studied little outside primates but three patterns have been described. Some species, such as elephants, show the macaque, baboon and vervet pattern of graded and more affiliative relationships with closer kin. In others, cooperative relationships are confined to the closest kin. This occurs among Belding's ground squirrels, *Spermophilus beldingi*, where helping between adult females occurs only in mother–daughter and sister–sister pairs and not in aunt–niece or first-cousin pairs (Sherman 1980). Third, relationships within groups may show little evidence of being differentiated. For example among females within lion prides dominance interactions have unpredictable outcomes which are dependent on the context and state of each participant, and no systematic relationships have been described (Bertram 1978b; Schaller 1972). Caution is required here, however, because under appropriate conditions normally covert relationships might be openly expressed. There are few pedigree data available for group-living birds, though preliminary information on the Mexican jay suggest that individuals tend to aid relatives more than others (Brown & Brown 1981).

Complexity of social structure

Cooperation is a prerequisite for the evolution of groups but most relationships also have competitive elements with a consequent tension between cooperative and competitive strategies. This is one of the factors responsible for the complexity of primate social structure, not only in generating dominance relationships but also in thereby allowing the interweaving of dominance and other relationships, such as grooming (section 10.2). As yet, there are few studies which match the wealth of primate data in this respect.

In some species there are few or no dominance interactions. Reviewing studies of prairie dogs, Michener (1982) concluded that within coteries there are no dominance relationships among adult females, juveniles or between females and the harem male. Similarly, within closed groups of jungle babblers, *Turdoides striatus*, Gaston (1977) found so little aggressive behaviour that no dominance relationships could be described, despite frequent interactions of other types, such as allo-preening. In both cases, the lack of aggression within groups contrasted with the fact that group members actively defended their shared home-range boundary against neighbours. African wild dogs provide an opposite example. In this species separate linear hierarchies occur among adult males and adult females, with higher-ranking dogs tending to monopolize breeding, to have high rates of social interaction and to direct pack movements (Frame *et al.* 1979).

Groove-billed anis, *Crotophaga sulcirostris*, provide a rare case of birds living in closed groups with several breeding females: females share a single nest but dominants tend to lay later than subordinates, removing previously laid eggs (Vehrencamp 1978). Dominance hierarchies are also widespread in species which live in open groups, but because cooperation is not a necessary part of their social structure their effects can be simple. Dominance relationships within closed groups, by contrast, potentially interact with affiliative relationships. African wild dogs play together and beg from each other for food but how such relationships interact with each other or with dominance has not yet been described.

Perhaps the most important relationships in generating social complexity are alliances. In primates, temporary coalitions make the outcome of dominance interactions less predictable, while the need for long-term allies appears partly responsible for the pattern of affiliative relationships such as grooming (section 10.2). Alliances within groups are probably a universal feature of primate societies, but in other animals strikingly few cases have been described (Wilson 1975). Coalitions are best known from social carnivores but even here descriptions are largely anecdotal and their importance is little understood (Kruuk 1972; Rudnai 1973; Zimen 1975). As a result, no descriptions of other animal societies yet compare with primate groups with respect to the complexity with which different kinds of relationships interact or in the extent to which relationships with some individuals affect relationships with others.

Significance of species differences

This brief survey illustrates the diversity of animal social relationships and how much remains to be learned. Already, however, comparison with other species shows clearly that primates are unusual. Thus, macaques, savannah baboons and vervets live in groups of several breeding females, with strikingly differentiated relationships within groups, marked social boundaries, frequent matching of kinship and social relationships and a complex social network in which alliances play an important role. Though none of these features is unique, the combination is rare. By including other characteristics of social organization, in particular whether or not the social group forages as a unit, female-bonded groups of monkeys prove to share a type of social structure unknown in other animals. This is shown in Table 13.1, which summarizes information from a selected set of mammals and birds living in complex societies. No female ungulates, for example, have yet been reported to form alliances against other females, and though social carnivores, such as lions and spotted hyaenas, show evidence of female alliances, individuals in the same social network travel independently.

Why should a number of different primates share a common type of social

Table 13.1. Species are selected to show social systems similar to those of female-bonded primates such as rhesus macaques. The listed characteristics refer to closed bisexual groups for all species except impala and elephant, where female groups have only temporary associations with particular males.

	More than one breeding female	Females tend to be related to each other	Female alliances versus females in the same group	Males normally unrelated to each other	Group forages as a unit	Groups mutually hostile	References
Macaques, savannah baboons, vervets	Yes	Yes	Yes	Yes	Yes	Yes	This volume
Spotted hyaena*	Yes	Yes	Rare	Probably	No	Yes	Kruuk 1972
Lion	Yes	Yes	Rare	No	No	Yes	Rudnai 1973; Schaller 1972
Wolf, African wild dog	Rarely	No	No	No	Yes	Yes	Frame et al. 1979; Mech 1970; Zimen 1975
African buffalo	Yes	Yes	No	Uncertain	Yes	Avoidance	Sinclair 1977
Plains zebra	Yes	No	No	Only one male	Yes	No	Klingel 1972
Many ungulates (e.g. impala)	Yes	Probably	No	Males not in closed groups	Yes	No	Murray 1981

				Males not in closed groups			
Elephant	Yes	Yes	No	Yes	Yes	No	Moss & Poole section 13.4
Prairie dog	Yes	Yes	No	Only one male	No	Yes	Michener 1982
Most group-living birds	No	No	No	No/variable	Variable	Yes	Brown 1978
Groove-billed ani	Yes	Unknown	No	Unknown	No	Yes	Vehrencamp 1978
Mexican jay	Yes	Variable	No	Variable	No	Yes	Brown & Brown 1981

*Though spotted hyaenas live in female-bonded clans in areas of high population density, their social system appears to be variable: at low or medium density clans may not occur (Kruuk 1972).

system not found in other animals? Phylogeny and ecology are the two major sources of adaptive explanations for species differences in social evolution and each has its merits here. The phylogenetic argument would point to the importance of characteristics such as unusual cognitive abilities in allowing primate societies to evolve in special ways. Thus Eisenberg (1973) suggested that primate relationships are more complex than in most animals because individuals are long lived and have a highly developed capacity for individual recognition. Ecological explanations, on the other hand, would suggest that social similarities stem from a common set of foraging or antipredator adaptations absent in other animals. Thus, it has been argued that the ultimate factors determining primate social structure depend on the primates' peculiar food distribution, which is itself seen to be a result of idiosyncratic abilities to harvest particular food types (section 12.12).

Comparison between species in different taxonomic groups will contribute to refining and integrating such explanations. Effective comparisons depend on an appropriate choice of variables and in this respect the theoretical frameworks developed by Trivers (1972), Emlen and Oring (1977) and Bradbury and Vehrencamp (1977) are particularly helpful. They lead to the suggestion that analysis should seek to describe the three sets of adult relationships in turn: those between females, between females and males and between males. These relationships need to be classified as simply as possible, e.g. are they invariably aggressive, competitive within a closed social network or non-aggressive within fluid groups, and are they with kin? This approach can classify species in ways which reveal common functional principles underlying the evolution of their social relationships: for instance the fact that lions, hyaenas and many monkeys form female-bonded kin groups with hostile intergroup relationships suggests that female relationships in these species are adapted to cooperative defence of large food sources (Wrangham 1982). When such hypotheses have been properly tested, subtler characteristics of social relationships will be amenable to comparative analysis.

13.6 The Human Species
ROBERT A. HINDE

It is proper to ask whether concepts and methods devised for the study of non-human species could be useful also in the human case. Our capacities for cognitive functioning, particularly for language, introduce new dimensions into behaviour, including cultural diversity of a different order from that encountered among other animals. We can plan ahead, talk about what we will do, assess whether we did what we intended to do and we are aware of the facts of birth and death. Of course, the limits of the cognitive capacities of animals are not yet known. It is now recognized that reaction against the excessive anthropormorphism of early workers and the misguided use of tests of cognitive functioning inappropriate for the environment to which the species had been adapted, has led scientists to underestimate the capacities animals can show in appropriate contexts and the flexibility of their responses in different contexts (sections 9.1 and 10.2). Nevertheless, even though students of primate behaviour are finding it necessary constantly to guard against the over-simplistic bias thus introduced, the principle of economy of hypothesis has required us not to invoke mechanisms more complex than the facts demand. The conceptual scheme used in this book has intentionally been rooted in observable behaviour: while the interpretations of the data may in due course be found to have erred in the direction of simplicity, yet the stripping down to basics which the objective analysis of behaviour requires provides a scheme which can then be transformed by the incorporation of specified complexities as they are found necessary.

Consider first the basic divisions between behaviour, interaction, relationships and social structure. The distinctions are clearly equally applicable to human behaviour but in each case cognizance of complexities known to be important in the human case must be taken. While much animal behaviour is clearly directed towards well-defined goals in a manner compatible with the current social situation, there are, so far as is known, no restrictions of the type imposed by group norms in human society. The interactions of non-human primates demonstrate great cognitive complexity (e.g. Bachman & Kummer 1980; Goodall 1968; Menzel 1971) but there is no reason to suppose that they contain simple parallels to the complex manner in which humans select from their repertoire of possible actions those most fitted to their assessment of whoever they are interacting with and their expectations of the encounter (McCall 1970; though see de Waal 1982).

As some of the contributions to this book have demonstrated, non-human primates form interindividual relationships of considerable durability. It has been argued elsewhere that their study can open our eyes to the nature of properties, such as 'controllingness' and 'permissiveness', that are emergent from the patterning of the interactions and are equally or more important in the human case (Hinde 1979a). However, there are also properties of human interpersonal relationships whose importance depends upon human capacities to reflect upon and evaluate relationships which are only doubtfully applicable to non-human species. In considering a human relationship, it can properly be asked whether A sees B as B really is, whether A sees B as B sees B (i.e. does A understand B?) and whether B believes that A sees B as B sees B (i.e. does B feel understood?). While the first of these questions is clearly applicable to non-human species, the following ones are more doubtfully so. Similarly, while the degree of intimacy

(e.g. Altman & Taylor 1973) shared by a human couple, the degree of commitment each feels towards the other (e.g. Lund 1981) and the extent to which each
believes in the others' commitment are all important properties of human
relationships, their applicability to animals remains to be explored (Hinde
1979a).

The social structures of human societies, like those of animals, involve the
patterning of dyadic (and higher-order) relationships. Yet while monkeys may
have some conception of relative dominance rank and behave to other individuals
in ways that differ according to their degree of relatedness, there is no evidence
that they are conscious of their roles in society and of the rights and duties
imposed upon them by virtue of their position within it, in the way in which human
beings are. Thus, humans strive to behave as spouses, parents, children, students
and teachers are supposed to behave in the particular group and subgroup to
which they belong. It is extremely doubtful whether the females represented in
Fig. 1.1 have any conception of the overall structure: indeed some people would
argue that it does not even make sense to ask the question (but see pp. 189–90).
Yet, as members of a society, human beings not only have some idea of the
manner in which the several roles they fill contribute towards the nature of their
society but they also have views about what that society ought to be like which they
try to realize.

The principles that have so far been used to account for the frequency and
patterning of interactions in the relationships of non-human primates are limited
in number, though those that have been used (related to age, sex, status,
blood-relatedness, familiarity) are also clearly applicable to the human species. It
seems likely that further understanding may be gained if primatologists are
willing to borrow from the social scientists and apply the approaches of exchange
and interdependence theorists (e.g. Homans 1961; Kelley 1979; Thibaut &
Kelley 1959). Thus, in the human being 'social approval' is given in the
expectation of future rewards, and grooming appears to play an analagous role in
some primate relationships (Cheney 1978b; Hinde & Stevenson-Hinde 1976;
Seyfarth 1977, 1980; see sections 12.1 and 12.2). The role of a *conscious*
calculation of costs relative to expected gains is almost equally dubious in both
cases. If this line of approach proves profitable, there may even be some returns
for the social sciences. For instance the difficult problem of what is considered to
be fair in interpersonal exchange (e.g. Lerner *et al.* 1976) also arises in non-
human species: does dominant A groom subordinate B less than vice versa
because A's 'investments' (i.e. what he is invested with—age, maleness,
dominance) are greater?

Some other principles (e.g. attribution, balance) useful for understanding the
dynamics of human relationships are perhaps less applicable to non-human
species but the utility of the concept of feedback in describing the dynamics of
relationships is as clearly shown in the relatively simple rhesus mother–infant

relationship as in any human case (sections 6.2 and 6.3).

In understanding human social structure the concept of institution is of course crucial. The monarchy is an institution, the king and queen roles which are filled by different incumbents from time to time. Marriage, the Church, the city council and so on are all institutions containing roles filled by different individuals as time goes on. Social scientists sometimes use the term 'institution' to refer to something actually observed, such as the common features of a number of marital relationships or a number of village councils, and sometimes to refer to an idealized pattern of relationships which the incumbents strive to achieve, when it becomes a principle with which to explain observed data. While the concept is dubiously applicable in the data-language sense to non-human primates, it is extremely unlikely that it is useful in the latter, theory-language sense.

There is no need here to spell out the manner in which studies of the dialectics between individual characteristics and relationships or between relationships and social structure in monkeys complement those in people. It will be apparent that many of the methods and concepts used in Chapters 4–7 have been taken from human work and studies of monkeys have been used to throw light on the dynamics of human relationships (e.g. Hinde 1979a, 1983). In particular, mother–infant and peer–peer relationships in non-human primates have many of the same properties (e.g. reciprocity versus complementarity, meshing) as early human relationships, and studies of non-human primates can illuminate the complex ways in which the motivations of an individual are modified and directed during development through harmonization and conflict with those of others (e.g. Bowlby 1969, 1973; Hinde 1983; Konner 1976; Chapter 8). Of course, in understanding the focuses of human socialization in the narrow sense of the acquisition of the norms and values of the group, studies of non-human primates are unlikely to be of use. However, therein lies their strength: it is precisely because the relationships and social structure of monkeys and apes are unclouded by the complexity of the human case that they provide such good material for studying some of the basic principles in the dynamics of social behaviour. Similar considerations apply to the study of triadic and higher-order relationships (Chapter 9) and to the study of social groups (Chapter 10). There are many problems in common between monkey and human, but to us the knowledge that we do or do not belong to a group has ramifying consequences many of which are unlikely to be important in non-verbal species (Tajfel 1978).

The issue about which most controversy has arisen concerns the relevance of questions of ultimate causation for the human case. It is often argued that our social environment has changed so much more rapidly than our genetic constitution, that our cultural environment gives us an independence from our biological environment and that the plasticity of our behaviour confers on us the ability to flourish in almost any environment. While all this is true, the conclusions drawn are often based on the simplistic assumption that behaviour is either immutably determined or infinitely modifiable by learning. The fallacy of this view has become apparent during the last few decades. All animals, including ourselves, are constrained in what they can learn: they have predispositions to learn some things and not others (Hinde & Stevenson-Hinde 1973; Seligman & Hager 1972). These constraints and predispositions, which differ between species, have presumably been shaped by adaptive forces in the course of evolution.

It is not known exactly what those adaptive forces were because there is no direct evidence about the physical and social conditions under which human behavioural propensities were shaped by evolutionary forces. However, anthropological, archaeological (e.g. Campbell 1979; Isaac 1978; Lee & DeVore 1976) and comparative anatomical and physiological (Alexander & Noonan 1979; Short 1979) evidence point to the importance of the long period when our ancestors were hunter-gatherers, probably living in a society in which males competed for females, where there was a polygynous or serially monogamous mating system and where both parents played some role in the nurturing of the offspring, with sex contributing to the maintenance of pair bonds. This probably constituted our 'environment of evolutionary adaptedness'.

The value of these studies must not be overemphasized. They provide considerable scope for unscientific speculation and post hoc rationalization and little solid basis for prediction: for example at present, they seem unlikely to be useful in predicting the relevant factors in an aetiological enquiry. Yet they do have a value in the perspective they provide on our nature and on the necessity for recognizing that there are limits to our plasticity. There is space here for only a few examples to illustrate this view.

1   Children tend to show fear responses when alone, if they feel themselves falling or are in the dark. Since fear was shown even when such events had not previously been followed by harmful sequelae, these were formerly labelled 'irrational fears of childhood'. However, such fears would have made good sense in an environment in which predators abounded and separation from the parent would have meant almost certain death (Bowlby 1969). They are, in fact, closely similar to those shown by the young of all non-human primates.

2   Mammalian milk varies in composition between species. It tends to be more concentrated in those species whose young suckle infrequently. Human milk is rather dilute, implying that human newborns are adapted to be fed more often than on the four-hourly schedules imposed in many maternity hospitals (Blurton Jones 1972). Breast milk is also less concentrated than most bottle milk and the behaviour of breast- and bottle-fed babies confirms this view (Bernal 1972).

3   It has been shown that, in animal species, natural selection operates so that mothers are likely to promote their infant's independence before, other things being equal, the latter would seek it on their own (section 6.1). This is in harmony with the way in which human mothers do in fact tend to move their baby from

breast to solid food, from cot to bed or from home to school before the child, left to its own devices, would make these moves. However, it must be remembered that natural selection must have acted both on mothers and on babies: if mothers are adapted to promote the independence of their offspring, the latter will be adapted to cope with mothers who promote their independence.

4   The biological data bring no clear recipe for 'good' mothering but merely support the view that mothers should be sensitive to the baby's needs. Indeed, one of the lessons that studies of adaptation clearly show is that what is 'best' usually depends on circumstances. It may be better to be at the top of a dominance hierarchy, but an individual who finds himself at the bottom may do better to bide his time than to challenge the leader against insuperable odds. By a similar token, prescriptions for how mothers 'should' behave under different circumstances are unlikely to be ubiquitously applicable (see also Chapter 12).

5   Many of the characteristics of human marriage systems, which have considerable cross-cultural generality, can be understood in terms of inclusive fitness given a basically polygynous mating system (e.g. Wilson 1978). Thus, in all societies outbreeding is ensured by the movement of one or other sex. Unlike most non-human primates (but see section 13.3), it is usually the women who move, the males controlling their movement by kinship systems. The double standard about virginity on marriage and adultery during marriage can be related to the fact that a woman must know that a child is hers but a man can be cuckolded. The fact that men generally marry at an older age than women is presumably related to the difference in reproductive potential: a young virgin has the capacity for producing more offspring than an older woman but an older man may have greater protective expertise to offer.

Examples such as these could be multiplied and show that an evolutionary approach can be valuable in integrating diverse facts about the near-universals of human behaviour. A second stage involves the search for evidence that human propensities, as deployed in *particular* cultures, are adaptive to the individuals concerned in those cultures. For example Irons (1979) has shown that the conscious goal of achieving wealth is, among the Turkman of Persia, associated with inclusive fitness in a manner which appears not to be due solely to the immediate effects of wealth on individual health and survival. Again Chagnon and Bugos (1979) showed that who helped whom in an axe fight in a Yanamamö village was closely associated with their genetic relatedness. Individuals were willing to risk injury to help their kin.

Yet another stage in this line of thought involves considerations of how cultures themselves may evolve in ways that are conducive to increases in the inclusive fitness of at least some of the participating individuals and of how such cultural evolution may in turn have genetic repercussions. These issues have been considered in some detail by Wilson and Lumsden (1981) with the help of mathematical models which have at least intermittent contact with reality.

References

Abegglen J-J. (1976) On Socialization in Hamadryas Baboons. PhD Thesis, University of Zurich.

Alcock J. (1982) Post-copulatory mate-guarding by males of the damselfly *Hetaerina vulnerata* Selys (Odonata: Calopterygidae). *Anim. Behav.* **30**, 99–107.

Aldrich-Blake F.P.G. (1970) Problems of social structure in forest monkeys. In *Social Behaviour in Birds and Mammals*, ed. Crook J.H. Academic Press, New York.

Aldrich-Blake F.P.G., Dunn T.K., Dunbar R.I.M. & Headley P.M. (1971) Observations on baboons, *Papio anubis*, in an arid region in Ethiopia. *Folia primatol.* **15**, 1–35.

Alexander R.D. & Borgia G. (1978) Group selection, altruism and the levels of organization of life. *Ann. Rev. ecol. Syst.* **9**, 449–74.

Alexander R.D. & Noonan K.M. (1979) Concealment of ovulation, parental care and human social evolution. In *Evolutionary Biology and Human Social Behaviour: an Anthropological Perspective*, eds Chagnon N.A. & Irons W. Duxbury Press, North Scituate, Massachusetts.

Ali R. (1981) The Ecology and Social Behaviour of the Agastyamalai Bonnet Macaque *(Macaca radiata diluta)*. PhD Thesis, University of Bristol.

Altman I. & Taylor D.A. (1973) *Social Penetration*. Holt, Rinehart & Winston, New York.

Altmann J. (1974) Observational study of behaviour: sampling methods. *Behavior*, **49**, 227–65.

Altmann J. (1978) Infant development in yellow baboons. In *The Development of Behaviour*, eds Burghardt G. & Beckoff M. Garland Publishing, New York.

Altmann J. (1979) Age-cohorts as paternal sibships. *Behav. Ecol. Sociobiol.* **6**, 161–9.

Altmann·J. (1980) *Baboon Mothers and Infants*. Harvard University Press, Cambridge, Massachusetts.

Altmann J., Altmann S.A., Hausfater G. & McCuskey S.A. (1977) Life history of yellow baboons: Physical development, reproductive parameters, and infant mortality. *Primates*, **18**(2) 315–30.

Altmann S.A. (1962) A field study of the sociobiology of rhesus monkeys, *Macaca mulatta*. *Ann. N.Y. Acad. Sci.* **102**, 338–435.

Altmann S.A. (1979) Baboon progressions: order or chaos? A study of one-dimensional group geometry. *Anim. Behav.* **27**, 46–80.

Altmann S.A. (1981) Dominance relations: the cheshire cat's grin? *Behav. Brain Sci.* **4**, 430–1.

Altmann S.A. & Altmann J. (1970) *Baboon Ecology: African Field Research*. University of Chicago Press, Chicago.

Altmann S.A. & Altmann J. (1979) Demographic constraints on behaviour and social organisation. In *Primate Ecology and Human Origins*, eds Bernstein I.S. & Smith E.O. Garland Publishing, New York.

Altmann S.A. & Walters J. (1978) Book review. *Man*, **13**, 324–5.

Anderson C.M. (1981) Subtrooping in a chacma baboon *(Papio ursinus)* population. *Primates*, **22**, 445–8.

Anderson C.O. & Mason W.A. (1974) Early experience and complexity of social organization in groups of young rhesus monkeys *(Macaca mulatta)*. *J. Comp. Physiol. Psychol.* **87**(4), 681–90.

Anderson D.M. & Simpson M.J.A. (1979) Breeding performance of a captive colony of rhesus macaques. *Lab. Anim.* **13**, 275–81.

Angst W. & Thommen D. (1977) New data and a discussion of infant killing in Old World monkeys and apes. *Folia primatol.* **27**(3), 198–229.

Arabie P., Boorman S.A. & Levitt P.R. (1978) Constructing blockmodels: how and why. *J. math. Psychol.* **17**, 21–63.

de Assumpcao T. & Deag J.M. (1979) Attention structure in monkeys. *Folia primatol.* **31**, 285–300.

Axelrod R. & Hamilton W.D. (1981) The evolution of cooperation. *Science, N.Y.* **211**, 1390–6.

Azuma S. (1973) Acquisition and propagation of food habits in a troop of Japanese monkeys. In *Behavioral Regulators of Behavior in Primates*, ed. Carpenter C.R. Bucknell University Press, Lewisburg, Pensylvania.

Bachmann C. & Kummer H. (1980) Male assessment of female choice in hamadryas baboons. *Behav. Ecol. Sociobiol.* **6**, 315–21.

Baker M.C., Belcher C.S., Deutsch L.C., Sherman G.L. & Thompson D.B. (1981) Foraging success in junco flocks and the effects of social hierarchy. *Anim. Behav.* **29**, 137–42.

Baker R.R. (1978) *The Evolutionary Ecology of Animal Migration*. Hodder & Stoughton, London.

Baldwin J.D. & Baldwin J.I. (1973) The role of play in social organisation: comparative observations on squirrel monkeys *(Saimiri)*. *Primates*, **14**, 369–81.

Baldwin J.D. & Baldwin J.I. (1976) Effects on ecology in social play. A laboratory simulation. *Z. Tierpsychol.* **40**, 1–14.

Baldwin J.D. & Baldwin J.I. (1977) The role of learning phenomena in the ontogeny of exploration and play. In *Primate Bio-social Development. Biological, Social and Ecological Determinants*, eds Chevalier-Skolnikofs S. & Poirier F.E. Garland Publishing, New York.

Barnes J.A. (1972) Social networks. *Addison-Welsley Module Anthrop.* **26**, 1–29.

Barth F. (1966) *Models of Social Organization*, Occasional Paper of the Royal Anthropological Institute 23. Royal Anthropological Institute, London.

Bateson P.P.G. (1981a) Discontinuities in development and changes in the organization of play in cats. In *Behavioural Development*, eds Immelmann K., Barlow G., Main M. & Petrinovich L. Cambridge University Press, Cambridge.

Bateson P.P.G. (1981b) Ontogeny of behaviour. *Br. med. Bull.* **37**(2), 159–64.

Beckoff M. (1981) Mammalian sibling interactions: Insightful genes, facilitative environments, and the coefficient of familiarity. In *Parental Care in Mammals*, eds Gubermick D.J. & Klopfer P.H. Plenum Press, London.

Beckoff M. & Byers J.A. (1981) A critical reanalysis of the ontogeny and phylogeny of mammalian social and locomotor play: an ethological hornet's nest. In *Behavioural Development*, eds Immelmann K., Barlow G., Main M. & Petrinovich L. Cambridge University Press, Cambridge.

Beer C.G. (1970) Individual recognition of voice in the social behaviour of birds. *Adv. Stud. Behav.* **3**, 27–74.

Bem D.J. & Funder D.C. (1978) Predicting more of the people more of the time: assessing the personality of situations. *Psychol. Rev.* **85**, 485–501.

Bengtsson B.O. (1978) Avoiding inbreeding: at what cost? *J. theor. Biol.* **73**, 439–44.

Berger J. (1979a) Social ontogeny and behavioural diversity: consequences for bighorn sheep *Ovis canadensis* inhabiting desert and mountain environments. *J. Zool. Lond.* **188**, 252–66.

Berger J. (1979b) Weaning conflict in desert and mountain bighorn sheep *(Ovis canadensis)*: an ecological interpretation. *Z. Tierpsychol.* **50**, 188–200.

Berger J. (1980) Ecology, structure and function of social play in bighorn sheep. *J. Zool. Lond.* **192**, 531–42.

Berman C.M. (1978a) Social Relationships among Free-ranging Infant Rhesus Monkeys. PhD Thesis, University of Cambridge.

Berman C.M. (1978b) Analysis of mother–infant interactions in groups: possible influences of yearling sibling. In *Recent Advances in Primatology*, vol. I, eds Chivers D.J. & Herbert J. Academic Press, London.

Berman C.M. (1980a) Mother–infant relationships among free-ranging rhesus monkeys on Cayo Santiago: a comparison with captive pairs. *Anim. Behav.* **28**, 860–73.

Berman C.M. (1980b) Early agonistic experience and rank acquisition among free-ranging infant rhesus monkeys. *Int. J. Primatol.* **1**, 153–70.

Berman C.M. (1982a) Ontogeny of social relationships with group companions among free-ranging companies among free-ranging infant rhesus monkeys. I. Social networks and differentiation. *Anim. Behav.* **30**, 149–62.

Berman C.M. (1982b) The ontogeny of social relationships with group companions among free-ranging infant rhesus monkeys. II. Differentiation and attractiveness. *Anim. Behav.* **30**, 163–70.

Berman C.M. (1982c) The social development of an orphaned rhesus infant on Cayo Santiago: Male care, foster mother–orphan interaction and peer interaction. *Am. J. Primatol.* **3** (in press).

Bernal J. (1972) Crying during the first 10 days of life, and maternal responses. *Devl. Med. Child Neurol.* **14**, 362–72.

Bernstein I.S. (1963) Social activities related to rhesus monkey consort behaviour. *Psychol. Rep.* **13**, 375–9.

Bernstein I.S. (1969) Stability of the status hierarchy in a pigtail monkey group *(Macaca nemestrina)*. *Anim. Behav.* **17**, 452–8.

Bernstein I.S. (1970) Primate status hierarchies. In *Primate Behaviour*, vol. 1, ed. Rosenblum L.A. Academic Press, New York.

Bernstein I.S. (1972) Daily activity cycles and weather influences on a pigtail monkey group. *Folia Primatol.* **18**, 390–415.

Bernstein I.S (1976a) Dominance, aggression, and reproduction in primate societies. *J. theor. Biol.* **60**, 459–72.

Bernstein I.S. (1976b) Activity patterns in a sooty mangabey group. *Folia primatol.* **26**, 185–206.

Bernstein I.S. (1981) Dominance: the baby and the bathwater. *Behav. Brain Sci.* **3**, 419–58.

Bernstein I.S. & Dobrofsky M. (1980) Compensatory social responses of older pigtailed monkeys to maternal separation. *Devl. Psychobiol.* **14**, 163–8.

Bernstein I.S. & Gordon T.P. (1974) The function of aggression in primate societies. *Am. Scient.*, **62**, 304–311.

Bernstein I.S., Rose R.M. & Gordon T.P. (1974) Behavioural and environmental events influencing primate testosterone levels. *J. hum. Evol.* 3, 517–25.

Bernstein I.S. & Sharpe L.G. (1966) Social roles in a rhesus monkey group. *Behaviour,* 26, 91–104.

Bernstein L., Rodman P.S. & Smith D.G. (1981) Social relations between fathers and offspring in a captive group of rhesus monkeys. *Anim. Behav.* 29, 1057–63.

von Bertalanffy L. (1968) *General System Theory.* George Braziller, New York.

Bertram B.C.R. (1976) Kin selection in lions and in evolution. In *Growing Points in Ethology,* eds Bateson P.P.G. & Hinde R.A. Cambridge University Press, Cambridge.

Bertram B.C.R. (1978a) Living in groups: predators and prey. In *Behavioural Ecology: an Evolutionary Approach,* eds Davies N.B. & Krebs J.R. Blackwell Scientific Publications, Oxford.

Bertram B.C.R. (1978b) *Pride of Lions.* J.M. Dent & Sons, London.

Bertram B.C.R. (1982) Problems with altruism. In *Current Problems in Sociobiology,* eds King's College Sociobiology Group. Cambridge University Press, Cambridge.

Biernoff A., Leary R.W. & Littman R.A. (1964) Dominance behaviour of paired primates in two settings. *J. abnorm soc. Psychol.* 68, 109–113

Birkhoff G. (1967) *Lattice Theory.* American Mathematical Society, Providence, Rhode Island.

Block J. (1977) Advancing the psychology of personality: Paradigmatic shift or improving the quality of research. In *Personality at the Crossroads,* eds Endler N.S. & Magnusson D. John Wiley & Sons, New York.

Blurton Jones N. (1972) Comparative aspects of mother–child contact. In *Ethological Studies of Child Behaviour,* ed. Blurton Jones N. Cambridge University Press, Cambridge.

Boelkins R.C. & Wilson A.P. (1972) Intergroup social dynamics of the Cayo Santiago rhesus *(Macaca mulatta)* with special reference to changes in group membership by males. *Primates,* 13(2), 125–40.

Boesch C. & Boesch H. (1981) Sex differences in the use of natural hammers by wild chimpanzees: A preliminary report. *J. hum. Evol.* 10, 585–93.

Boggess J. (1979) Troop male membership changes and infant killing in langurs *(Presbytis entellus). Folia primatol.* 32, 65–107.

Bolwig N. (1980) Early social development and emancipation of *Macaca nemistrina* and species of Papio. *Primates,* 21(3), 357–75.

Boorman S.A. & Arabie P. (1980) Algebraic approaches to the comparison of concrete social structures represented as networks. Reply to Professor Bonacich. *Am. J. Sociol.* 86(1), 166–74.

Boorman S.A. & White H.C. (1976) Social structure from multiple networks. II. Role structures. *Am. J. Sociol.* 81, 1384–446.

Bowlby J. (1969) *Attachment and Loss. Vol. 1 Attachment.* Hogarth Press, London.

Bowlby J. (1973) *Attachment and Loss. Vol. II Separation.* Hogarth Press, London.

Bradbury J.W. & Vehrencamp S.L. (1976) Social organization and foraging in emballonurid bats. I. Field studies. *Behav. Ecol. Sociobiol.* 1, 337–81.

Bradbury J.W. & Vehrencamp S.L. (1977) Social organization and foraging in emballonurid bats. III. Mating systems. *Behav. Ecol. Sociobiol.* 2, 1–17.

Bramblett C.A. (1970) Coalitions among gelada baboons. *Primates,* 11, 327–34.

Bramblett C.A. (1978) Sex differences in the acquisition of play among juvenile vervet monkeys. In *Social Play in Primates,* ed. Smith E.O. Academic Press, London.

Breiger R.L., Boorman S.A. & Arabie P. (1975) An algorithm for clustering relational data, with applications to social network analysis and comparison with multidimensional scaling. *J. math. Psychol.* **12**, 328–83.

Brown J.L. (1978) Avian communal breeding systems. *Ann. Rev., ecol. Syst.* **9**, 123–55.

Brown J.L. & Brown E.R. (1981) Extended family system in a communal bird. *Science, N.Y.* **211**, 959–60.

Bruner J.S. & Tagiuri R. (1954) The perception of people. In *Handbook of Social Psychology*, vol. 1. ed. Lindzey I.G. Addison-Wesley, Reading, Massachusetts.

Buirski P., Kellerman H., Plutchik R., Weininger R. & Buirski N. (1973) A field study of emotions, dominance, and social behaviour in a group of baboons *(Papio anubis)*. *Primates*, **14**(1), 67–78.

Buirski P., Plutchik R. & Kellerman H. (1978) Sex differences, dominance, and personality in the chimpanzee. *Anim. Behav.* **26**, 123–9.

Bunnell B.N. & Perkins M.N. (1980) Performance correlates of social behavior and organization. *Primates*, **21**, 515–23.

Burton J.J. (1980) Changes in Social Integration Following the Natural Leader Male Take-over in a Free-ranging Group of *Macaca fuscata*. Paper presented at the Eighth Congress of the International Primatology Society, Florence.

Buskirk W.H., Buskirk R.E. & Hamilton W.J. (1974) Troop-mobilizing behaviour of adult male chacma baboons. *Folia primatol.* **22**, 9–18.

Buss I.O. (1961) Some observations on food habits and behavior of the African elephant. *J. Wildl. Mgmt.* **25**(2), 131–48.

Busse C. (1981) Infanticide and Paternal Care by Male Chacma Baboons, *Papio ursinis*. PhD Thesis, University of California, Davis.

Busse C. & Hamilton W.J. (1981) Infant carrying by male chacma baboons. *Science, N.Y.* **212**, 1281–3.

Bygott J.D. (1974) Agonistic Behaviour and Dominance in Wild Chimpanzees. PhD Thesis, Cambridge University.

Bygott J.D. (1979) Agonistic behaviour, dominance, and social structure in wild chimpanzees of the Gombe National Park. In *The Great Apes*, eds Hamburg D.A. & McCown E.R. Benjamin Cummings, Menlo Park, California.

Bygott J.D., Bertram B.C.R. & Hanby J.P. (1979) Male lions in large coalitions gain reproductive advantage. *Nature, Lond.* **282**, 839–41.

Caine N.G. & Mitchell G.D. (1979) The relationship between maternal rank and companion choice in immature macaques *(Macaca mulatta* and *M. radiata)*. *Primates*, **20**, 583–90.

Campbell B. (1979) Ecological factors and social organization in human evolution. In *Primate Ecology and Human Origins*, eds Bernstein I.S. & Smith E.O. Garland Publishing, New York.

Cant J.H.G. (1980) What limits primates? *Primates*, **21**, 538–44.

Capitanio J.P. (1982). Early experience and social responsiveness in rhesus monkeys. *Int. J. Primatol.* **3**(3), 268.

Caplow T. (1968) *Two Against One.* Prentice-Hall, Englewood Cliffs, New Jersey.

Carpenter C.R. (1942) Sexual behavior of free-ranging rhesus monkeys, *Macaca mulatta*. I. Specimens, procedures and behavioural characteristics of estrus. *J. comp. Psychol.* **33**, 113–42.

Catchpole H.R. & van Wagenen G. (1978) Reproduction in rhesus monkeys, *Macaca mulatta*. In *The Rhesus Monkey*, vol. 2, ed. Bourne G.H.

Cavalli-Sforza L.L. & Bodmer W.F. (1971) *The Genetics of Human Populations*. W.H. Freeman, San Francisco.

Chagnon N.A. & Bugos P.E. (1979) Kin selection and conflict: an analysis of a Yanomamö axe fight. In *Evolutionary Biology and Human Social Behaviour: an Anthropological Perspective*, eds Chagnon N.A. & Irons W. Duxbury Press, North Scituate, Massachusetts.

Chalmers N.R. (1968) The social behaviour of free-living mangabeys in Uganda. *Folia primatol.* 8, 263–81.

Chalmers N.R. (1980) The ontogeny of play in feral olive baboons *(Papio anubis)*. *Anim. Behav.* 28, 570–85.

Chance M.R.A. (1967) Attention structure as the basis of primate rank orders. *Man*, 2, 503–18.

Chance M.R.A., Emory G.R. & Payne R.G. (1977) Status referents in long-tailed macaques *(Macaca fascicularis)*. *Primates*, 18, 611–32.

Chapais B. (1981) The Adaptiveness of Social Relationships Among Adult Rhesus Monkeys. PhD Thesis, University of Cambridge.

Chapais B. (1983) Reproductive activity in relation to male dominance and the likelihood of ovulation in rhesus monkeys. *Behav. Ecol. Sociobiol.*

Chapais B. & Schulman S. (1980) An evolutionary model of female dominance relations in primates. *J. theor. Biol.* 82, 47–89.

Chapman M. & Hausfater G. (1979) The reproductive consequences of infanticide in langurs: a mathematical model. *Behav. Ecol. Sociobiol.* 5, 227–40.

Charlesworth B. & Charnov E.L. (1981). Kin selection in age-structured populations. *J. theor. Biol.* 88, 103–19.

Cheney D.L. (1977) The acquisition of rank and the development of reciprocal alliances among free-ranging immature baboons. *Behav. Ecol. Sociobiol.* 2, 303–18.

Cheney D.L. (1978a) The play partners of immature baboons. *Anim. Behav.* 26, 1038–50.

Cheney D.L. (1978b) Interactions of immature male and female baboons with adult females. *Anim. Behav.* 26, 389–408.

Cheney D.L. (1981) Intergroup encounters among free-ranging vervet monkeys. *Folia primatol.* 35, 124–46.

Cheney D.L., Lee P.C. & Seyfarth R.M. (1981) Behavioural correlates of non-random mortality among free-ranging female vervet monkeys. *Behav. Ecol. Sociobiol.* 9, 153–61.

Cheney D.L. & Seyfarth R.M. (1977) Behaviour of adult and immature baboons during intergroup encounters. *Nature, Lond.* 269, 404–6.

Cheney D.L. & Seyfarth R.M. (1980) Vocal recognition in free-ranging vervet monkeys. *Anim. Behav.* 28, 362–7.

Cheney D.L. & Seyfarth R.M. (1981) Selective forces affecting the predator alarm calls of vervet monkeys. *Behaviour*, 76, 25–61.

Cheney D.L. & Seyfarth R.M. (1982) Recognition of individuals within and between free-ranging groups of vervet monkeys. *Am. Zool.* 22, 519–29.

Chepko-Sade B.D. & Olivier T.J. (1979) Coefficients of genetic relationship and the probability of intergenealogical fission in *Macaca mulatta. Behav. Ecol. Sociobiol.* 5, 263–78.

Chepko-Sade B.D. & Sade D.S. (1979) Patterns of group splitting within matrilineal kinship groups. *Behav. Ecol. Sociobiol.* 5, 67–86.

Chism J. (1978) Relationships between Patas infants and group members other than the mother. In *Recent Advances in Primatology*, vol. 1, eds Chivers D.J. & Herbert J. Academic Press, London.

Clark D.L. & Dillon J.E. (1973) Evaluation of the water incentive method of social dominance measurement in primates. *Folia primatol.* **19**, 293–311.

Clutton-Brock T.H. (1977) Some aspects of intraspecific variation in feeding and ranging behaviour in primates. In *Primate Ecology*, ed. Clutton-Brock T.H. Academic Press, London.

Clutton-Brock T.H., Guiness F.E. & Albon S.D. (1982) *Red Deer: The Behaviour and Ecology of Two Sexes.* University of Chicago Press, Chicago.

Clutton-Brock T.H. & Harvey P.H. (1976) Evolutionary rules and primate societies. In *Growing Points in Ethology*, eds Bateson P.P.G. & Hinde R.A. Cambridge University Press, Cambridge.

Clutton-Brock T.H. & Harvey P.H. (1977) Primate ecology and social organisation. *J. Zool. Lond.* **183**, 1–39.

Clutton-Brock T.H. & Harvey P.H. (1979) Home range size, population density and phylogeny in primates. In *Primate Ecology and Human Origins*, eds Bernstein I.S. & Smith E.O. Garland Publishing, New York.

Collins D.A. (1981) Social Behaviour and Patterns of Mating in Adult Yellow Baboons *(Papio c. cynocephalus* L. 1766). PhD Thesis, University of Edinburgh.

Colvin J.D. (1982) Social Integration and Emigration of Immature Male Rhesus Macaques. PhD Thesis, University of Cambridge.

Colvin J.D. & Tissier G. Social relationships with male peers and siblings among free-ranging immature male rhesus monkeys: Intimacy, friendship and altruism. (In prep.)

Crook J.H. (1966) Gelada baboon herd structure and movement: a comparative report. *Symp. zool. Soc. Lond.* **18**, 237–58.

Crook J.H. (1972) Sexual selection, dimorphism, and social organization in primates. In *Sexual Selection and the Descent of Man*, ed. Campbell B. Aldine, Chicago.

Crook J.H. (1975) Primate social structure and dynamics—conspectus 1974. In *Proceedings from the Symposia of the Fifth Congress of the International Primatological Society*, eds Kondo S., Kawai M., Ehara A. & Kawamura S. Japan Science Press, Tokyo.

Crook J.H. & Aldrich–Blake P. (1968) Ecological and behavioural contrasts between sympatric ground-dwelling primates in Ethiopia. *Folia primatol.* **8**, 192–227.

Crook J.H., Ellis J.E. & Goss-Custard J.D. (1976) Mammalian social systems: structure and function. *Anim. Behav.* **24**, 261–74.

Crook J.H. & Gartlan (1966) On the evolution of primate societies. *Nature*, **210**, 1200–3.

Croze H. (1974) The Seronera bull problem, I: The elephants. *E. Afr. Wildl. J.* **12**, 1–27.

Cubicciotti D.D. III & Mason W.A. (1975) Comparative studies of social behavior in *Callicebus* and *Saimiri:* male–female emotional attachments. *Behav. Biol.* **16**, 185–97.

Cubicciotti D.D. III & Mason W.A. (1978) Comparative studies of social behaviour in *Callicebus* and *Saimoi:* heterosexual jealousy behavior. *Behav. Ecol. Sociobiol.* **3**, 311–22.

Cummins M.S. & Suomi S.J. (1976) Long-term effects of social rehabilitation in rhesus monkeys. *Primates*, **17**, 43–51.

Czaja J.A., Eisele S.G. & Goy R.W. (1975) Cyclical changes in the sexual skin of female rhesus: relationship to mating behavior and successful artificial insemination. *Fedn Proc. Fedn Ann. Socs exp. Biol.* **34**, 1680–4.

Datta S.B. (1981) Dynamics of Dominance Among Rhesus Females. PhD Thesis, Cambridge University.

Dawkins R. (1976) Hierarchical organization: a candidate principle for ethology. In *Growing Points in Ethology*, eds Bateson P.P.G. & Hinde R.A. Cambridge University Press, Cambridge.

Dawkins R. (1982) Replicators and vehicles. In *Current problems in Sociobiology*, eds King's College Sociobiology Group. Cambridge University Press, Cambridge.

Deag J.M. (1974) A Study of the Social Behaviour and Ecology of the Wild Barbary Macaque, *Macaca sylvanus* L. PhD Thesis, University of Bristol.

Deag J.M. (1977) Aggression and submission in monkey societies. *Anim. Behav.* **25**, 465–74.

Deag J.M. (1980) Interactions between males and unweaned Barbary macaques: testing the agonistic buffering hypothesis. *Behaviour*, **75**, 54–81.

Deag J.M. & Crook J.H. (1971) Social behaviour and "Agonistic Buffering" in the wild barbary macaque *Macaca sylvana* L. *Folia primatol.* **15**, 183–200.

Defler T.R. (1978) Allogrooming in two species of macaque *(Macaca nemestrina* and *Macca radiata)*. *Primates*, **19**, 153–67.

Delvoye P., Badawi M., Demaegd M. & Robyn C. (1978) Long-lasting lactation is associated with hyperprolactinemia and amenorrhea. *Progress in Prolactin Physiology and Pathology*, eds Robyn C. & Harter M. Elsevier-North Holland, New York.

DeVore I. (1963a) Mother–infant relations in free-ranging baboons. In *Maternal Behavior in Mammals*, ed. Rheingold H. John Wiley & Sons, New York.

DeVore I. (1963b) A comparison of the ecology and behavior of monkeys and apes. In *Classification and Human Evolution*, ed. Washburn S.L. Aldine, Chicago.

DeVore I. & Hall K.R.L. (1965) Baboon ecology. In *Primate Behaviour. Field Studies of Monkeys and Apes*, ed. DeVore I. Holt, Rinehart & Winston, New York.

Dittus W.P.J. (1977) The social regulation of population density and age–sex distribution in the toque monkey. *Behaviour*, **63**, 281–322.

Dittus W.P.J. (1979) The evolution of behaviours regulating density and age-specific sex ratios in a primate population. *Behaviour*, **69**, 265–302.

Dittus W.P.J. (1980) The social regulation of primate populations: a synthesis. In *The Macaques: Studies in Ecology, Behavior, and Evolution*, ed. Lindburg D.G. Van Nostrand Reinhold, New York.

Dolhinow P.J. (1980) An experimental study in mother loss in the Indian langur monkey *(Presbytis entellus)*. *Folia primatol.* **33**, 77–128.

Dolhinow P.J. (1982) Primate strategies of early attachment to caregivers: a case study. *Int. J. Primatol.* 3(3), 277.

Dolhinow P.J. & Bishop N. (1970) The development of motor skills and social relationships among primates through play. In *Minnesota Symposia in Child Psychology*, vol. 4, ed. Hill J.P. University of Minnesota Press, Minneapolis.

Dolhinow P.J., McKenna J.J. & Vonder Haar Laws J. (1979) Rank and reproduction among female langur monkeys. *Aggress. Behav.* **5**, 19–30.

Douglas-Hamilton I. (1972) On the Ecology and Behaviour of the African Elephant. DPhil Thesis, University of Oxford.

Douglas-Hamilton I. (1973) On the ecology and behaviour of the Lake Manyara elephants. *E. Afr. wildl. J.* **11**, 401–3.

Douglas-Hamilton I. & Douglas-Hamilton O. (1975) *Among the Elephants.* Collins, London.

Drickamer L.C. (1974a) A ten year summary of reproductive data for free-ranging *Macaca mulatta. Folia primatol.* 21, 61–80.

Drickamer L.C. (1974b) Social rank, observability, and sexual behaviour of rhesus monkeys *(Macaca mulatta). J. Reprod. Fert.* 37, 117–20.

Drickamer L.C. & Vessey S.H. (1973) Group changing in free-ranging male rhesus monkeys. *Primates,* 14(4), 359–68.

Dunbar R.I.M. (1976) Some aspects of research design and their implications in the observational study of behaviour. *Behaviour,* 58, 79–98.

Dunbar R.I.M. (1978) Sexual behaviour and social relationships among gelada baboons. *Anim. Behav.* 26, 167–78.

Dunbar R.I.M. (1979a) Population demography, social organisation and mating strategies. In *Primate Ecology and Human Evolution,* eds Bernstein I. & Smith E.O. Garland Publishing, New York.

Dunbar R.I.M. (1979b) Structure of gelada baboon reproductive units. I. Stability of social relationships. *Behaviour,* 69, 72–87.

Dunbar R.I.M. (1980a) Demographic and life history variables of a population of gelada baboons *(Theropithecus gelada). J. Anim. Ecol.* 49, 485–506.

Dunbar R.I.M. (1980b) Determinants and evolutionary consequences of dominance among female gelada baboons. *Behav. Ecol. Sociobiol.* 7, 253–65.

Dunbar R.I.M. (1982) Adaptation, fitness and the evolutionary tautology. In *Current Problems in Sociobiology,* eds King's College Sociobiology Group. Cambridge University Press, Cambridge.

Dunbar R.I.M. (1983) *Reproductive Strategies of Gelada Baboons: Economics of Decision-Making.* Harvard University Press, Cambridge, Massachusetts.

Dunbar R.I.M. & Dunbar E.P. (1974) Behaviour related to birth in wild gelada baboons *(Theropithecus gelada). Behaviour,* 50, 185–91.

Dunbar R.I.M. & Dunbar E.P. (1975) *Social Dynamics of Gelada Baboons.* Karger, Basel.

Dunbar R.I.M. & Dunbar E.P. (1976) Contrasts in social structure among black-and-white colobus monkey groups. *Anim. Behav.* 24, 84–92.

Dunbar R.I.M. & Dunbar E.P. (1977) Dominance and reproductive success among female gelada baboons. *Nature, Lond.* 266, 351–2.

Dunn J. & Kendrick C. (1980) The arrival of a sibling: changes in patterns of interaction between mother and first born child. *J. Child Psychol. Psychiat.* 21, 119–32.

Eisenberg J.F. (1973) Mammalian social systems: are primates social systems unique? In *Preceedings from the Symposia of the Fourth Congress of the International Primatological Society,* vol. 1, ed. Menzei E.W. Karger, Basel.

Eisenberg J.F., McKay G.M. & Jainundeen M.R. (1971) Reproductive behaviour of the Asiatic elephant (*Elephas maximus maximus* L.). *Behaviour,* 38, 193–225.

Eisenberg J.F., Muckenhirn N.A. & Rudran R. (1972) The relation between ecology and social structure in primates. *Science, N.Y.* 176, 863–74.

Emerson R.M. (1962) Power-dependence relations. *Am. Soc. Rev.* 27, 31–41.

Emlen S.T. (1982) The evolution of helping behaviour. I. An ecological constraints model. *Am. Nat.* 119, 29–39.

Emlen S.T. & Oring L.W. (1977) Ecology, sexual selection and the evolution of mating systems. *Science, N.Y.* 197, 215–23.

Endler N.S. & Magnusson D. (1976) *Interactional Psychology and Personality.* John Wiley & Sons, New York.

Enomoto T. (1974) The sexual behavior of Japanese monkeys. *J. hum. Evol.* 3, 351–72.

Epple G. (1981) Effect of pair-bonding with adults on the ontogenetic manifestation of aggressive behavior in a primate, *Sanguinus fuscicollis. Behav. Ecol. Sociobiol.* 8, 117–24.

Estrada A., Estrada R. & Ervin F. (1977) Establishment of a free-ranging colony of stumptail macaques *(Macaca arctoides).* Social relations I. *Primates,* 18(3), 647–76.

Estrada A. and Sandeval J.M. (1977) Social relations in a free-ranging troop of stumptail macaques *(Macaca arctoides)*: male care behaviour I. *Primates,* 18, 793–813.

Fady J-C. (1969) Les jeux sociaux: le compagnon de jeux chez les jeunes. Observations chez *Macaca iris. Folia primatol.* 11, 134–43.

Fagen R.M. (1981) *Animal Play Behaviour.* Oxford University Press, Oxford.

Fairbanks L.A. (1975) Communication of food quality in captive *Macaca nemestrina* and free-ranging *Ateles geoffroyi. Primates,* 16(2), 181–90.

Fairbanks L.A. (1980) Relationships among adult females in captive vervet monkeys: testing a model of rank-related attractiveness. *Anim. Behav.* 28, 853–9.

Fairbanks L.A. & Bird J. (1978) Ecological correlates of interindividual distance in the St. Kitts vervets *(Cercopithecus aethiops sabaeus). Primates,* 19, 605–14.

Fedigan L.M. (1972) Social and solitary play in a colony of vervet monkeys. *Primates,* 13, 347–64.

Festinger L. (1949) The analysis of sociograms using matrix algebra. *Hum. Relat.* 2, 153–8.

Fletemeyer J.R. (1978) Communication about potentially harmful foods in free-ranging chacma baboons, *Papio ursinus. Primates,* 19(1), 223–6.

Forsyth E. & Katz L. (1946) A matrix approach to the analysis of sociometry data. *Sociometry,* 9, 340–7.

Frame L.H., Malcolm J.R., Frame G.W. & van Lawick H. (1979) Social organization of African wild dogs *(Lycaon pictus)* on the Serengeti Plains, Tanzania 1967–1978. *Z. Tierpsychol.* 50, 225–49.

Fricke H.W. (1980) Control of different mating systems in a coral reef fish by one environmental factor. *Anim. Behav.* 28, 561–9.

Furuya F. (1968) On the fission of troops of Japanese monkeys. *Primates,* 9, 323–49.

Furuya Y. (1969) On the fission of troops of Japanese monkeys. II. General view of troop fission of Japanese monkeys. *Primates,* 10, 47–69.

Gadgil M. (1982) Changes with age in the strategy of social behaviour. *Perspect. Ethol.* 5, 489–502.

Galdikas B.M.F. (1979) Orangutan adaptation at Tanjung Puting Reserve: Mating and ecology. In *The Great Apes,* eds Hamburg D.A. & McCown E.R. Benjamin-Cummings, Menlo Park, California.

Gartlan J.S. (1969) Sexual and maternal behaviour of the vervet monkey, *Cercopithecus aethiops. J. Reprod. Fert.* 6, 137–50.

Gartlan J.S. & Brain C.K. (1968) Ecology and social variability in *Cercopithecus aethiops* and *C. mitis.* In *Primates: Studies in Adaptation and Variability,* ed. Jay P. Holt, Rinehart & Winston, New York.

Gaston A.J. (1977) Social behaviour within groups of jungle babblers. *Anim. Behav.* 25, 828–48.

Gericke H. (1966) *Lattice Theory.* Frederick Ungar, New York.

Ginsburg B. & Allee W.C. (1942) Some effects of conditioning on social dominance and subordination in inbred strains of mice. *Physiol. Zoöl.* 15(4), 485–506.

Glick B.B. (1980) Ontogenetic and psychobiological aspects of the mating activities of male *Macaca radiata.* In *The Macaques: Studies in Ecology, Behavior and Evolution.* ed. Lindburg D.G. Van Nostrand Reinhold, New York.

Goldfoot D.A. (1971) Hormonal and social determinants of sexual behavior in the pigtail monkey *(Macaca nemestrina).* In *Normal and Abnormal Development of Brain and Behavior,* eds Stoelinger G.B.A. & van der Werff ten Bosch J.J. University of Leiden Press, Leiden.

Goodall J. (1968) The behaviour of free-living chimpanzees in the Gombe Stream Reserve. *Anim. Behav. Monogr.* 3, 165–311.

Goodall J. van Lawick (1975) The behaviour of the chimpanzee. In *Hominisation und Verhalten,* eds Kurth G. & Eibl-Eibesfeldt I. Gustav Fischer Verlag, Stuttgart.

Goodall J. (1978) *The Shadow of Man.* William Collins Sons & Co., London.

Gordon T.P., Rose R.M., Grady C.L. & Berstein I.S. (1979) Effects of increased testosterone secretion on the behavior of adult male rhesus living in a social group. *Folia primatol.* 32, 149–60.

Gouzoules H. (1980) A description of genealogical rank changes in a troop of Japanese monkeys *(Macaca fuscata). Primates,* 21, 262–7.

Goy R.W. & Goldfoot D.A. (1973) Experimental and hormonal factors influencing development of sexual behavior in the male rhesus monkey. In *The Neurosciences: Third Study Program,* vol. 3, eds Schmitt F.O. & Worden F.G. MIT Press, Cambridge, Massachusetts.

Greenwood P.J. (1980) Mating systems, philopatry and dispersal in birds and mammals. *Anim. Behav.* 28, 1140–62.

Greenwood P.J., Harvey P.H. & Perrins C.M. (1979) The role dispersal in the great tit *(Parus major):* the causes, consequences and heritability of natal dispersal. *J. Anim. Ecol.* 48, 123–42.

Grewal B.S. (1980a) Social relationships between adult central males and kinship groups of Japanese macaques at Arashiyama with some aspects of troop organization. *Primates,* 21, 161–80.

Grewal B.S. (1980b) Changes in relationship of multiparous and parous females of Japanese monkeys at Arashiyama with some aspects of troop organization. *Primates,* 21, 330–9.

Gutstein J. (1978) Behavioural correlates of male dispersal in patas monkeys. In *Recent Advances in Primatology,* vol. 1, eds Chivers D.J. & Herbert J. Academic Press, London.

Hall K.R.L. (1963) Variations in the ecology of the chacma baboon *Papio ursinus. Symp. zool. Soc. Lond.* 10, 1–28.

Hall K.R.L. (1967) Social interactions of the adult male and adult females of a patas monkey group. In *Social Communication Among Primates,* ed. Altmann S.A. University of Chicago Press, Chicago.

Hall K.R.L. & DeVore I. (1965) Baboon social behaviour. In *Primate Behaviour: Field Studies of Monkeys and Apes,* ed. DeVore I., pp. 53–110. Holt, Rinehart & Winston, New York.

Hamilton W.D. (1964) The genetical evolution of social behaviour I, II. *J. theor. Biol.* 7, 1–52.

Hamilton W.J., Buskirk R.E. & Buskirk W.H. (1975) Chacma baboon tactics during inter-troop encounters. *J. Mammal.* **56**, 857–70.

Hamilton W.J., Buskirk R.E. & Buskirk W.H. (1976) Defense of space and resources by chacma *(Papio ursinus)* baboon troops in an African desert and swamp. *Ecology,* **57**, 1264–72.

Hamilton W.J. & Busse C. (1980) Male transfer and offspring protection in chacma baboons. In *Proceedings from the Symposia of the Eighth Congress of the International Primatological Society,* eds Chiarelli A.B. & Corruccini R.S.

Hamilton W.J., Busse C. & Smith K.S. (1982) Adoption of infant orphan chacma baboons. *Anim. Behav.* **30**, 29–34.

Hanby J.P. (1980a) Relationships in six groups of rhesus monkeys I. Networks. *Am. J. phys. Anthrop.* **52**, 549–64.

Hanby J.P. (1980b) Relationships in six groups of rhesus monkeys II. Dyads. *Am. J. phys. Anthrop.* **52**, 565–75.

Hanby J.P., Robertson L.T. & Phoenix C.H. (1971) The sexual behavior of a confined troop of Japanese macaques. *Folia primatol.* **16**, 123–43.

Harcourt A.H. (1978a) Activity periods and patterns of social interaction: a neglected problem. *Behaviour,* **66**, 121–35.

Harcourt A.H. (1978b) Strategies of emigration and transfer by primates with particular reference to gorillas. *Z. Tierpsychol.* **48**, 401–20.

Harcourt A.H. (1979a) Social relationships between adult male and female mountain gorillas in the wild. *Anim. Behav.* **27**, 325–42.

Harcourt A.H. (1979b) Social relationships among adult female mountain gorillas. *Anim. Behav.* **27**, 251–61.

Harcourt A.H. (1979c) Contrasts between male relationships in wild gorilla groups. *Behav. Ecol. Sociobiol.* **5**, 39–49.

Harcourt A.H., Fossey D. & Sabater Pi J. (1981) Demography of *Gorilla gorilla. J. Zool. Lond.* **195**(2), 215–34.

Harcourt A.H., Fossey D., Stewart K.J. & Watts D. (1980) Reproduction in wild gorillas and some comparisons with chimpanzees. *J. Reprod. Fert.* **28**(suppl.), 59–70.

Harcourt A.H. & Stewart K.J. (1981) Gorilla male relationships: can differences during immaturity lead to contrasting reproductive tactics in adulthood? *Anim. Behav.,* **29**, 206–10.

Harcourt A.H., Stewart K.J. & Fossey D. (1976) Male emigration and female transfer in wild mountain gorillas. *Nature, Lond.* **263**, 226–7.

Harding R.S.O. (1973a) Range utilization by a troop of olive baboons *(Papio anubis).* PhD Thesis, University of California, Berkeley.

Harding R.S.O. (1973b) Predation by a troop of olive baboons *(Papio anubis). Am. J. phys. Anthrop.* **38**, 587–92.

Harding R.S.O. (1976) Ranging patterns of a troop of baboons *(Papio anubis)* in Kenya. *Folia primatol.* **25**, 143–85.

Harlow H.F. (1965) Sexual behavior in the rhesus monkey. In *Sex and Behavior,* ed. Beach F.A. John Wiley & Sons, New York.

Harlow H.F. & Harlow M.K. (1965) The affectional system. In *Behaviour of Non-human Primates 2,* eds Schrier A.M., Harlow H.F. & Stollnitz F. Academic Press, New York.

Harrington J.E. (1976) Discrimination between individuals by scent in *Lemur fulvus. Anim. Behav.* **24**, 207–212.

Hartl D.L. (1900). *Principles of Population Genetics.* Sinauer Press, Sunderland, Massachusetts.

Harvey P.H. & Clutton-Brock T. (1981) Primate home range size and metabolic needs. *Behav. Ecol. Sociobiol.* 8(2), 151–6.

Harvey P.H., Kavanagh M. & Clutton-Brock T.H. (1978) Sexual dimorphism in primate teeth. *J. Zool. Lond.* 186, 475–85.

Hasegawa T. & Hiraiwa M. (1980) Social interactions of orphans observed in a free-ranging troop of Japanese monkeys. *Folia primatol.* 33, 129–58.

Haude R.H., Graber J.G. & Farres A.G. (1976) Visual observing by rhesus monkeys: some relationships with social dominance rank. *Anim. Learn. Behav.* 4, 163–6.

Hausfater G. (1972) Intergroup behaviour of free-ranging rhesus monkeys *(Macaca mulatta). Folia primatol.* 18, 78–107.

Hausfater G. (1975) *Dominance and Reproduction in Baboons* (Papio cynocephalus). Karger, Basel.

Hausfater G., Altmann S.A. & Altmann J. (1982) Long-term Consistency of Dominance Relationships in Baboons. Paper presented at the Ninth Congress of the International Primatology Society, Atlanta.

Hendrichs H. (1971) Freilandbeibachtungen zum Sozialsystem ses Afrikanischen Elefanten, *Loxodonta africana (Blumenbach,* 1797). In *Dikdik und Elefanten,* Ethologische Studien. Verlag Piper, Munich.

Hendricks A.G. & Kraemer D.C. (1969) Observation of the menstrual cycle, optimal mating time, and preimplantation embryos of the baboon, *Papio anubis* and *Papio cynocephalus. J. Reprod. Fert.* 6(suppl.), 119–28.

Henzi S.P. & Lucas J.W. (1980) Observations on the intertroop movement of adult vervet monkeys *(Cercopithecus aethiops). Folia primatol.* 33, 220–35.

Herbert J. (1969) Neural and hormonal factors concerned in sexual attraction between rhesus monkeys. In *Proceedings from the Symposia of the Second Congress of the International Primatological Society,* vol. 2. Karger, Basel and New York.

Hill W.C.O. (1966) *Primates: Comparative Anatomy and Taxonomy, Vol. 6, Cercopithicoidea.* Edinburgh University Press, Edinburgh.

Hill W.C.O. (1970) *Primates: Comparative Anatomy and Taxonomy, Vol. 8, Cynopithecinae.* Edinburgh University Press, Edinburgh.

Hinde R.A. (1968) Dichotomies in the study of development. In *Genetic and Environment Influences on Behavior,* eds Thoday J.M. & Parkes A.S. Oliver & Boyd, Edinburgh.

Hinde R.A. (1969) Analyzing the roles of the partners in a behavioral interaction— mother–infant relations in rhesus macaques. *Ann. N.Y. Acad. Sci.* 159, 651–67.

Hinde R.A. (1970) *Animal Behaviour: a Synthesis of Ethology and Comparative Psychology,* 2nd edn. McGraw Hill, London.

Hinde R.A. (1973) On the design of check-sheets. *Primates,* 14(4), 393–406.

Hinde R.A. (1974) *Biological Bases of Human Social Behaviour.* McGraw Hill, New York.

Hinde R.A. (1976) Interactions, relationships and social structure. *Man,* 11, 1–17.

Hinde R.A. (1977a) On assessing the basis of partner preferences. *Behaviour,* 62, 1–9.

Hinde R.A. (1977b) Mother–infant separation and the nature of inter-individual relationships: experiments with rhesus monkeys. *Proc. R. Soc. Lond.* 196, 29–50.

Hinde R.A. (1978) Dominance and role—two concepts with dual meaning. *J. soc. Biol. Struct.* 1, 27–38.

Hinde R.A. (1979a) *Towards Understanding Relationships.* Academic Press, London.

Hinde R.A. (1979b) The nature of social structure. In *The Great Apes,* eds Hamburg D.A. & McCann C.R. Benjamin–Cummings, Menlo Park, California.

Hinde R.A. (1981a) Animal signals: ethological and games-theory approaches are not incompatible. *Anim. Behav.* 29, 535–42.

Hinde R.A. (1981b) The bases of a science of interpersonal relationships. In *Personal Relationships 1: Studying Personal Relationships,* eds Duck S.W. & Gilmour R. Academic Press, London.

Hinde R.A. (1982) *Ethology.* Fontana, London.

Hinde R.A. (1983) Ethological contributions to the study of child development. In *Carmichael's Manual of Child Psychology,* ed. Mussen P. John Wiley & Sons, New York.

Hinde R.A. & Atkinson S. (1970) Assessing the roles of social partners in maintaining mutual proximity as exemplified by mother–infant relations in rhesus monkeys. *Anim. Behav.* 18, 169–76.

Hinde R.A. & Datta S.B. (1981) Dominance — an intervening variable. *Behav. Brain Sci.* 4, 442.

Hinde R.A. & Davies L. (1972) Removing infant rhesus from mother for 13 days compared with removing mother from infant. *J. Child Psychol. Psychiat.* 13, 227–37.

Hinde R.A. & Herrmann J. (1977) Frequencies, durations, derived measures and their correlations in studying dyadic and triadic relationships. In *Studies in Mother–Infant Interactions,* ed. Schaffer H.R. Academic Press, London.

Hinde R.A., Leighton-Shapiro M.E. & McGinnis L. (1978) Effects of various types of separation experience on rhesus monkeys five months later. *J. Child Psychol. Psychiat.* 19, 199–211.

Hinde R.A. & McGinnis L. (1977) Some factors influencing the effects of temporary mother–infant separation — some experiments with rhesus monkeys. *Psychol. Med.* 7, 197–212.

Hinde R.A. & Simpson M.J.A. (1975) Qualities of mother–infant relationships in monkeys. *Parent–Infant Interactions,* CIBA Foundation Symposium 33, pp. 39–68. Elsevier, Amsterdam.

Hinde R.A. & Spencer-Booth Y. (1967) The behaviour of socially living rhesus monkeys in their first two and a half years. *Anim. Behav.* 15, 169–96.

Hinde R.A. & Spencer-Booth Y. (1970) Individual differences in the responses of rhesus monkeys to a period of separation from their mothers. *J. Child Psychol. Psychiat.* 11, 159–76.

Hinde R.A. & Spencer-Booth Y. (1971a) Towards understanding individual differences in rhesus mother–infant interactions. *Anim. Behav.* 19, 165–73.

Hinde R.A. & Spencer-Booth Y. (1971b) Effects of brief separation from mothers on rhesus monkeys. *Science, N.Y.* 173, 111–18.

Hinde R.A. & Stevenson-Hinde J. (eds) (1973) *Constraints on Learning: Limitations and Predispositions.* Academic Press, London.

Hinde R.A. & Stevenson-Hinde J. (1976) Towards understanding relationships: dynamic stability. In *Growing Points in Ethology,* eds Bateson P.P.G. & Hinde R.A. Cambridge University Press, Cambridge.

Hiraiwa M. (1981) Maternal and alloparental care in a troop of free-ranging Japanese monkeys. *Primates,* 22, 309–29.

Hoff M.P. (1982) A reassessment of developing infant independence in captive rhesus monkeys. *Int. J. Primatol.* 3(3), 297.

Hölldobler B. (1976) Tournaments and slavery in a desert ant. *Science, N.Y.,* 192, 912–14.

Hölldobler B. & Michener C.D. (1980) Mechanisms of identification and discrimination in social Hymenoptera. In *Evolution of Social Behavior: Hypotheses and Empirical Tests,* ed. Markl H. Verlag Chemie, Weinheim.

Hölldobler B. & Wilson E.O. (1977) The number of queens: an important trait in ant evolution. *Naturwissenschaften,* 64, 8–15.

Holman S.D., Yoshihara D.L. & Goy R.W. (1982) Effects of sibling birth on the social development of rhesus yearlings. *Int. J. Primatol.* 3(3), 298.

Holmes W.G. & Sherman P.W. (1982) The ontogeny of kin recognition in two species of ground squirrels. *Am. Zool.* (in press).

Homans G.C. (1951) *The Human Group.* Routledge & Kegan Paul, London.

Homans G.C. (1961) *Social Behaviour: its Elementary Forms.* Routledge & Kegan Paul, London.

Homewood K.M. (1978) Feeding strategy of the Tana mangabey— *Cercocebus galeritus galeritus. J. Zool. Lond.* 186, 375–91.

van Hooff J.A.R.A.M. & de Waal F. (1975) Aspects of an ethological analysis of polyadic agonistic interactions in a captive group of *Macaca fascicularis.* In *Proceedings from the Symposia of the Fifth Congress of the International Primatological Society,* eds Kondo S., Kawai M., Ehara A. & Kawamura S. Japan Science Press, Tokyo.

Hooley J.M. & Simpson M.J.A. (1981) A comparison of primiparous and multiparous mother–infant dyads in *Macaca mulatta. Primates,* 22, 379–92.

Hooley J.M. & Simpson M.J.A. The influence of siblings on the early social development of infant rhesus monkeys *(Macaca mulatta).* (In prep.)

Hopf S. (1981) Conditions of failure and recovery of maternal behavior in captive squirrel monkeys *(Saimiri). Int. J. Primatol.* 2, 335–49.

Horn H.S. (1968) The adaptive significance of colonial nesting in the Brewer's blackbird *(Euphagus cyanocephalus). Ecology,* 49, 682–94.

Horvat J.R., Coe C.L. & Levine S. (1980) Infant development and maternal behavior in captive chimpanzees. In *Maternal Influences and Early Behavior,* eds Beur W. & Smotherman W.P. Spectrum Publications, New York.

Hrdy S.B. (1976) Care and exploitation of non-human primate infants by conspecifics other than the mother. In *Advances in the Study of Behaviour 6,* eds Rosenblatt J.S., Hinde R.A., Shaw E. & Beer C. Academic Press, London.

Hrdy S.B. (1977a) Infanticide as a primate reproductive strategy. *Am. Scient.* 65, 40–9.

Hrdy S.B. (1977b) *Langurs of Abu: Female and Male Strategies of Reproduction.* Harvard University Press, Cambridge, Massachusetts.

Hrdy S.B. (1979) Infanticide among animals: a review, classification and examination of the implications for the reproductive strategies of females. *Ethol. Sociobiol.* 1, 13–40.

Hrdy S.B. & Hrdy D.B. (1976) Hierarchical relations among female hanuman langurs (Primates: Colubinae, *Presbytis entellus). Science, N.Y.* 193, 913–15.

Humphrey N.K. (1976) The social function of intellect. In *Growing Points in Ethology,* eds Bateson P.P.G. & Hinde R.A. Cambridge University Press, Cambridge.

Humphrey N.K. (1980) Nature's psychologists. In *Consciousness and the Physical World,* eds Josephson B.D. & Ramachandran V.S. Pergamon Press, Oxford. Also in *The Exercise of Intelligence,* eds Sunderland E. & Smith M.T. Garland Publishing, New York.

Imanishi K. (1960) Social organisation of subhuman primates in their natural habitat. *Curr. Anthrop.* 1, 393–407.

Irons W. (1979) Natural selection, adaptation and human social behaviour. In *Evolutionary Biology and Human Social Behaviour: an Anthropological Perspective,* eds Chagnon N.A. & Irons W. Duxbury Press, North Scituate, Massachusetts.

Isaac G.L. (1978) The archaelogical evidence for the activities of early African hominids. In *African Hominidae of the Plio-Pleistocene,* ed. Jolly C. Duckworth, London.

Itani J. (1963) Paternal care in the wild Japanese monkey. *Primate Social Behavior,* ed. Southwick C.H. Van Nostrand, Princeton.

Itani J. (1972) A preliminary essay on the relationship between social organization and incest avoidance in non-human primates. In *Primate Socialization,* ed. Poirier F.E. Random House, New York.

Itani J. (1980) Social structure of African great apes. *J. Reprod. Fert.* **28**(suppl.), 33–41.

Itoigawa N. (1975) Variables in male leaving a group of Japanese macaques. In *Proceedings from the Symposia of the Fifth Congress of the International Primatological Society,* eds Kondo S., Kawai M., Ehara A. & Kawamura S. Japan Science Press, Tokyo.

Itoigawa N., Negayama K. & Kondo K. (1981) Experimental study on sexual behaviour between mother and son in Japanese monkeys *(Macaca fuscata). Primates,* **22,** 494–502.

Iwamoto T. & Dunbar R.I.M. (1983) Thermoregulation, habitat quality and the behavioural ecology of gelada baboons. *J. anim. Ecol.* **53,** 357–66.

Izawa K. (1980) Social behaviour of the wild black-capped capuchin *(Cebus apella). Primates,* **21,** 443–67.

Jay P. (1968) Mother–infant relations in langurs. In *Maternal Behavior in Mammals,* ed. Rheingold H. John Wiley & Sons, New York.

Johnson C.K., Koerner C., Esrin M. & Duoos D. (1980) Alloparental care and kinship in captive social groups of vervet monkeys *(Cercopithecus aethiops sabaeus). Primates,* **21**(3), 406–15.

Jolly A. (1972) *The Evolution of Primate Social Behaviour.* Macmillan, London.

Jones C.B. (1979) Grooming in the mantled howler monkey *(Alouatta palliata).* Grey. *Primates,* **20,** 289–92.

Jones C.B. (1980) The functions of status in the mantled howler monkey *(Alouatta palliata). Primates,* **21,** 389–405.

Jones C.B. (1981) The evolution and socioecology of dominance in primate groups: a theoretical formulation, classification and assessment. *Primates,* **22,** 70–83.

Kaplan J. (1977). Patterns of fight interference in free-ranging rhesus monkeys. *Am. J. phys. Anthrop.* **47,** 279–88.

Kaplan J.R. (1978) Fight interference and altruism in rhesus monkeys. *Am. J. phys. Anthropol.* **49,** 241–50.

Katz L. (1947) On the matrix analysis of sociometric data. *Sociometry,* **10,** 233–41.

Kaufman I.C. & Rosenblum L.A. (1967) The reaction to separation in infant monkeys: Anaclitic depression and conservation-withdrawal. *Psychsom. Med.* **29,** 648–75.

Kaufmann J.H. (1965) A three-year study of mating behavior in a free-ranging band of rhesus monkeys. *Ecology,* **46,** 500–12.

Kaufmann J.H. (1967) Social relations of adult males in a free-ranging band of rhesus monkeys. In *Social Communication Among Primates,* ed. Altmann S.A. University of Chicago Press, Chicago.

Kawai M. (1958) On the rank system in a natural group of Japanese monkeys I & II. *Primates,* **1,** 111–48.

Kawamura S. (1965) Matriarchal social ranks in the Minoo B troop: a study of the rank system of Japanese monkeys. In *Japanese monkeys*, eds Imanishi K. & Altmann S.A. Published by the editors.

Kawanaka K. (1973) Intertroop relations among Japanese monkeys. *Primates*, **14**, 113–59.

Kawanaka K. (1977) Division of males in a Japanese monkey troop on the basis of numerical data. *Bull. Hiruzen Res. Inst.* **3**, 11–44.

Kawanaka K. & Nishida T. (1975) Recent advances in the study of inter-unit relationships and social structure of wild chimpanzees of the Mahali Mountains. In *Proceedings from the Symposia of the Fifth Congress of the International Primatological Society*, eds Kondo S., Kawai M., Ehara A. & Kawamura S. Japan Science Press, Tokyo.

Kelley H.H. (1979) *Personal Relationships*. Erlbaum, Hillsdale, New Jersey.

Kenrick D.T. & Stringfield D.O. (1980) Personality traits and the eye of the beholder: crossing some traditional philosophical boundaries in the search for consistency in all of the people. *Psychol. Rev.* **87**, 88–104.

Keverne E.B., Miller R.E. & Eberhart A. (1982) Dominance and subordination: concepts or physiological states? In *Proceedings from the Symposia of the Eighth Congress of the International Primatological Society*, eds Chiarelli A.B. & Corruccini R.S.

Kleiman D.G. (1977) Monogamy in mammals. *Q. Rev. Biol.* **52**, 36–9.

Kleiman D.G. & Eisenberg J.F. (1973) Comparisons of canid and felid social organizations from an evolutionary perspective. *Anim. Behav.* **21**, 637–59.

Klein D.J. (1978) The Diet and Reproductive Cycles of a Population of Vervet Monkeys *Cercopithecus aethiops*. PhD Thesis, New York University.

Klingel H. (1972) Social behaviour of African Equidae. *Zool. Afr.* **7**, 175–86.

Koford C.B. (1963) Rank of mothers and sons in bands of rhesus monkeys. *Science, N.Y.* **141**, 356–7.

Koford C.B. (1965) Population dynamics of rhesus monkeys on Cayo Santiago. In *Primate Behavior: Field Studies of Monkeys and Apes*, ed. DeVore I. Holt, Rinehart & Winston, New York.

Konner M.J. (1976) Maternal care, infant behavior and development among the !Kung. In *Kalahari Hunter-Gatherers*, eds Lee R.B. & DeVore I. Harvard University Press, Cambridge, Massachusetts.

Koyama N. (1967) On dominance rank and kinship of a wild Japanese monkey troop in Arashujama. *Primates*, **8**, 189–216.

Koyama N. (1970) Changes in dominance rank and division of a wild Japanese monkey troop at Arashujama. *Primates*, **11**, 335–90.

Kraemer H.C. (1979) One-zero sampling in the study of primate behaviour. *Primates*, **20**(2), 237–44.

Krebs J.R. (1979) Foraging strategies and their social significance. In *Handbook of Behavioral Neurobiology*, vol. 3, eds Marler P. & Vandenbergh J.G. Plenum Press, New York.

Kruuk H. (1972) *The Spotted Hyaena*. University of Chicago Press, Chicago.

Kummer H. (1967) Tripartite relations in hamadryas baboons. In *Social Communication Among Primates*, ed. Altmann S.A. University of Chicago Press, Chicago.

Kummer H. (1968) *Social Organisation of Hamadryas Baboons*. Karger, Basel.

Kummer H. (1971) *Primate Societies: Group Techniques of Ecological Adaptation*. Aldine Atherton, Chicago.

Kummer H. (1975) Rules of dyad and group formation among captive gelada baboons *(Theropithecus gelada)*. In *Proceedings from the Symposia of the Fifth Congress of the*

International Primatological Society, eds Kondo S., Kawai M., Ehara A. & Kawamura S. Japan Science Press, Tokyo.

Kummer H. (1978) On the value of social relationships to non-human primates: a heuristic scheme. *Soc. Sci. Inform.* **17**, 687–705.

Kummer H. (1981) Social knowledge in free-ranging primates. In *Animal Mind–Human Mind*, ed. Griffin D.R. Springer-Verlag, Berlin.

Kummer H. Abegglen J.J., Bachman Ch., Falett J. & Sigg H. (1978) Grooming relationship and object competition among Hamadryas baboons. In *Recent Advances in Primatology*, vol. 1, eds Chivers D.J. & Herbert J. Academic Press, London.

Kummer H., Götz W. & Angst W. (1974) Triadic differentiation: an inhibitory process protecting pair bonds in baboons. *Behaviour*, **49**, 62–87.

Kurland J.A. (1977) Kin selection in the Japanese monkey. In *Contributions to Primatology*, vol. 12, ed. Szalay F.S. Karger, Basel.

Kuroda S. (1979) Grouping of the pygmy chimpanzees. *Primates*, **20**, 161–83.

Kuroda S. (1980) Social behavior of the pygmy chimpanzees. *Primates*, **21**, 181–97.

Lack D. (1954) *The Natural Regulation of Animal Numbers.* Clarendon Press, Oxford.

Lack D. (1968) *Ecological Adaptations for Breeding in Birds.* Methuen, London.

Lancaster J.B. (1972) Play-mothering: the relations between juvenile females and young infants among free ranging vervet monkeys. In *Primate Socialization*, ed. Poirier F.E. Random House, New York.

Lauer C. (1980) Seasonal variability in spatial defense by free-ranging rhesus monkeys *(Macaca mulatta). Anim. Behav.* **28**, 476–82.

Laws R.M. & Parker I.S.C. (1968) Recent studies on elephant populations in East Africa. *Symp. zool. Soc. Lond.* **21**, 319–59.

Lee P.C. Early infant development and maternal care in free-ranging vervet monkeys. (In prep.)

Lee P.C. (1981) Ecological and Social Influences on Development in Vervet Monkeys *(Cercopithecus aethiops).* PhD Thesis, Cambridge University.

Lee P.C. Early infant development and maternal care in free-ranging vervet monkeys. *Primates* (in press).

Lee P.C. & Oliver J.I. (1979) Competition, dominance and the acquisition of rank in juvenile yellow baboons *(Papio cynocephalus). Anim. Behav.* **27**, 576–85.

Lee R.B. & DeVore I. (1976) *Kalahari Hunter-Gatherers.* Harvard University Press, Cambridge, Massachusetts.

Lehrman D.S. (1953) A critique of Konrad Lorenz's theory of instinctive behaviour. *Q. Rev. Biol.* **28**, 337–63.

Lerner M.J., Miller D.T. & Holmes J.G. (1976) Deserving and the emergence of forms of justice. In *Advances in Experimental Social Psychology*, vol. 9, eds Berkowitz L. & Walster E.

Leuthold W. (1976) Group size in elephants of Tsavo National Park and possible factors influencing it. *J. Anim. Ecol.* **45**, 425–39.

Leuthold W. (1977a) Spatial organization and strategy of habitat utilization of elephants in Tsavo National Park, Kenya. *Z. Säugetierk.* **42**, 358–79.

Leuthold W. (1977b) *African Ungulates: A Comparative Review of their Ethology and Behavioral Ecology.* Springer-Verlag, Berlin.

Leuthold W. & Sale J.B. (1973) Movements and patterns of habitat utilization in Tsavo National Park, Kenya. *E. Afr. wildl. J.* **11**, 369–84.

Lindburg D.G. (1969) Rhesus monkeys: mating season mobility of adult males. *Science, N.Y.* **166**, 1176–8.

Lindburg D.G. (1971) The rhesus monkey in North India: an ecological and behavioral study. In *Primate Behavior*, vol. 2, ed. Rosenblum L.A. Academic Press, New York.

Lindsay W.K. (1982) Habitat Selection and Social Group Dynamics of African Elephants in Amboseli, Kenya. MSc Thesis, University of British Columbia.

Loizos C. (1967) Play behaviour in higher primates: a review. In *Primate Ethology*, ed. Morris D. Weidenfeld & Nicolson, London.

Lorenz K. (1935) Der Kumpan in der Umwelt des Vogels, *J.f. Ornith.* **83**, 137–213 and 289–413. (Translated in Martin R.B. (ed.) *Studies in Animal and Human Behaviour*, vol. 1. Harvard University Press, Cambridge, Massachusetts.)

Lorenz K. (1970) *Studies in Animal and Human Behavior*, vol. I. Harvard University Press, Cambridge, Massachusetts.

Loy J. (1970) Behavioural responses of free-ranging rhesus monkeys to food shortage. *Am. J. phys. Anthrop.* **33**, 263–72.

Loy J. & Loy K. (1974) Behavior of an all juvenile group of rhesus monkeys. *Am. J. phys. Anthrop.* **40**, 83–96.

Luce R.D. (1950) Connectivity and generalized cliques in sociometric group structure. *Psychometrika*, **15**, 169–90.

Luce R.D. & Perry A.D. (1949) A method of matrix analysis of group structure. *Psychometrika*, **14**, 95–116.

Lund M. (1981) The Development of Commitment to a Close Relationship. PhD Thesis, University of California, Los Angeles. (University Microfilm A/C 313130.)

McArthur P.D. (1982) Mechanisms and development of parent–young vocal recognition in the pinon jay *(Gymnorhinus cyanocephalus). Anim. Behav.* **30**, 62–74.

McCall G.J. (1970) The social organization of relationships. In *Social Relationships*, eds McCall G.J., McCall M., Denzin N.K., Suttles G.D. & Kurth S.B. Aldine Publishing, Hawthorne, New York.

McCann T.S. (1982) Aggressive and maternal activities of female southern elephant seals *(Mirounga leonina). Anim. Behav.* **30**, 268–76.

McCracken G.F. & Bradbury J.W. (1977) Paternity and genetic heterogeneity in the polygynous bat, *Phyllostomus hastatus. Science, N.Y.* **198**, 303–6.

Macdonald D.W. (1979a) The flexible social system of the golden jackal, *Canis aureus. Behav. Ecol. Sociobiol.* **5**, 17–38.

Macdonald D.W. (1979b) Helpers' in fox society. *Nature, Lond.* **282**, 69–71.

McGinnis L.M. (1980) Maternal separation studies in children and non-human primates. In *Maternal Influences and Early Behavior*, eds Bell R.W. & Smotherman W.P. Spectrum Publications, Jamaica, New York.

McGrew W.C., Baldwin P.J. & Tutin C.E.G. (1981) Chimpanzees in a hot, dry, and open habitat, Mt. Assirik, Senegal, West Africa. *J. hum. Evol.* **10**, 227–44.

McGrew W.C., Tutin C. & Baldwin P.J. (1979) Chimpanzees, tools and termites: cross-cultural comparisons of Senegal, Tanzania and Rio Muni. *Man*, **14**, 185–214.

Mackinnon J.R. (1974) The behaviour and ecology of wild orang-utans *(Pongo pygmaeus). Anim. Behav.* **22**, 3–74.

McMillan C.A. (1982) Factors Affecting Mating Success Among Rhesus Macaque Males on Cayo Santiago. PhD Thesis, State University of New York, Buffalo.

Marsden H.M. (1968) Behavior between two social groups of rhesus monkeys within two tunnel-connected enclosures. *Folia primatol.* **8**, 240–6.

Marsden H.M. (1973) Aggression within social groups of rhesus monkeys *(Macaca mulatta):* effect of contact between groups. *Anim. Behav.* **21**, 247–9.

Marsh C.W. (1979) Comparative aspects of social organization in the Tana River red Colobus, *Colobus badius rufonitratus. Z. Tierpsychol.* **51**, 337–62.

Martin P. (1982) The energetic costs of play: definition and estimation. *Anim. Behav.* **30**, 292.

Maslow A.H. (1936) The role of dominance in the social and sexual behavior of infra-human primates IV. *J. gen. Psychol.* **49**, 161–98.

Mason W.A. (1963) The effects of environmental restriction on the social development of rhesus monkeys. In *Primate Social Behaviour,* ed. Southwick C.H. D. Van Nostrand, Princeton.

Mason W.A. (1965) The social development of monkeys and apes. In *Primate Behaviour. Field Studies of Monkeys and Apes,* ed. DeVore I. Holt, Rinehart & Winston, New York.

Mason W.A. (1973) Field and laboratory studies of social organization in *Saimiri* and *Callicebus.* In *Primate Behavior,* vol. 2, ed. Rosenblum L.A. Academic Press, New York.

Mason W.A. (1974) Comparative studies of social behavior in *Callicebus* and *Saimiri*: Behavior of Male–Female pairs. *Folia primatol.* **22**, 1–8.

Mason W.A. (1975) Comparative studies of social behavior in *Callicebus* and *Saimiri:* Strength and specificity of attraction between male–female cagemates. *Folia primatol.* **23**, 113–23.

Massey A. (1977) Agonistic aids and kinship in a group of pigtail macaques. *Behav. Ecol. Sociobiol.* **2**(1), 31–40.

Maxim P.E. & Buettner-Janisch J. (1963) A field study of the Kenya baboon. *Am. J. phys. Anthrop.* **21**, 165–80.

May R.M. (1982) Mutualistic interactions among species. *Nature, Lond.* **296**, 803–4.

Maynard Smith J. (1964) Group selection and kin selection. *Nature, Lond.* **201**, 1145–7.

Maynard Smith J. (1974) The theory of games and the evolution of animal conflict. *J. theor. Biol.* **47**, 209–21.

Maynard Smith J. (1976) Group selection. *Q. Rev. Biol.* **51**, 277–83.

Maynard Smith J. (1979) Game theory and the evolution of behaviour. *Proc. R. Soc. Lond. B.* **205**, 475–88.

Maynard Smith J. (1982) The evolution of social behaviour—a classification of models. In *Current problems in Sociobiology,* eds King's College Sociobiology Group. Cambridge University Press, Cambridge.

Maynard Smith J. & Parker G.A. (1976) The logic of asymetric contests. *Anim. Behav.* **24**, 159–75.

Mech L.D. (1970) *The Wolf: The Ecology and Behaviour of an Endangered Species.* Natural History Press, Garden City, New York.

Meikle D.B. & Vessey S.H. (1981) Nepotism among rhesus monkey brothers. *Nature, Lond.* **294**, 160–1.

Melnick D. (1981) Microevolution in a Population of Himalayan Rhesus Monkeys *(Macaca mulatta).* PhD Thesis, Yale University.

Melnick D. & Pearl M. (1982) Intergroup migration and the incidence of inbreeding in a wild population of rhesus monkeys. *Int. J. Primatol.* **3**(3), 313.

Menzel E.W. (1971) Communication about the environment in a group of young chimpanzees. *Folia primatol.* **15**, 220–32.

Michener G.R. (1982) Kin identification, matriarchies, and the evolution of sociality in ground-dwelling sciurids. In *Recent Advances in the Study of Mammalian Behavior*, eds Eisenberg J.F. & Kleiman D.G. American Society of Mammalogists, Special Publication No. 7.

Mineka S. & Suomi S.J. (1978) Social separation in monkeys. *Psychol. Bull.* **85**, 1376–400.

Mischel W. (1973) Toward a cognitive social learning reconceptualization of personality. *Psychol. Rev.* **80**, 252–83.

Missakian E.A. (1972) Genealogical and cross-genealogical dominance relations in a group of free-ranging rhesus monkeys *(Macaca mulatta)* on Cayo Santiago. *Primates*, **13**, 169–80.

Missakien E.A. (1973) Genealogical mating activity in free-ranging groups of rhesus monkeys *(Macaca mulatta)* in Cayo Santiago. *Behaviour*, **45**, 225–41.

Missakian E.A. (1976) Dominance relations in groups of free-ranging rhesus monkeys on Cayo Santiago. Talk given at Animal Behavior Society meeting, Boulder, Colorado, June 20–25.

Mitani J.C. & Rodman P.S. (1979) Territoriality: the relation of ranging patterns and home range size to defendability, with an analysis of territoriality among primate species. *Behav. Ecol. Sociobiol.* **5**, 214–51.

Mitchell G.D. (1968) Attachment differences in male and female infant monkeys. *Child Dev.* **39**, 611–20.

Mitchell G.D. (1979) *Behavioral Sex Differences in Non-human Primates.* Van Nostrand Reinhold, New York.

Moehlman P.D. (1979) Jackal helpers and pup survival. *Nature, Lond.* **277**, 382–3.

Moore J. (1978) Dominance relations among free-ranging female baboons in Gombe National Park, Tanzania. In *Recent Advances in Primatology I*, eds Chivers D.J. & Herbert J. Academic Press, London.

Morgan B.J.T., Simpson M.J.A., Hanby J.P. & Hall-Craggs J. (1976) Visualizing interactions and sequential data in animal behaviour: theory and application of cluster-analysis methods. *Behaviour*, **56**, 1–43.

Mori A. (1975) Signals found in the grooming interactions of wild Japanese monkeys of the Koshima troop. *Primates*, **16**, 107–40.

Mori A. (1979) Analysis of population changes by measurement of body weight in the Koshima troop of Japanese monkeys. *Primates*, **20**, 371–98.

Mori U. (1979) Unit formation and the emergence of a new leader. In *Ecological and Sociological Studies of Gelada Baboons*, ed. Kawai M. Karger, Basel and Kodansha, Tokyo.

Moss C.J. (1977) The Amboseli elephants. *Wildl. News*, **12**(2), 9–12.

Moss C.J. (1981) Social circles. *Wildl. News*, **16**(1), 2–7.

Moss C.J. Social relationships among female elephants in Amboseli National Park, Kenya. (In prep.)

Murdock G.P. (1957) World ethnographic sample. *Am. Anthrop.* **59**, 664–87.

Murray M.G. (1981) Structure of association in impala, *Aepyceros melampus. Behav. Ecol. Sociobiol.* **9**, 23–33.

Nagel U. (1971) Social organisation in a baboon hybrid zone. In *Proceedings from the Symposia of the Second Congress of the International Primatological Society*, ed. Kummer H. Karger, Basel.

Napier J.R. & Napier P.H. (1967) *A Handbook of the Living Primates*. Academic Press, New York.

Nash L.T. (1974) Parturition in a feral baboon. *Primates*, **15**, 279–85.

Nash L.T. (1976) Troop fission in free-ranging baboons in the Gombe Stream National Park, Tanzania. *Am. J. phys. Anthrop.* **44**, 63–77.

Nash L.T. (1978a). The development of the mother–infant relationship in wild baboons *(Papio anubis)*. *Anim Behav.* **26**, 746–59.

Nash L.T. (1978b) Kin preference in the behavior of young baboons. In *Recent Advances in Primatology I*, eds Chivers D.J. & Herbert J. Academic Press, London.

Negayama K. (1981) Maternal aggression to its offspring in Japanese monkeys. *J. hum. Evol.* **10**, 523–7.

Nicolson N. (1982) Weaning and the Development of Independence in Olive Baboons. PhD Thesis, Harvard University.

Nie N.H., Hull C.H., Jenkins J.G., Steinbrenner K. & Bent D.H. (1975) *Statistical Package for the Social Sciences*. McGraw-Hill, New York.

Nishida T. (1979) The social structure of chimpanzees of the Mahale Mountains. In *The Great Apes*, eds Hamburg D.A. & McCown E.R.

Norikoshi K. & Koyama N. (1975) Group shifting and social organization among Japanese monkeys. In *Proceedings from the Symposia of the Fifth Congress of the International Primatological Society*, eds Kondo S., Kawai M., Ehara A. & Kawamura S. Japan Science Press, Tokyo.

O'Hara R.K. & Blaustein A.R. (1981) An investigation of sibling recognition in *Rana cascadae* tadpoles. *Anim. Behav.* **29**, 1121–6.

Ohsawa H. & Dunbar R.I.M. Environmental and demographic determinants of demographic structure in gelada baboons. (In prep.)

Oki J. & Maeda Y. (1973) Grooming as a regulator of behavior in Japanese macaques. In *Behavioral Regulators of Behavior in Primates*, ed. Carpenter C.R. Bucknell University Press, Cranbury, New Jersey.

Oliver J.I. & Lee P.C. (1978) Comparative aspects of the behaviour of juveniles in two species of baboon in Tanzania. In *Recent Advances in Primatology. Vol. 1. Behaviour*, eds Chivers D.J. & Herbert J. Academic Press, London.

Olivier T.J., Ober C., Buettner-Janisch J. & Sade D.S. (1981) Genetic differentiation among matrilines in social groups of rhesus monkeys. *Behav. Ecol. Sociobiol.* **8**, 279–86.

Owens N.W. (1975a) Social play behaviour in free-living baboons, *Papio anubis*. *Anim. Behav.* **23**, 387–408.

Owens N.W. (1975b) A comparison of aggressive play and aggression in free-ranging baboons, *Papio anubis*. *Anim. Behav.* **23**, 757–65.

Packer C. (1977) Reciprocal altruism in *Papio anubis*. *Nature, Lond.* **265**, 441–3.

Packer C. (1979a) Inter-troop transfer and inbreeding avoidance in *Papio anubis*. *Anim. Behav.* **27**, 1–36.

Packer C. (1979b) Male dominance and reproductive activity in *Papio anubis*. *Anim. Behav.* **24**, 37–45.

Packer C. (1900) Male care and exploitation of infants in *Papio anubis*. *Anim. Behav.* **28**, 512–20.

362 References

Packer C. & Pusey A.E. (1979) Female aggression and male membership in troops of Japanese macaques and olive baboons. *Folia primatol.* **31**, 212–18.

Packer C. & Pusey A.E. (1982) Cooperation and competition within coalitions of male lions: kin selection or game theory? *Nature, Lond.* **296**, 740–2.

Parker G.A. (1974) Assessment strategy and the evolution of fighting behaviour. *J. theor. Biol.* **47**, 223–43.

Pearl M.C. & Schulman S.R. (1983) Techniques for the analysis of social structure in animal societies. In *Advances in the Study of Behaviour*, eds Rosenblatt J.S., Hinde R.A., Beer C. & Busnel M.C. Academic Press, New York. (In press.)

Phillips M.J. & Mason W.A. (1976) Comparative studies of social behaviour in *Callicebus* and *Saimiri:* Social looking in male–female pairs. *Bull. Psychon. Soc.* **7**(1), 55–6.

Pilbeam D. (1972) *The Ascent of Man.* Macmillan, New York.

Poirier F.E. (1968) The Nilgiri langur *(Presbytis johnii)* mother–infant dyad. *Primates,* **9**, 45–68.

Poirier F.E. (1969) The Nilgiri langur troop: its composition, structure, function and change. *Folia primatol.* **10**, 20–47.

Poole J.H. (1982) Musth and Male–Male Competition in the African Elephant. PhD Thesis, University of Cambridge.

Poole J.H., Lasley W. & Kasman L. Musth and urinary testosterone in the African elephant, *Loxodonta africana.* (In prep.)

Poole J.H. & Moss C.J. (1981) Musth in the African elephant, *Loxodonta africana. Nature, Lond.* **292**, 830–1.

Poole J.H. & Moss C.J. Association patterns of adult male elephants in Amboseli National Park, Kenya. (In prep.)

Popp J.L. & DeVore I. (1979) Aggressive competition and social dominance theory: synopsis. In *The Great Apes*, eds Hamburg D.A. & McCown E.K. Benjamin-Cummings, Menlo Park, California.

Porter R.H. & Wyrick M. (1979) Sibling recognition in spiny mice *(Acomys cahirinus)*: influence of age and isolation. *Anim. Behav.* **27**, 761–6.

Post D.G., Hausfater G. & McCuskey S.A. (1980) Feeding behaviour of yellow baboons *(Papio cynocephalus):* relationships to age, gender and dominance rank. *Folia primatol.* **34**, 170–95.

Pusey A.E. (1978a) The Physical and Social Development of Wild Adolescent Chimpanzees. PhD Thesis, Stanford University.

Pusey A.E. (1978b) Age changes in the mother–offspring association of wild chimpanzees. In *Recent Advances in Primatology*, vol. 1, eds Chivers D.J. & Herbert J. Academic Press, London.

Pusey A.E. (1980) Inbreeding avoidance in chimpanzees. *Anim. Behav.* **28**, 543–52.

Quiatt D. (1966) *Social Dynamics of Rhesus Monkey Groups.* University Microfilms, Ann Arbor.

Radesäter T. (1976) Individual sibling recognition in juvenile Canada geese *(Branta canadensis). Can. J. Zool.* **54**, 1069–72.

Raleigh M.J., Flannery J.W. & Ervin F.R. (1979) Sex differences in behavior among juvenile vervet monkeys *(Cercopithecus aethiops sabaeus). Behav. neural Biol.* **26**(4), 455–65.

Ransom T.W. (1981) *Beach Troops of the Gombe.* Bucknell University Press, Lewisburg, Pennsylvania.

Ransom T.W. & Ransom B.S. (1971) Adult male–infant relations among baboons *(Papio anubis). Folia primatol.* **16**, 179–95.

Ransom T.W. & Rowell T.E. (1972) Early social development of feral baboons. In *Primate Socialization,* ed. Poirer F.E. Random House, New York.

Rasmussen D.R. (1979) Correlates of patterns of range use of a troop of yellow baboons *(Papio cynocephalus)* I. Sleeping sites, impregnable females, births, and male emigrations and immigrations. *Anim. Behav.* **27**, 1098–1112.

Rasmussen D.R. (1981) Communities of baboon troops *(Papio cynocephalus)* in Mikumi National Park, Tanzania. *Folia primatol.* **36**, 232–42.

Rasmussen K.L.R. (1980) Consort Behaviour and Mate Selection in Yellow Baboons *(Papio cynocephalus).* PhD Thesis, University of Cambridge.

Redican W.K. (1976) Adult male–infant interactions in non-human primates. In *The Role of the Father in Child Development,* ed. Lamb M.E. John Wiley & Sons, New York.

Redican W.K. & Mitchell G. (1973) The social behavior of adult male–infant pairs of rhesus macaques in a laboratory environment. *Am. J. phys. Anthrop.* **38**, 523–6.

van Rhijn J.G. & Vodegel R. (1980) Being honest about one's intentions: An evolutionary stable strategy for animal conflicts. *J. theor. Biol.* **85**(4), 623–41.

Rhine R.J. (1972) Changes in the social structure of two groups of stump-tail macaques *(Macaca arctoides). Primates,* **13**, 181–94.

Rhine R.J. & Flanigon M. (1978) An empirical comparison of one-zero, focal-animal, and instantaneous methods of sampling spontaneous primate social behaviour. *Primates,* **19**(2), 353–61.

Rhine R.J., Forthman D.L., Stillwell-Barnes R., Westlund B.J. & Westlund H.D. (1979) *Folia primatol.* **32**, 241–51.

Richard A. (1974) Intra-specific variation in the social organization and ecology of *Propithecus verreauxi. Folia primatol.* **22**, 178–207.

Richards S.M. (1974) The concept of dominance and methods of assessment. *Anim. Behav.* **22**, 914–30.

Rijksen H.D. (1978) *A Field Study on Sumatran Orangutans.* Veenman & Zonen, Wageningen.

Riss D. & Goodall J. (1977) The recent rise to the alpha-rank in a population of free-living chimpanzees. *Folia primatol.* **27**, 134–51.

Robinson J.G. (1981) Vocal regulation of inter- and intragroup spacing during boundary encounters in the Titi monkey, *Callicebus moloch. Primates,* **22**, 161–72.

Rodman P.S. (1973) Population composition and adaptive organization among orangutans of the Kutai Nature Reserve, East Kalimantan. In *Comparative Ecology and Behaviour of Primates,* eds Michael R.P. & Crook J.H. Academic Press, London.

Rodman P.S. (1977) Feeding behaviour of orang-utans of the Kutai Nature Reserve, East Kalimantan. In *Primate Ecology: Studies of Feeding and Ranging Behaviour in Lemurs, Monkeys and Apes,* ed. Clutton-Brock T.H. Academic Press, London.

Rood J.P. (1978) Dwarf mongoose helpers at the den. *Z. Tierpsychol.* **48**, 277–87.

Rose R.M., Bernstein I.S. & Gordon T.P. (1975) Consequences of social conflict on plasma testosterone levels in rhesus monkeys. *Psychosom. Med.* **37**, 50–61.

Rosenblum L.A. & Harlow H.F. (1963) Approach–avoidance in the mother-surrogate situation. *Psychol. Rep.* **12**, 83 5.

Ross N.M. & Gamboa G.J. (1981) Nestmate discrimination in social wasps *(Polistes metricus,* Hymenoptera: Vespidae). *Behav. Ecol. Sociobiol.* **9**, 163–5.

Rowell T.E. (1966) Hierarchy in the organisation of a captive baboon group. *Anim. Behav.* **14**, 430–43.

Rowell T.E. (1969) Intra-sexual behaviour and female reproductive cycles of baboons *(Papio anubis)*. *Anim. Behav.* **17**, 159–67.

Rowell T.E. (1970) Baboon menstrual cycles affected by social environment. *J. Reprod. Fert.* **21**, 133–41.

Rowell T.E. (1972a) *Social Behaviour of Monkeys.* Penguin Books, Harmondsworth.

Rowell T.E. (1972b) Female reproductive cycles and social behaviour in primates. In *Advances in the Study of Behaviour,* vol. 4, eds Lehrman D.S., Hinde R.A. & Shaw E. Academic Press, London.

Rowell T.E. (1974) The concept of social dominance. *Behav. Biol.* **11**, 131–54.

Rowell T.E. (1978) How female reproductive cycles affect interaction patterns in groups of patas monkeys. In *Recent Advances in Primatology VI,* eds Chivers D.J. & Herbert J. Academic Press, London.

Rowell T.E., Hinde R.A. & Spencer Booth Y. (1964) "Aunt"–infant interactions in captive rhesus monkeys. *Anim. Behav.* **12**, 219–26.

Rowell T.E. & Richards S.M. (1979) Reproductive strategies of some African monkeys. *J. Mammal.* **60**, 58–69.

Rubenstein D.I. (1980) On the evolution of alternative mating strategies. In *Limits to Action: The Allocation of Individual Behaviour,* ed. Staddon J.E.R. Academic Press, New York.

Rubenstein D.I. (1982) Reproductive value, and behavioral strategies: coming of age in monkeys and horses. *Perspect. Ethol.* **5**, 469–88.

Rubin Z. (1980) *Children's Friendships.* Collins (William), Sons (Fontana), London.

Rudnai J.A. (1973) *The Social Life of the Lion.* MTP Press, Lancaster.

Ruppenthal G.C., Arling G.L., Harlow H.F., Sackett G.P. & Suomi S.J. (1976) A 10 year perspective of motherless-mother monkey behavior. *J. Abnorm. Psychol.* **85**, 341–9.

Ruppenthal G.C., Harlow M.K., Eisele C.D., Harlow H.F. & Suomi S. (1974) Development of peer interactions of monkeys reared in a nuclear family environment. *Child Dev.* **45**, 670–82.

Saayman G.S. (1971) Behaviour of the adult males in a troop of free-ranging chacma baboons *(Papio ursinus)*. *Folia primatol.* **15**, 36–57.

Sackett G.P. (1982) Can single processes explain effects of postnatal influences on primate development? In *The Development of Attachment and Affiliative Systems,* eds Emde R.N. & Harman R.J. Plenum Press, New York.

Sackett G.P., Holm R.A., Davis A.E. & Farhenbruck E.E. (1975) Prematurity and low birth weight in pigtail macaques: incidence, prediction and effects on infant development. In *Proceedings from the Symposia of the Fifth Congress of the International Primatological Society,* eds Kondo S., Kawai M., Ehara A. & Kawamura S. Japan Science Press, Tokyo.

Sackett G.P., Ruppenthal G.C. & Gluck J. (1978) An overview of methodological and statistical problems in observational research. In *Observing Behaviour II,* ed. Sackett G.P. University Park Press, Baltimore.

Sade D.S. (1965) Some aspects of parent–offspring and sibling relationships in a group of rhesus monkeys, with a discussion of grooming. *Am. J. phys. Anthrop.* **23**(1), 1–18.

Sade D.S. (1967) Determinants of dominance in a group of free-ranging rhesus monkeys.

In *Social Communication Among Primates,* ed. Altmann S.A. University of Chicago Press, Chicago.

Sade D.S. (1968) Inhibition of son–mother mating in free-ranging rhesus monkeys. *Sci. Psychoanal.* **12**, 18–38.

Sade D.S. (1969) An algorithm for dominance relations: Rules for adult females and sisters. *Am. J. phys. Anthrop.* **31**, 271.

Sade D.S. (1971) Life cycle and social organization among free-ranging rhesus monkeys. Paper read at American Association of Anthropologists meeting, November 1971, New York.

Sade D.S. (1972a) Sociometrics of *Macaca mulatta* I: Linkages and cliques in grooming matrices. *Folia primatol.* **196**, 223.

Sade D.S. (1972b) A longitudinal study of social behavior of rhesus monkeys. In *The Functional and Evolutionary Biology of Primates,* ed. Tuttle R. Aldine, Chicago.

Sade D.S., Cushing K., Cushing P., Dunale J., Figueroa A., Kaplan J.R., Lauer C., Rhodes D. & Schneider J. (1976) Population dynamics in relation to social structure on Cayo Santiago. *Yearbook phys. Anthrop.* **20**, 253–62.

Sassenrath E.N. (1970) Increased adrenal responsiveness related to social stress in rhesus monkeys. *Horm. Behav.* **1**, 283–398.

Saunders C. & Hausfater G.H. (1978) Sexual selection in baboons *(Papio cynocephalus):* a computer simulation of differential reproduction with respect to dominance rank in males. In *Recent Advances in Primatology,* vol. I, eds Chivers D. & Herbert J. Academic Press, New York.

Schaller G.B. (1963) *The Mountain Gorilla: Ecology and Behavior.* University of Chicago Press, Chicago.

Schaller G.B. (1972) *The Serengeti Lion.* University of Chicago Press, Chicago.

Schulman S.R. (1980) Intragroup Spacing and Multiple Social Networks in *Macaca mulatta.* PhD Thesis, School of Forestry and Environmental Studies, Yale University.

Schulman S.R. & Boorman S.A. New methods for the identification and interspecific comparison of social structure in non-human vertebrate populations: Blockmodels of multiple networks in a free-ranging population of *Macaca mulatta.* (In prep.)

Schulman S.R. & Chapais B. (1980) Reproductive value and rank relations among macaque sisters. *Am. Nat.* **115**, 580–93.

Schulman S.R. & Rubenstein D.I. (1983) Kinship, need, and the distribution of altruism. *Am. Nat.* **121** (in press).

Seligman M.E.P. & Hager J.L. (1972) *Biological Boundaries of Learning.* Appleton-Century-Crofts, New York.

Selman R. & Jaquette D. (1978) Stability and oscillation in interpersonal awareness: A clinical–developmental analysis. In *Nebraska Symposium on Motivation,* ed. Keasey C.B. University of Nebraska Press, Lincoln.

Seyfarth R.M. (1975) The Social Relationships among Adults in a Troop of Free-ranging Baboons *(Papio cynocephalus ursinus).* PhD Thesis, University of Cambridge.

Seyfarth R.M. (1976) Social relationships among adult female baboons. *Anim. Behav.* **24**, 917–38.

Seyfarth R.M. (1977) A model of social grooming among adult female monkeys., *J. theor. Biol.* **65**, 671–98.

Seyfarth R.M. (1978a) Social relationships among adult male and female baboons. I. Behaviour during sexual consortships. *Behaviour,* **64**, 204–26.

Seyfarth R.M. (1978b) Social relationships among adult male and female baboons. II. Behaviour throughout the female reproductive cycle. *Behaviour,* **64**, 227–47.

Seyfarth R.M. (1980) The distribution of grooming and related behaviours among adult female vervet monkeys. *Anim. Behav.* **28**, 798–813.

Seyfarth R.M. (1981) Do monkeys rank each other? *Behav. Brain Sci.* **4**, 447–8.

Seyfarth R.M., Cheney D.L. & Hinde R.A. (1978) Some principles relating social interactions and social structure among primates. In *Recent Advances in Primatology*, vol. 1, eds Chivers D.J. & Herbert J. Academic Press, London.

Sharman M.J. (1981) Feeding, Ranging and Social Organisation of the Guinea Baboon, *Papio Papio*. PhD Thesis, University of St Andrews.

Sharman M.J. & Dunbar R.I.M. Observer bias in the selection of study groups in baboon field studies. *Primates* (in press).

Sherman P.W. (1980) The limits of ground squirrel nepotism. In *Sociobiology: Beyond Nature/Nurture?*, eds Barlow G.W. & Silverberg J. Westview Press, Boulder, Colorado.

Short R. (1979) Sexual selection and its component parts, somatic and genital selection, as illustrated by man and the great apes. *Adv. Stud. Behav.* **9**, 131–58.

Sigg H. (1980) Differentiation of female positions in hamadryas one-male units. *Z. Tierpsychol.* **53**, 265–302.

Sigg H. & Stolba A. (1981) Home range and daily march in a hamadryas baboon troop. *Folia primatol.* **36**, 40–75.

Silk J.B. (1979) Feeding, foraging and food sharing behavior of immature chimpanzees. *Folia primatol.* **31**, 123–42.

Silk J.B. (1980) Kidnapping and female competition among captive bonnet macaques. *Primates*, **21**(1), 100–10.

Silk J.B. (1981) Social Behavior of Female *Macaca radiata:* the Influence of Kinship and Rank on Cooperation and Competition. PhD Thesis, University of California.

Silk J.B. (1982) Altruism among female *Macaca radiata:* explanations and analysis of patterns of grooming and coalition formation. *Behaviour*, **79**, 162–88.

Silk J.B., Clark-Wheatley C.B., Rodman P.S. & Samuels A. (1981a) Differential reproductive success and facultative adjustment of sex ratios among captive female bonnet macaques. *Anim. Behav.* **29**, 1106–20.

Silk J.B., Samuels S.A. & Rodman P.S. (1981b) The influence of kinship, rank and sex on affiliation and aggression between adult female and immature bonnet macaques *(Macaca radiata). Behaviour*, **78**, 111–37.

Simonds P.E. (1973) Outcast males and social structure among bonnet macaques *(Macaca radiata). Am. J. phys. Anthrop.* **38**, 599–604.

Simpson M.J.A. (1973) The social grooming of male chimpanzees. In *Comparative Ecology and Behaviour of Primates*, eds Michael R.P. & Crook J.H. Academic Press, London.

Simpson M.J.A. (1979a) Daytime rest and activity in socially living rhesus monkey infants. *Anim. Behav.* **27**, 602–12.

Simpson M.J.A. (1979b) Problems of recording behavioural data by keyboard. In *Social Interaction Analysis: Methodological Issues*, eds Lamb M.E., Suomi S.J. & Stephenson G.R. University of Wisconsin Press, Madison.

Simpson M.J.A. & Howe S. (1980) The interpretation of individual differences in rhesus monkey infants. *Behaviour*, **72**, 127–55.

Simpson M.J.A. & Simpson A.E. (1977) One-zero and scan methods for sampling behaviour. *Anim. Behav.* **25**, 726–31.

Simpson M.J.A. & Simpson A.E. Birth/sex ratios and social rank in rhesus monkey mothers. *Nature, Lond.* **300**, 440–1.

Simpson M.J.A., Simpson A.E., Hooley J. & Zunz M. (1981) Infant-related influences on birth intervals in rhesus monkeys. *Nature, Lond.* **290**, 49–51.

Slatkin M. & Hausfater G. (1976) A note on the activities of a solitary male baboon. *Primates*, **17**(3), 311–22.

Small M.F. & Smith D.G. (1981) Interactions with infants by full siblings, paternal half siblings and non-relatives in a captive group of rhesus macaques *(Macaca mulatta)*. *Am. J. Primatol.* **1**, 91–4.

Smith D.G. (1981) The association between rank and reproductive success of male rhesus monkeys. *Am. J. Primatol.* **1**, 83–90.

Smith S.M. (1976) Ecological aspects of dominance hierarchies in black-capped chickadees. *Auk*, **93**, 95–107.

Smuts B.B. (1982) Special Relationships between Adult Male and Female Olive Baboons *(Papio anubis)*. PhD Thesis, Stanford University.

Smuts B.B. (1983) Male–male competition, female choice, and mating activity among savannah baboons: a review. In *Evolution, Adaptation and Behavior*, eds Tooby J. & DeVore I. Aldine, Hawthorne, New York.

Snowdon C.T. & Cleveland J. (1980) Individual recognition of contact calls by pygmy marmosets. *Anim. Behav.* **28**, 717–27.

Southwick C.H., Beg M.A. & Siddiqi M.R. (1965) Rhesus monkeys in North India. In *Primate Behavior: Field Studies of Monkeys and Apes*, ed. DeVore I. Holt, Rinehart & Winston, New York.

Spencer-Booth Y. (1968) The behaviour of group companions towards rhesus monkey infants. *Anim. Behav.* **16**, 541–57.

Spencer-Booth Y. (1970) The relationship between mammalian young and conspecifics other than mothers and peers. In *Advances in the Study of Behaviour 3*, eds Lehrman D.S., Hinde R.A. & Shaw E. Academic Press, London.

Spencer-Booth Y. & Hinde R.A. (1967) The effects of separating rhesus monkey infants from their mothers for six days. *J. Child Psychol. Psychiat.* **7**, 179–97.

Spencer-Booth Y. & Hinde R.A. (1969) Tests of behavioural characteristics for rhesus monkeys. *Behaviour*, **33**, 179–211.

Spencer-Booth Y. & Hinde R.A. (1971a) Effects of 6 days separation from mother on 18- to 32-week old rhesus monkeys. *Anim. Behav.* **19**, 174–91.

Spencer-Booth Y. & Hinde R.A. (1971b) Effects of brief separation from mothers during infancy on behaviour of rhesus monkeys 6–24 months later. *J. Child Psychol. Psychiat.* **12**, 157–72.

Stacey P.B. & Bock C.E. (1978) Social plasticity in the acorn woodpecker. *Science, N.Y.* **202**, 1298–1300.

Stammbach E. (1978) On social differentiation in groups of captive female hamadryas baboons. *Behaviour*, **67**, 322–38.

Stein D.M. (1981) The Nature and Function of Social Interactions between Infant and Adult Male Yellow Baboons *(Papio cynocephalus)*. PhD Thesis, University of Chicago.

Stephenson G.R. (1975) Social structure of mating activity in Japanese macaques. In *Proceedings from the Symposia of the Fifth Congress of the International Primatological Society*, eds Kondo S., Kawai M., Ehara A. & Kawamura S. Japan Science Press, Tokyo.

Stevenson-Hinde J. & Simpson M.J.A. (1981) Mothers' characteristics, interactions, and infants' characteristics. *Child Dev.* **52**, 1246–54.

Stevenson-Hinde J., Stillwell-Barnes R. & Zunz M. (1980a) Subjective assessment of rhesus monkeys over four successive years. *Primates*, **21**, 66–82.

Stevenson-Hinde J., Stillwell-Barnes R. & Zunz M. (1980b) Individual differences in young rhesus monkeys: consistency and change. *Primates*, **21**, 498–509.

Stevenson-Hinde J. & Zunz M. (1978) Subjective assessment of individual rhesus monkeys. *Primates*, **19**, 473–82.

Stevenson-Hinde J., Zunz M. & Stillwell-Barnes R. (1980) Behaviour of one-year-old rhesus monkeys in a strange situation. *Anim. Behav.* **28**, 266–77.

Stewart K.J. (1981) Social Development of Wild Mountain Gorillas PhD. Thesis, University of Cambridge.

Stolba A. (1979) Entscheidungsfindung in Verbänden von *Papio hamadryas*. PhD Thesis, University of Zurich.

Stolte L.A.M. (1978) Pregnancy in the rhesus monkey. In *The Rhesus Monkey*, vol. 2, ed. Bourne G.H. Academic Press, New York.

Stoltz L.P. & Saayman G.S. (1970) Ecology and behaviour of baboons in the Northern Transvaal. *Ann. Transv. Mus.* **26**, 99–143.

Strayer F.F. (1976) Learning and imitation as a function of social status in macaque monkeys *(Macaca nemestrina)*. *Anim. Behav.* **24**, 835–48.

Strayer F.F. & Harris P.J. (1979) Social cohesion among captive squirrel monkeys *(Saimiri scuireus)*. *Behav. Ecol. Sociobiol.* **5**(1), 93–109.

Struhsaker T.T. (1967a) Behaviour of vervet monkeys *(Cercopithecus aethiops)*. *Univ. Calif. Publs Zool.* **82**, 1–64.

Struhsaker T.T. (1967b) Social structure among vervet monkeys *(Cercopithecus aethiops)*. *Behaviour*, **29**, 6–121.

Struhsaker T.T. (1967c) Ecology of vervet monkeys *(Cercopithecus aethiops)* in the Masai–Amboseli Game Reserve, Kenya. *Ecology*, **48**, 891–904.

Struhsaker T.T. (1967d) Auditory communication among vervet monkeys *(Cercopithecus aethiops)*. In *Social Communication Among Primates*, ed. Altmann S.A. University of Chicago Press, Chicago.

Struhsaker T.T. (1971) Social behaviour of mother and infant vervet monkeys *(Cercopithecus aethiops)*. *Anim. Behav.* **19**, 233–50.

Struhsaker T.T. (1977) Infanticide and social organisation in the redtail monkey *(Cercopithecus ascanius schmidti)* in the Kibale Forest, Uganda. *Z. Tierpsychol.* **45**, 75–84.

Strum S.C. (1975) Life with the Pumphouse Gang: New insights into baboon behavior. *Natn. geogr. Mag.* **147**, 672–91.

Strum S.C. (1982) Agonistic dominance in male baboons: An alternative view. *Int. J. Primatol.* **3**, 175–202.

Strum S.C. Why do males use infants? In *Primate Paternalism*, ed. Taub D. Van Nostrand Reinhold, New York. (In press a.)

Strum S.C. Why males use females among olive baboons. *Am. J. Primatol.* (in press b).

Sugiyama Y. (1960) On the division of a natural troop of Japanese monkeys at Takasaki-yama. *Primates*, **2**, 109–48.

Sugiyama Y. (1976) Life history of male Japanese monkeys. In *Advances in the Study of Behavior 7*, eds Rosenblatt J.S., Hinde R.A., Shaw E. & Beer C. Academic Press, New York.

Suomi S.J. (1976) Factors affecting responses to social separation in rhesus monkeys. In *Animal Models in Human Psychobiology*, eds Serban G. & Kling A. Plenum, New York.

Suomi S.J. (1977) Adult male–infant interactions among monkeys living in nuclear families. *Child Dev.* **48**, 1255–70.

Suomi S.J., Collins M.L. & Harlow H.F. (1973) Effects of permanent separation from mother on infant monkeys. *Devl. Psychol.* **9**, 376–84.

Suomi S.J. & Harlow H.F. (1978) Early experience and social development in rhesus monkeys. In *Social and Personality Development*, ed. Lamb M. Holt, Rinehart & Winston, New York.

Symons D. (1974) Aggressive play and communication in rhesus monkeys *(Macaca mulatta). Am. Zool.* **14**, 317–22.

Symons D. (1978) The question of function: dominance and play. In *Social Play in Primates*, ed. Smith E.O. Academic Press, London.

Szalay F.S. & Delson E. (1979) *The Evolutionary History of the Primates*. Academic Press, New York.

Tajfel H. (1978) The psychological structure of intergroup relations. In *Differentiation between Social Groups*, ed. Tajfel H. Academic Press, London.

Takahata Y. (1982) Social relations between adult males and females of Japanese monkeys in the Arashiyama B troop. *Primates*, **23**, 1–23.

Tartabini A. & Dienske H. (1979) Social play and rank order in rhesus monkeys *(Macaca mulatta). Behav. Process*, **4**, 375–83.

Taub D.M. (1980a) Female choice and mating strategies among wild Barbary macaques *(Macaca sylvanus L.)*. In *The Macaques. Studies in Ecology, Behavior and Evolution*, ed. Lindburg D.A. Van Nostrand Reinhold, New York.

Taub D.M. (1980b) Testing the 'agonistic buffering' hypothesis. *Behav. Ecol. Sociobiol.* **6**, 187–98.

Tenaza R.R. (1976) Songs, choruses and countersinging of Kloss' gibbon *(Hylobates klossii)* in Siberut Island, Indonesia. *Z. Tierpsychol.* **40**, 37–52.

Thibaut J.W. & Kelley H.H. (1959) *The Social Psychology of Groups*. John Wiley & Sons, New York.

Tinbergen N. (1959) Comparative studies of the behaviour of gulls (Laridae): a progress report. *Behaviour*, **15**, 1–70.

Tinbergen N. (1963) On aims and methods of ethology. *Z. Tierpsychol.* **20**, 410–33.

Tissier G., Colvin J.D. & Barton R.A. Models of peer networks among immature male rhesus monkeys. (In prep.)

Trivers R.L. (1971) The evolution of reciprocal altruism. *Q. Rev. Biol.* **46**, 35–57.

Trivers R.L. (1972) Parental investment and sexual selection. In *Sexual Selection and the Descent of Man*, ed. Campbell B. Aldine, Chicago.

Trivers R.L. (1974) Parent–offspring conflict. *Am. Zool.* **14**, 249–64.

Tsumori A. (1967) Newly acquired behavior and social interactions of Japanese monkeys. In *Social Communication among Primates*, ed. Altmann S.A. The University of Chicago Press, Chicago.

Tutin C.E.G. (1979) Mating patterns and reproductive strategies in a community of wild chimpanzees *(Pan troglodytes schweinfurthii). Behav. Ecol. Sociobiol.* **6**, 29–38.

Tutin C.E.G. (1980) Reproductive behaviour of wild chimpanzees in the Gombe National Park, Tanzania. *Reprod. Fert. Suppl.* **28**, 43–57.

Vaitl E.A. (1977) Social context as a structuring mechanism in captive groups of squirrel monkeys. *Primates*, **18**, 861–74.

Vaitl E. (1978) Nature and implications of complexly organised social systems in non-human primates. In *Recent Advances in Primatology 1*, eds Chivers D.J. & Herbert J. Academic Press, London.

Vaitl E.A., Mason W.A., Taub D.M. & Anderson C.O. (1978) Contrasting effects of living in heterosexual pairs and mixed groups on the structure of social attraction in squirrel monkeys. *Anim. Behav.* **26**, 358–67.

Vandell D.L. & Mueller E.C. (1980) Peer play and friendships during the first two years. In *Friendship and Social Relations in Children*, eds Foot H.C., Chapman A.J. & Smith J.R. John Wiley & Sons, London.

Varley M. & Symmes D. (1966) The hierarchy of dominance in a group of macaques. *Behaviour*, **27**, 54–75.

Vehrencamp S.L. (1978) The adaptive significance of communal nesting in groove-billed anis *(Crotophaga sulcirostris). Behav. Ecol. Sociobiol.* **4**, 1–33.

Vehrencamp S.L. (1979) The roles of individuals, kin and group selection in the evolution of sociality. In *Handbook of Neurobiology, Vol. 3: Social Behavior and Communication*, eds Marler P. & Vandenburgh J.G. Plenum, New York.

Vessey S.H. (1968) Interactions between free-ranging groups of rhesus monkeys. *Folia primatol.* **8**, 228–39.

Vogt J.L. (1978) The social behavior of a marmoset group. *Folia primatol.* **29**, 250–67.

de Waal F.B.M. (1975) The wounded leader. A spontaneous temporary change in the structure of agonistic relations among captive java-monkeys *(Macaca fascicularis). Neth. J. Zool.* **25**(4), 529–49.

de Waal F.B.M. (1977) The organisation of agonistic relations within two captive groups of Java monkeys *(Macaca fascicularis). Z. Tierpsychol.* **44**, 225–82.

de Waal F.B.M. (1978) Exploitative and familiarity dependent support strategies in a colony of semi-free living chimpanzees. *Behaviour*, **66**, 268–312.

de Waal F.B.M. (1982) *Chimpanzee Politics.* Jonathan Cape, London.

de Waal F.B.M., van Hooff J.A.R.A.M. & Nelto W.J. (1976) An ethological analysis of types of agonistic interaction in a captive group of Java monkeys *(Macaca fascicularis). Primates*, **17**, 257–90.

de Waal F.B.M. & van Roosmalen A. (1979) Reconciliation and consolation among chimpanzees. *Behav. Ecol. Sociobiol.* **5**, 55–66.

Wade T.D. (1977) Complementarity and symmetry in social relationships of non-human primates. *Primates*, **18**, 835–47.

Wade T.D. (1979) Inbreeding and kin selection in primate social evolution. *Primates*, **20**, 355–70.

Walters J. (1980) Interventions and the development of dominance relationships in female baboons. *Folia primatol.* **34**, 61–89.

Walters J. (1981) Inferring kinship from behaviour: maternity determinations in yellow baboons. *Anim. Behav.* **29**, 126–36.

Waser P.M. (1976) *Cercocebus albigena:* site attachment, avoidance and intergroup spacing. *Am. Nat.* **110**, 911–35.

Waser P.M. & Homewood K. (1979) Cost-benefit approaches to territoriality: A test with forest primates. *Behav. Ecol. Sociobiol.* **6**, 115–19.

Washburn S.L. & DeVore I. (1961) The social life of baboons. *Scient. Am.* **204**(6), 62–71.

Watanabe K. (1979) Alliance formation in a free-ranging troop of Japanese macaques. *Primates*, **20**(4), 459–74.

Weigel R.M. (1981) The distribution of altruism among kin: a mathematical model. *Am. Nat.* **118**, 191–201.

Weisbard C. & Goy R. (1976) Effect of parturition and group composition on competitive drinking order in stumptail macaques. *Folia primatol.* **25**, 95–121.

Welker C., Lührmann B. & Meinel W. (1980) Behavioural sequences and strategies of female crab-eating monkeys. *Behaviour,* **73**, 219–37.

Western D. (1975) Water availability and its influence on the structure and dynamics of a savannah large mammal community. *E. Afr. wildl. J.* **13**, 265–86.

White H.C., Boorman S.A. & Breiger R. (1976) Social structure from multiple networks. I. Blockmodels of roles and positions. *Am. J. Sociol.* **81**, 730–80.

White L.E. (1977) The Nature of Social Play and the Development of the Rhesus Monkey. PhD Thesis, University of Cambridge.

White R.E.C. (1971) WRATS: a computer compatible system for automatically recording and transcribing behavioral data. *Behaviour,* **40**, 135–61.

Whitten P. (1982) Female Reproductive Strategies among Vervet Monkeys. PhD Thesis, Harvard University.

Williams G.C. (1966) *Adaptation and Natural Selection.* Princeton University Press, New Jersey.

Wilson E.O. (1971) *The Insect Societies.* Harvard University Press, Cambridge, Massachusetts.

Wilson E.O. (1975) *Sociobiology: the New Synthesis.* Harvard University Press, Cambridge, Massachusetts.

Wilson E.O. (1978) *On Human Nature.* Harvard University Press, Cambridge, Massachusetts.

Wilson E.O. & Lumsden C. (1981) *Genes, Mind and Culture.* Harvard University Press, Cambridge, Massachusetts.

Witt R., Schmidt C. & Schmitt J. (1981) Social rank and Darwinian fitness in a multi-male group of Barbary macaques *(Macaca sylvana* Linnaeus, 1758). *Folia primatol.* **36**, 201–11.

Wittenberger J.F. & Tilson R.L. (1980) The evolution of monogamy: hypotheses and evidence. *Ann. Rev. ecol. Syst.* **11**, 197–232.

Wrangham R.W. (1974) Artificial feeding of chimpanzees and baboons in their natural habitat. *Anim. Behav.* **22**, 83–93.

Wrangham R.W. (1977) Feeding behaviour of chimpanzees in Gombe National Park, Tanzania. In *Primate Ecology,* ed. Clutton-Brock T.H. Academic Press, London.

Wrangham R.W. (1979) On the evolution of ape social systems. *Soc. Sci. Inf.* **18**, 335–68.

Wrangham R.W. (1980) An ecological model of female-bonded primate groups. *Behaviour,* **75**, 262–300.

Wrangham R.W. (1981) Drinking competition among vervet monkeys. *Anim. Behav.* **29**, 904–10.

Wrangham R.W. (1982) Mutualism, kinship and social evolution. In *Current Problems in Sociobiology,* eds Bertram B.C.R., Clutton-Brock T.H., Dunbar R.I.M., Rubenstein D.I. & Wrangham R.W. Cambridge University Press, Cambridge.

Wrangham R.W. & Smuts B.B. (1980) Sex differences in the behavioural ecology of chimpanzees in Gombe National Park, Tanzania. *J. Reprod. Fert.* **28**(suppl.), 1–20.

Wrangham R.W. & Waterman P.G. (1981) Feeding behaviour of vervet monkeys on *Acacia tortilis* and *Acacia xanthophloea:* with special reference to reproductive strategies and condensed tannin production. *J. Anim. Ecol.* **50**, 715–31.

Wu H.M.H., Holmes W.G., Medina S.R. & Sackett G.P. (1980) Kin preferences in infant *Macaca nemistrina. Nature, Lond.* **203**, 225–77.

Wynne-Edwards V.C. (1962) *Animal Dispersion in Relation to Social Behaviour.* Oliver & Boyd, Edinburgh.

Yamada M. (1963) A study of blood relationships in the natural society of the Japanese monkey. *Primates,* **4,** 43–65.

Yamada M. (1971) Five natural troops of Japanese monkeys on Shodoshima Island, II. A Comparison of social structure. *Primates,* **12,** 125–50.

Yarrow L.J. & Yarrow M.R. (1964) Personality and change in the family context. In *Personality Change,* eds Worchel P. & Byrne D. John Wiley & Sons, New York.

Yoshiba K. (1968) Local and intertroop variability in ecology and social behavior of common Indian langurs. In *Primates: Studies in Adaptation and Variability,* ed. Jay P.C. Holt, Rinehart & Winston, New York.

Zimen E. (1975) Social dynamics of the wolf pack. In *The Wild Canids,* ed. Fox M.W. Van Nostrand Reinhold, New York.

Zucker E.L. (1982) Comparative behavior of African monkeys: are primate models applicable? *Int. J. Primatol.* 3(3), 253.

Author index

Page numbers shown in italics refer to the list of references.

Abegglen J-J.   302, (63, 195, Kummer *et al.* 1978), *340, 357*
Albon S.D.   (328, Clutton-Brock *et al.* 1982), *346*
Alcock J.   326, *340*
Aldrich-Blake F.P.G.   177, 181, *340, 346*
Alexander R.D.   338, *340*
Ali R.   169, *340*
Allee W.C.   103, 110, *350*
Altman I.   336, *340*
Altmann J.   9, 10, 14, 15, 19, 37, 112, 134, 146, 154, 159, 165, 166, 176, 181, 188, 200, 225, 232, 237, 253, 254, 258, 262, 263, 303, 308, (234, Hausfater *et al.* 1982), *340, 352*
Altmann S.A.   9, 10, 152, 165, 166, 181, 188, 189, 232, 242, 237, 253, 255, 270, 303, 308, (234, Hausfater *et al.* 1982), *340, 352*
Anderson C.M.   162, 181, 233, *340*
Anderson C.O.   122, (128, Vaitl *et al.* 1978), *341, 370*
Anderson D.M.   54, 180, *341*
Angst W.   252, (52, 153, 195, Kummer *et al.* 1974), *341, 357*
Arabie P.   222, 223, (222, 223, Breiger *et al.* 1975), *341, 343, 344*
Arling G.L.   (54, 122, Ruppenthal *et al.* 1976), *364*
de Assumpcao T.   179, *341*
Atkinson S.   16, *353*
Axelrod R.   261, *341*
Azuma S.   252, *341*

Bachmann C.   63, 68, 153, 195, 335, (63, 195, Kummer *et al.* 1978), *341, 357*
Badawi M.   (56, Delvoye *et al.* 1978), *347*
Baker M.C.   182, *341*
Baker R.R.   160, *341*
Baldwin J.D.   82, 225, *341*

Baldwin J.I.   82, 225, *341*
Baldwin P.J.   (68, 309, McGrew *et al.* 1979, 1981), *358*
Barnes J.A.   168, *341*
Barth F.   188, *341*
Barton R.A.   (58, 191, 195, 196–7, 199, Tissier *et al.* in prep.), *369*
Bateson P.P.G.   82, 88, *341*
Beckoff M.   82, *341*
Beer C.G.   326, *341*
Beg M.A.   (9, 221, Southwick *et al.* 1965), *367*
Belcher C.S.   (182, Baker *et al.* 1981), *341*
Bem D.J.   29, *342*
Bengtsson B.O.   248, *342*
Bent D.H.   (31, Nie *et al.* 1975), *361*
Berger J.   82, 86, 227, *342*
Berman C.M.   10, 79–81, 89–93, 132–4, 146, 154–9, 188, 275, 295, *342*
Bernal J.   338, *342*
Bernstein I.S.   10, 35, 37, 82, 106, 110, 122, 181, 289, 294, (122, Gordon *et al.* 1979; Rose *et al.* 1975), *342–3, 350, 363*
Bernstein L.   263, *343*
von Bertalanffy L.   199, *343*
Bertram B.C.R.   259, 329, 330, (329, Bygott *et al.* 1979), *343, 344*
Biernoff A.   106, *343*
Bird J.   181, *349*
Birkhoff G.   224, *343*
Bishop N.   82, *347*
Blaustein A.R.   327, *361*
Block J.   29, 30, 35, *343*
Blurton Jones N.   338, *343*
Bock C.E.   259, 326, *367*
Bodmer W.F.   248, *345*
Boelkins R.C.   162, 181, 236, 238, 240, 243, 283, *343*
Boesch C.   68, *343*
Boesch H.   68, *343*
Boggess J.   252, *343*

373

Bolwig N.   134, *343*
Boorman S.A.   222, 223, 224, (222, 223, Arabie *et al.* 1978; Breiger *et al.* 1975), *341, 343, 344, 365*
Borgia G.   *340*
Bowlby J.   337, 338, *343*
Bradbury J.W.   233, 327, 329, 334, *343, 358*
Brain C.K.   9, 231, *349*
Bramblett C.A.   83, 86, 104, 183, *343*
Breiger R.L.   222, 223, *344*
Brown E.R.   329, 330, *344*
Brown J.L.   328, 329, 330, *344*
Bruner J.S.   30, *344*
Buettner-Janisch J.   180, (180, Olivier *et al.* 1981), *359, 361*
Bugos P.E.   339, *345*
Buirski N.   29, *344*
Buirski P.   29, *344*
Bunnell B.N.   180, *344*
Burton J.J.   112, *344*
Buskirk R.E.   180, 235, (10, 232, 236 237, Hamilton *et al.* 1975, 1976), *344, 351*
Buskirk W.H.   180, 235, (10, 232, 236, 237, Hamilton *et al.* 1975, 1976), *344, 351*
Buss I.O.   315, 316, *344*
Busse C.   146, 165, 261, 263, (260, Hamilton *et al.* 1982), *344, 351*
Byers J.A.   82, *341*
Bygott J.D.   177, 180, 181, 309, 310, 329, *344*

Caine N.G.   61, 89, *344*
Campbell B.   338, *344*
Cant J.H.G.   258, *344*
Capitanio J.P.   122, *344*
Caplow T.   262, *344*
Carpenter C.R.   10, 294, *344·*
Catchpole H.R.   268, *345*
Cavalli-Sforza L.L.   248, *345*
Chagnon N.A.   339, *345*
Chalmers N.R.   82, 225, *345*
Chance M.R.A.   110, 179, *345*
Chapais B.   9, 101, 112, 171–5, 200–21, 224, 234, 266, 267–78, 281–2, 286–9, *345, 365*
Chapman M.   252, *345*
Charlesworth B.   274, 277, *345*
Charnov E.L.   274, 277, *345*
Cheney D.L.   10, 12, 22, 36, 47, 58, 63, 69, 73, 74, 83, 84, 85, 86, 88, 92, 96, 108, 110, 111, 128, 132, 146, 153, 154, 157, 162, 163, 169, 179, 184, 186, 188, 189–90, 228, 232, 233–49, 251, 252, 253, 259, 277, 278–86, 290, 291, 292, 293–4, 295, 297, 312, 327, 336, (160, Seyfarth *et al.* 1978), *345, 366*
Chepko-Sade B.D.   132, 181, 221, 224, 232, 240, 261, *345*
Chism J.   146, *346*
Clark D.L.   93, *346*
Clark-Wheatley C.B.   (63, 164, 168, 251, 272, 273, Silk *et al.* 1981a), *366*
Cleveland J.   *367*
Clutton-Brock T.H.   9, 10, 200, 229, 231, 259, 273, 328, (259, Harvey *et al.* 1978), *346, 352*
Coe C.L.   *354*
Collins D.A.   229, *346*
Collins M.L.   (121, Suomi *et al.* 1973), *368*
Colvin J.D.   19, 20–7, 57–64, 160–71, 190–200, (58, 191, 195, 196, 197, 199, Tissier *et al.* in prep.), *346, 369*
Crook J.H.   146, 153, 177, 181, 232, 252, 259, 300, 308, *346*
Croze H.   315, 323, *346*
Cubicciotti D.D.III   45, *346*
Cummins M.S.   122, *346*
Cushing K.   (132, 157, 251, 272, 273, Sade *et al.* 1976), *365*
Cushing P.   (132, 157, 251, 272, 273, Sade *et al.* 1976), *365*
Czaja J.A.   10, *347*

Datta S.B.   2, 10, 20, 35, 36, 63, 93–112, 140, 145, 178, 179, 275, 289–97, *347, 353*
Davies L.   73, 78, *353*
Davis A.E.   (272, Sackett *et al.* 1975), *364*
Dawkins R.   36, *347*
Deag J.M.   10, 37, 46, 104, 105, 146, 153, 177, 178, 179, *341, 347*
Defler T.R.   181, *347*
Delson E.   9, *369*
Delvoye P.   56, *347*
Demaegd M.   (56, Delvoye *et al.* 1978), *347*
Deutsch L.C.   (182, Baker *et al.* 1981), *341*
DeVore I.   9, 104, 259, 263, 270, 283, 338, *347, 350, 357, 362, 370*

Dienske H. 83, 154, *369*
Dillon J.E. 93, *346*
Dittus W.P.J. 93, 165, 225, 238, 258, 259, 260, 272, 284, *347*
Dobrofsky M. *342*
Dolhinow W. 73, 82, 273, 299, *347*
Douglas-Hamilton I. 315, 316, 329, *347*
Douglas-Hamilton O. 329, *347*
Drickamer L.C. 94, 132, 160, 163, 169, 171, 238, 239, 241, 243, 244, 258, 267, 269, 272, 273, *348*
Duggleby 79
Dunale J. (132, 157, 251, 272, 273, Sade *et al.* 1976), *365*
Dunbar E.P. 94, 177, 231, 232, 300, 301, 302, 304, 307, *348*
Dunbar R.I.M. 13, 15, 37, 47, 65, 94, 98, 165, 168, 177, 179, 181, 183, 187, 188, 192, 198, 225, 226, 231, 232, 251, 253, 272, 299–307, (177, 181, Aldrich-Blake *et al.* 1971), *340, 348, 355, 361, 366*
Dunn J. 134, 138, *348*
Dunn T.K. (177, 181, Aldrich-Blake *et al.* 1971), *340*
Duoos D. (146, Johnson *et al.* 1980), *355*

Eberhart A. (94, Keverne *et al.* 1982), *356*
Eisele S.G. (10, Czaja *et al.* 1975), (54, 121, Ruppenthal *et al.* 1974), *346, 364*
Eisenberg J.F. 177, 181, 200, 322, 325, 327, 334, *348, 356*
Ellis J.E. (252, Crook *et al.* 1976), *346*
Emlen S.T. 233, 326, 334, *348*
Emory G.R. (110, 179, Chance *et al.* 1977), *345*
Endler N.S. 28, *348*
Enomoto T. 241, *349*
Epple G. 122, *349*
Ervin F. (180, Estrada *et al.* 1977), *349*
Ervin F.R. (83, Raleigh *et al.* 1979), *362*
Esrin M. (146, Johnson *et al.* 1980), *355*
Estrada A. 46, 180, *349*
Estrada R. 180, *349*

Fady J-C. 154, 188, 194, *349*
Fagen R.M. 82, *349*
Fairbanks L.A. 179, 181, 183, 185–6, 192, 252, *349*
Falett J. (63, 195, Kummer *et al.* 1978), *357*

Farhenbruck E.E. (272, Sackett *et al.* 1975), *364*
Farres A.G. (178, Haude *et al.* 1976), *352*
Fedigan L.M. 36, 83, 86, *349*
Festinger L. 221, *349*
Figueroa A. (132, 157, 251, 272, 273, Sade *et al.* 1976), *365*
Flanigon M. 15, *363*
Flannery J.W. (83, Raleigh *et al.* 1979), *362*
Fletemeyer J.R. 180, *349*
Forsyth E. 221, *349*
Forthman D.L. (176, Rhine *et al.* 1979), *363*
Fossey D. (311, 314, Harcourt *et al.* 1976, 1980, 1981), *351*
Frame G.W. 329, 330, *349*
Frame L.H. 329, 330, *349*
Fricke H.W. 328, *349*
Funder D.C. 29, *342*
Furuya F. 181, 232, *349*
Furuya Y. 232, *349*

Gadgil M. 277, *349*
Galdikas B.M.F. 68, 308, *349*
Gamboa G.J. 327, *363*
Gartlan J.S. 9, 146, 231, 300, *346, 349*
Gaston A.J. 330, *349*
Gericke H. 224, *350*
Ginsburg B. 103, 110, *350*
Glick B.B. 63, *350*
Gluck J. (15, Sackett *et al.* 1978), *364*
Goldfoot D.A. 47, 53, *350*
Goodall J. 68, 69, 181, 267, 309, 310, 335, *350, 363*
Gordon T.P. 122, 289, (122, Bernstein *et al.* 1974; Rose *et al.* 1975), *342–3, 350, 363*
Goss-Custard J.D. (252, Crook *et al.* 1976), *346*
Götz W. (52, 153, 195, Kummer *et al.* 1974), *357*
Gouzoules H. 112, 179, 281, *350*
Goy R.W. 53, 179, (10, Czaja *et al.* 1975), (140, Holman *et al.* 1982), *346, 350, 354, 371*
Graber J.G. (178, Haude *et al.* 1976), *352*
Grady C.L. (122, Gordon *et al.* 1979), *350*
Greenwood P.J. 160, 241, 310, 329, *350*
Grewal D.S. 46, 180, 200, *350*

Guiatt 188
Guiness F.E. (328, Clutton-Brock *et al.*
    1982), *346*
Gutstein J. 165, *351*

Hager J.L. 338, *365*
Hall K.R.L. 9, 82, 165, 259, 270, 283,
    347, 350
Hall-Craggs J. (201, 209, 303, Morgan *et
    al.* 1976), *360*
Hamilton W.D. 183, 252, 261, 289, 294,
    341, 350
Hamilton W.J. 10, 73, 146, 165, 232,
    235, 236, 237, 260, 261, 263, (180,
    Buskirk *et al.* 1974), *344, 351*
Hanby J.P. 47, 48, 168, (329, Bygott *et al.*
    1979; 201, 209, 303, Morgan *et al.*
    1976), *344, 351, 360*
Harcourt A.H. 9, 18, 46, 160, 165, 176,
    299, 307–14, 329, *351*
Harding R.S.O. 13, *351*
Harlow H.F. 53, 56, 70, 121, 122, (54,
    121, 122, Ruppenthal *et al.* 1974, 1976),
    (121, Suomi *et al.* 1973), *351, 363, 364,
    368–9*
Harlow M.K. 70, (54, 121, Ruppenthal *et
    al.* 1974), *351, 364*
Harrington J.E. 326, *351*
Harris P.J. 176, *368*
Hartl D.L. 248, *351*
Harvey P.H. 9, 10, 200, 229, 231, 259,
    273, (160, Greenwood *et al.* 1979), *346,
    350, 352*
Hasegawa T. 73, 110, 169, *352*
Haude R.H. 178, *352*
Hausfater G.H. 10, 96, 116, 225, 232,
    234, 235, 236, 237, 238, 240, 244, 251,
    252, 261, 267, 270, 284, 305, 310, (10,
    Altmann *et al.* 1977), (258, Post *et al.*
    1980), *340, 352, 362, 365, 367*
Headley P.M. (177, 181, Aldrich-Blake
    et al. 1971), *340*
Hendrichs H. 315, 316, *352*
Hendricks A.G. 10, *352*
Henzi S.P. 235, 236, 241, *352*
Herbert J. 52, *353*
Herrmann J. 16, *353*
Hill W.C.O. 9, *352*
Hinde R.A. 1–7, 14, 16, 17–20, 21, 22,
    24, 26, 28, 35, 45–7, 55, 56, 63, 65–73,
    78, 79, 91, 113, 115, 121–2, 123, 128–

31, 152–4, 176–82, 250–5, 298–9,
    308, 334–9, (146, Rowell *et al.* 1964),
    (160, Seyfarth *et al.* 1978), *352–3, 364,
    366, 367*
Hiraiwa M. 46, 73, 110, 169, *352, 353*
Hoff M.P. 68, *353*
Hölldobler B. 327, 328, *353*
Holm R.A. (272, Sackett *et al.* 1975), *364*
Holman S.D. 140, *354*
Holmes J.G. (336, Lerner *et al.* 1976),
    357
Holmes W.G. 326, (181, Wu *et al.* 1980)
    354, 371
Homans G.C. 69, 179, 188, 336, *354*
Homewood K.M. 231, 233, *354, 370*
van Hooff J.A.R.A.M. 152, (105, 152, de
    Waal *et al.* 1976), *354, 370*
Hooley J.M. 134, 139–45, 146, (19, 53,
    54, 55, 56, Simpson *et al.* 1981), *354,
    367*
Hopf S. 70, *354*
Horn H.S. 259, *354*
Horvat J.R. *354*
Howe S. *366*
Hrdy D.B. 179, *354*
Hrdy S.B. 146, 162, 179, 252, *354*
Hull C.H. (31, Nie *et al.* 1975), *361*
Humphrey N.K. 68, 153, *354*

Imanishi K. 255, *354*
Irons W. 339, *354*
Isaac G.L. 338, *354*
Itani J. 153, 241, 309, *355*
Itoigawa N. 163, 169, 236, 238, 243, *355*
Iwamoto T. 226, *355*
Izawa K. 180, *355*

Jainundeen M.R. (322, Eisenberg *et al.*
    1971), *348*
Jaquette D. 27, *365*
Jay P. 146, *355*
Jenkins J.G. (31, Nie *et al.* 1975), *361*
Johnson C.K. 146, *355*
Jolly A. 9, 300, *355*
Jones C.B. 178, 179, *355*

Kaplan J.R. 106, 110, 180, 289, 290,
    291, 293, 294, (132, 157, 251, 272, 273,
    Sade *et al.* 1976), *355, 365*

Katz L.   221, *349, 355*
Kaufman I.C.   79, *355*
Kaufmann J.H.   37, 47, 171, 174, 235, 267, *355*
Kavanagh M.   (259, Harvey *et al.* 1978), *352*
Kawai M.   36, 94, 234, 290, 291, 310, *355*
Kawamura S.   94, 295, *356*
Kawanaka K.   160–1, 163, 283, 310, *356*
Kellerman H.   (29, Buirski *et al.* 1973, 1978), *344*
Kelley H.H.   309, 336, *356, 369*
Kendrick C.   134, 138, *348*
Kenrick D.T.   29, *356*
Keverne E.B.   94, *356*
Kleiman D.G.   327, *356*
Klein D.J.   39, *356*
Klingel H.   329, *356*
Koerner C.   (146, Johnson *et al.* 1980), *355*
Koford C.B.   9, 170, 171, *356*
Konner M.J.   337, *356*
Koyama N.   94, 110, 112, 234, 236, 238, 240, 281, *356, 361*
Kraemer D.C.   10, *352*
Kraemer H.C.   15, *356*
Krebs J.R.   186, *356*
Kruuk H.   329, 331, 333, *356*
Kummer H.   52, 63, 66, 68, 128, 145, 152–3, 176, 177, 180, 181, 183, 186, 188, 193, 195, 232, 298, 300–1, 302, 304, 305, 306, 335, *341, 356–7*
Kurland J.A.   180, 185, 235, 253, 275, 278, 295, *357*
Kuroda S.   309, *357*

Lack D.   251, 328, *357*
Lancaster J.B.   86, 146, 151, *357*
Lauer C.   232, (132, 157, 251, 272, 273, Sade *et al.* 1976), *357, 365*
van Lawick H.   (329, 330, Frame *et al.* 1979), *349*
Laws R.M.   315, 323, *357*
Leary R.W.   (106, Biernoff *et al.* 1964), *343*
Lee P.C.   8–16, 35–44, 73–9, 82–9, 110, 111, 134–9, 145–51, 179, 225–9, 231–3, 295, (169, 186, 234, 239, 240, 251, 259, 277, 279, 283, 284, Cheney *et al.* 1981), *345, 357, 361*

Lee R.B.   338, *357*
Lehrman D.S.   28, *357*
Leighton-Shapiro M.E.   (122, Hinde *et al.* 1978), *353*
Lerner M.J.   336, *357*
Leuthold W.   316, 327, 328, *357–8*
Levine S.   *354*
Levitt P.R.   (222, 223, Arabie *et al.* 1978), *341*
Lindberg   241
Lindburg D.G.   55, 267, *358*
Lindsay W.K.   316, 318, *358*
Littman R.A.   (106, Biernoff *et al.* 1964), *343*
Loizos C.   82, 88, *358*
Lorenz K.   7, 88, *358*
Loy J.   58, 82, 154, 159, 166, *358*
Loy K.   58, 154, 159, 166, *358*
Lucas J.W.   235, 236, 241, *352*
Luce R.D.   221, *358*
Lumsden C.   339, *371*
Lund M.   336, *358*

McArthur P.D.   326, *358*
McCall G.J.   335, *358*
McCann T.S.   328, *358*
McCracken G.F.   327, 329, *358*
McCuskey S.A.   (10, Altmann *et al.* 1977), (258, Post *et al.* 1980), *340, 362*
Macdonald D.W.   259, 327, 329, 358
McGinnis L.   73, (122, Hinde *et al.* 1978), *353, 358*
McGrew W.C.   68, 309, *358*
McKay G.M.   (322, Eisenberg *et al.* 1971), *348*
McKenna J.J.   (273, Dolhinow *et al.* 1979), *347*
Mackinnon J.R.   308, *358*
McMillan C.A.   132, *359*
Maeda Y.   179, *361*
Magnusson D.   28, *348*
Malcolm J.R.   (329, 330, Frame *et al.* 1979), *349*
Marsden H.M.   110, 162, 232, *359*
Marsh C.W.   160, 165, *359*
Martin P.   88, 228, *359*
Maslow A.H.   179, *359*
Mason W.A.   45, 122, 255, (128, Vaitl *et al.* 1978), *341, 346, 359, 362, 370*
Massey A.   110, 180, 278, 289, 293, *359*

Maxim P.E.   180, *359*
May R.M.   277, *359*
Maynard Smith J.   251, 254, 255, 267, 273, 281, 282, *359*
Mech L.D.   *359*
Medina S.R.   (181, Wu *et al.* 1980), *371*
Meikle D.B.   163, 238, 239, 243, *359*
Melnick D.   238, 241, 277, *359*
Menzel E.W.   68, 335, *360*
Michener C.D.   327, *353*
Michener G.R.   327, 329, 330, *360*
Miller D.T.   (336, Lerner *et al.* 1976), *357*
Miller R.E.   (94, Keverne *et al.* 1982), *356*
Mineka S.   71, 73, 79, *360*
Mischel W.   28, *360*
Missakian E.A.   55, 94, 96, 110, 140, 174, 275, *360*
Mitani J.C.   232, *360*
Mitchell G.D.   46, 53, 61, 89, *344, 360, 363*
Moehlman P.D.   329, *360*
Moore J.   10, 229, *360*
Morgan B.J.T.   201, 209, 303, *360*
Mori A.   69, 258, 301, *360*
Mori U.   258, 301, *360*
Moss C.J.   315–25, *360, 362*
Muckenhirn N.A.   (177, 181, 200, Eisenberg *et al.* 1972), *348*
Mueller E.C.   26, *370*
Murdock G.P.   314, *360*
Murray M.G.   *360*

Nagel U.   178, *360*
Napier J.R.   9, 174, *361*
Napier P.H.   9, 174, *361*
Nash L.T.   13, 179, *361*
Negayama K.   46, *361*
Nelto W.J.   (105, 152, de Waal *et al.* 1976), *370*
Nicolson N.   262, 263, *361*
Nie N.H.   31, *361*
Nishida T.   309, 310, *356, 361*
Noonan K.M.   338, *340*
Norikoshi K.   236, 238, *361*

Ober C.   (180, Olivier *et al.* 1981), *361*
O'Hara R.K.   327, *361*
Ohsawa H.   225, *361*
Oki J.   179, *361*

Oliver J.I.   36, 82, 111, 179, 226, 228, 229, 295, *357, 361*
Olivier T.J.   132, 180, 181, 224, 232, 240, *345, 361*
Oring L.W.   233, 334, *348*
Owens N.W.   13, 83, 86, 194, *361*

Packer C.R.   9, 13, 36, 69, 93, 116, 146, 153, 160, 161, 162, 165, 180, 225, 232, 235, 236, 237, 238, 239, 241, 243, 244, 248, 251, 252, 253, 263, 266, 267, 270, 278, 283, 284, 305, *361–2*
Parker G.A.   98, 102, 110, 254, 267, 273 281, 282, *359, 362*
Parker I.S.C.   315, 323, *357*
Payne R.G.   (110, 179, Chance *et al.* 1977), *345*
Pearl M.   241, *359*
Pearl M.C.   223, 224, *362*
Perkins M.N.   180, *344*
Perrins C.M.   (160, Greenwood *et al.* 1979), *350*
Perry A.D.   221, *358*
Phillips M.J.   45, *362*
Phoenix C.H.   (47, 48, Hanby *et al.* 1971), *351*
Pilbeam D.   9, *362*
Plutchik R.   (29, Buirski *et al.* 1973, 1978), *344*
Poirier F.E.   146, 165, *362*
Poole J.H.   315–25, *362*
Popp J.L.   104, *362*
Porter R.H.   326, *362*
Post D.G.   258, *362*
Pusey A.E.   73, 162, 180, 181, 236, 237, 238, 243, 266, 284, 310, *362*

Quiatt D.   *362*

Radesäter T.   326, *362*
Raleigh   M.J.   83, *362*
Ransom B.S.   112, 146, 263, *363*
Ransom T.W.   112, 146, 263, *363*
Rasmussen D.R.   165, 231, 252, *363*
Rasmussen K.L.R.   9, 12, 47–53, 112, 116–20, 225, 305, *363*
Redican W.K.   46, *363*
van Rhijn J.G.   255, *363*
Rhine R.J.   15, 176, 183, *363*

Rhodes D. (132, 157, 251, 272, 273, Sade *et al.* 1976), *365*

Richard A. 82, 225, *363*

Richards S.M. 9, 36, *363, 364*

Rijksen H.D. 308, *363*

Riss D. 310, *363*

Robertson L.T. (47, 48, Hanby *et al.* 1971), *351*

Robinson J.G. 232, *363*

Robyn C. (56, Delvoye *et al.* 1978), *347*

Rodman P.S. 232, 308, (263, Bernstein *et al.* 1981), (55, 63, 145, 164, 168, 183, 186, 187, 239, 251, 272, 273, Silk *et al.* 1981a, 1981b), *343, 360, 363, 366*

Rood J.P. 327, 330, *363*

van Roosmalen A. 65, *370*

Rose R.M. 122, (122, Bernstein *et al.* 1974; Gordon *et al.* 1979), *343, 350, 363*

Rosenblum L.A. 56, 79, *355, 363*

Ross N.M. 327, *363*

Rowell T.E. 9, 35, 37, 47, 94, 104, 105, 110, 112, 146, 176, 177, 178, 181, *363, 364*

Rubenstein D.I. 277, 278, 326, *364, 365*

Rubin Z. 27, *364*

Rudnai J.A. 331, *364*

Rudran R. (177, 181, 200, Eisenberg *et al.* 1972), *348*

Ruppenthal G.C. 54, 121, 122, (15, Sackett *et al.* 1978), *364*

Saayman G.S. 235, 270, *364, 368*

Sabater P.J. (311, 314, Harcourt *et al.* 1981), *351*

Sackett G.P. 15, 55, 272, (54, 122, Ruppenthal *et al.* 1976), (181, Wu *et al.* 1980), *364, 371*

Sade D.S. 9, 36, 37, 58, 91, 94, 125, 132, 140, 155, 157, 166, 174, 178, 183, 185, 186, 187, 208, 221, 222, 224, 234, 240, 251, 261, 272, 273, 275, 295, 310, (180, Olivier *et al.* 1981), *345, 361, 364–5*

Sale J.B. 316, *358*

Samuels S.A. (55, 63, 145, 164, 168, 183, 187, 239, 251, 272, 273, Silk *et al.* 1981a,b), *366*

Sandeval J.M. 46, *349*

Sassenrath E.N. 94, 122, *365*

Saunders C. 251, *365*

Schaller G.B. 311, 313, 330, *365*

Schmidt C. (251, 267, Witt *et al.* 1981), *371*

Schmidt J. (251, 267, Witt *et al.* 1981), *371*

Schneider J. (132, 157, 251, 272, 273, Sade *et al.* 1976), *365*

Schulman S.R. 101, 177, 221–5, 234, 271–8, 281–2, *345, 362, 365*

Seligman M.E.P. 338, *365*

Selman R. 27, *365*

Seyfarth R.M. 12, 27, 36, 37, 47, 57, 58, 69, 93, 104, 106, 108, 110, 111, 112, 113, 132, 153, 160, 162, 179, 181, 182–90, 191–3, 195, 196–7, 200, 208, 211, 229, 232, 235, 236, 237, 238, 239, 240, 241, 242, 243, 244, 246, 252, 253, 284, 305, 312, 327, 336, (169, 186, 234, 239, 240, 251, 259, 277, 279, 283, 284, Cheney *et al.* 1981), *345, 365–6*

Sharman M.J. 13, 181, 232, *366*

Sharpe L.G. 37, 106, *343*

Sherman G.L. (182, Baker *et al.* 1981), *341*

Sherman P.W. 326, 330, *354, 366*

Short R. 338, *366*

Siddiqi M.R. (9, 221, Southwick *et al.* 1965), *367*

Sigg H. 68, 176, (63, 195, Kummer *et al.* 1978), *357, 366*

Silk J.B. 55, 57, 63, 145, 146, 150, 164, 168, 183, 186, 187, 192, 198, 239, 251, 253, 272, 273, 281, *366*

Simonds P.E. 165, 169, *366*

Simpson A.E. 15, 19, 53, 54, 55, 56, *366–7*

Simpson M.J.A. 14, 15, 18, 19, 26, 35, 46, 53–7, 79, 123, 125, 134, 139–42, 146, 178, 180, 310, 312, (201, 209, 303, Morgan *et al.* 1976), *341, 353, 354, 360, 366–7*

Slatkin M. 235, 237, 252, *367*

Small M.F. 146, 181, *367*

Smith D.G. 146, 181, 235, 244, 267, (263, Bernstein *et al.* 1981), *343, 367*

Smith K.S. (260, Hamilton *et al.* 1982), *351*

Smith S.M. 328, *367*

Smuts B.B. 110, 112–16, 181, 200, 231, 261, 262–6, 309, *367, 371*

Snowdon C.T. *367*

Southwick C.H. 9, 221, *367*

Spencer-Booth Y. 18, 21, 29, 57, 71, 73, 89, 91, 122, 142, 146, (146, Rowell *et al.* 1964), *353, 364, 367*

Stacey P.B. 259, 326, *367*

Stammbach E.   94, 183, 185, 192, *367*
Stein D.M.   263, 266, *367*
Steinbrenner K.   (31, Nie *et al.* 1975), *361*
Stephenson G.R.   267, *367*
Stevenson-Hinde J.   5, 28–35, 53, 55, 68, 122–7, 143, 179, 336, 338, *353, 367–8*
Stewart K.J.   46, 82, 165, 299, 307–14, 329, (311, 314, Harcourt *et al.* 1976, 1980), *351, 368*
Stillwell-Barnes R.   (29, 32, 33, 34, 35, 125, 143, Stevenson-Hinde *et al.* 1980a,b), (176, Rhine *et al.* 1979), *363, 367–8*
Stolba A.   68, 302, *366, 368*
Stolte L.A.M.   268, *368*
Stoltz L.P.   235, *368*
Strayer F.F.   176, 180, *368*
Stringfield D.O.   29, *356*
Strum S.C.   13, 112, 200, 263, 266, 287, *368*
Stuhsaker T.T.   9, 10, 37, 110, 146, 232, 235, 236, 252, 283, *368*
Sugiyama Y.   161, 163, 232, 235, 236, 238, 241, 243, 283, 284, 310, *368*
Suomi S.J.   71, 73, 79, 121, 122, 180, (54, 121, 122, Ruppenthal *et al.* 1974, 1976), *346, 360, 364, 368–9*
Symmes D.   110, *370*
Symons D.   82, 86, *369*
Szalay F.S.   9, *369*

Tagiuri R.   30, *344*
Tajfel H.   337, *369*
Takahata Y.   200, *369*
Tartabini A.   83, 154, *369*
Taub D.M.   169, 267, (129, Vaitl *et al.* 1978), *369, 370*
Taylor D.A.   336, *340*
Tenaza R.R.   232, *369*
Thibaut J.W.   336, *369*
Thommen D.   252, *341*
Thompson D.B.   (182, Baker *et al.* 1981), *341*
Tilson R.L.   326, 327, *371*
Tinbergen N.   3, 7, *369*
Tissier G.   21, 58, 61, 191, 195, 196–7, 199, *346, 369*
Trivers R.L.   134, 233, 253, 254, 256, 257, 294, 334, *369*
Tsumori A.   252, *369*

Tutin C.E.G.   266, (68, 309, McGrew *et al.* 1979, 1981), *358, 369*

Vaitl E.A.   129–30, 188, 310, 311, *369–70*
Vandell D.L.   26, *370*
Varley M.   110, *370*
Vehrencamp S.L.   233, 327, 330, 331, 334, *343, 370*
Vessey S.H.   160, 163, 169, 171, 235, 238, 239, 241, 243, 244, *348, 359, 370*
Vodegel R.   255, *363*
Vogt J.L.   176, *370*
Vonder Haar Laws J.   (273, Dolhinow *et al.* 1979), *347*

de Waal F.B.M.   65, 68, 69, 104, 105, 110, 152, 153, 179, 180, 181, 289, 290, 291, 293, 335, *354, 370*
Wade T.D.   19, 169, 181, *370*
van Wagenen G.   268, *345*
Walters J.   10, 14, 36, 37, 74, 78, 98, 110, 154, 159, 253, 290, 295, 297, *340, 370*
Waser P.M.   231, 233, *370*
Washburn S.L.   9, *370*
Watanabe K.   152, 180, *370*
Waterman P.G.   258, *371*
Watts D.   (314, Harcourt *et al.* 1980), *351*
Weigel R.M.   277, *370*
Weininger R.   (29, Buirski *et al.* 1973), *344*
Weisbard C.   179, *371*
Welker C.   66, *371*
Western D.   315, *371*
Westlund B.J.   (176, Rhine *et al.* 1979), *363*
Westlund H.D.   (176, Rhine *et al.* 1979), *363*
White H.C.   222, 223, 224, *343, 371*
White L.E.   82, 86, *371*
White R.E.C.   14, 123, *371*
Whitten P.   258, *371*
Williams G.C.   251, *371*
Wilson A.P.   162, 181, 236, 238, 240, 243, 283, *343*
Wilson E.O.   311, 325, 327, 328, 331, 339, *353, 371*
Witt R.   251, 267, *371*
Wittenberger J.F.   326, 327, *371*
Wrangham R.W.   160, 181, 184, 186, 200, 220, 228, 231, 234, 235, 236–7,

239, 252, 255–62, 272, 287, 309, 311,
313, 314, 325–34, *371*
Wu H.M.H.　181, *371*
Wynne-Edwards V.C.　251, *372*
Wyrick M.　326, *362*

Yamada M.　180, 188, *372*
Yarrow L.J.　29, *372*
Yarrow M.R.　29, *372*

Yoshiba K.　326, *372*
Yoshihara D.L.　(140, Holman *et al.*
1982), *354*

Zimen E.　331, *372*
Zucker E.L.　299, *372*
Zunz M.　29, (19, 53, 54, 55, 56, Simpson
et al. 1981), (29, 32, 33, 34, 35, 125, 143,
Stevenson-Hinde *et al.* 1980a,b), *367–8*

Subject index

Affiliative behaviour 23–7, 48–53, 57–62, 116–20, 209–16, 309–10, 319–20
Agonistic buffering 153, 266
Aids (*see* Alliances)
Alliances 22–7, 41–4, 57–61, 68, 98–112, 154–9, 179, 206–8, 214–19, 248, 278–97, 331
Allomothering 145–51
Alouatta 179
Alternative strategies 254, 339
Altruism 252–3, 289
Amboseli National Park 13, 315
Anemone fish 328
Anis 33
Approaches–leavings 16, 61, 67, 115, 144–5
Attention structure 179
Attribution 336

Babbler 329, 330
Baboon
  gelada 47, 66, 183, 187, 192, 198, 231–2, 272, 299–307
  hamadryas 52, 68, 128, 152, 177, 180, 183, 231–2, 299–307
  savanah 8–10, 47–53, 73, 112–20, 162, 179, 181, 183, 225–9, 232, 234–41, 243, 262–6, 278, 283, 305, 331
Balance theory 336
Birth interval 53–7
Blood-relatedness (*see also* Kin selection) 89–92, 180–2, 208–17, 247–9

Callicebus 45
Cayo Santiago 10–12
Cebus 180
Chikadees 328
Chimpanzee 68, 178, 180–1, 234, 307–14
Chimpanzee, Pigmy 308–9
Coalitions (*see* Alliances)
Cognitive capacities 68, 189–91, 242–3, 306, 327, 334–5

Competition 38–43, 77–9, 226–9, 232, 234–41, 258
  for grooming partners 182–200
Conflict, agonistic 254
  mother–infant 66–8, 253–4, 338
Consortship 48–53, 116–20, 262–6
Culture xi, 180, 334–9

Data and theory language 3–4, 19–20
Description
  behaviour 1, 17–19
  interactions 1, 18
  relationships 2, 18–19
  social structure 3, 176–8
Development, analysis of 65–70
Dialectic
  personality versus relationships 5, 28–30, 45–7, 66, 121–7, 337
  relationships versus social group 5–6, 47, 128–51, 337
Dominance 20, 35–44, 178–80
  acquisition of 36, 93–103, 154–9, 171–5
  and alliances 279–86, 289–97
  and attractiveness 148–51, 182–219
  and familiarity 190–200
  and female–female relationships 208–19
  and grooming 182–208, 312
  and intergroup encounters 234–41
  maintenance of 103–12
  and male emigration 160–71, 238–9
  and male–female relationships 200–8
  and mating success 267–78
  and matriline membership 132–4, 171–5
  in orphans 78
  and peer relationships 57–64
  reversals 41–3
  and sex of infant 56–7
  and social structure 178–200, 228–9
  stability of 310–11
Dragonfly 326

Elephant 315–25, 329

Elephant seal   327
Emergent properties   2–3, 19, 298, 335
Emigration, male   160–71, 238–9,
    241–9
Evolutionary stable strategy   254, 281
Exchange   68–9, 179–80, 336

Familiarity   190–200, 261
Feedback   69–73
Female-bonding   257
Female–female versus female–male
    bonding   300–2, 310–13
Female-resident versus male-resident
    species (*see also* Emigration, male)
    307–14
Four whys   xi, 3–7, 298–9
Fox   329
Friendship   26–7
Frogs   327
Fruit bat   329, 330
Function, concept of   250–1

Gibbon   232
Gilgil   13
Gombe Stream National Park   12
Gorilla   46, 165, 176, 234, 260, 307–14,
    329
Ground squirrel   330
Group-living, advantages of   255–62,
    313
Group selection   7, 252–3

Herding   118, 235, 300–2
Home range   231–49
Hyaena   329
Hybrids   178
Human species   334–9
Human versus non-human primates
    xi–xii, 189–90, 334–9

Implicit personality theories   30
Inbreeding avoidance   247–9
Individual characteristics   4, 28–44
    consistency across situations   34–5
        mother and offspring   53, 122–7
        parity and   143
    consistency over time   30–3
Infanticide   251–2
Interactions
    content of   18, 20–7
    quality of   18, 24–5
Interdependence theory   336

Intergroup relationships   6, 162, 231–49,
    310–13, 315–22
Inter-personal perception   25–6, 335–6
Institutions   337
Investment   180, 336
Irrational fears of childhood   338

Jackal   329
Jay, Mexican   329
Jealousy   45, 134, 139–42
Junco   182

Kin selection   183, 186–8, 252–3,
    261–2

Langur   146, 165, 179, 273
Leadership   180, 302
Lemur   225
Lion   329

Macaque
    Barbary   46, 178
    bonnet   164, 183, 186–7, 192, 198,
        253, 272–3
    Japanese   46, 48, 73, 162, 163, 169,
        180, 232, 235–6, 238, 273
    rhesus   8–10, 20–7, 30–5, 53–64,
        66–7, 70–3, 79–81, 89–112,
        121–7, 130–4, 139–45, 154–75,
        178, 181, 183, 200–21, 222–5, 232,
        234–41, 243, 267–78, 286–9,
        290–7
    stump-tail   46, 183
    toque   225, 272
Madingley   10
Mangabey   225, 233
Marriage   339
Mating success   165, 233–4, 239,
    247–9, 267–78, 294–5, 304
Maximum spanning tree   201–3, 209
Meshing   26
Mikumi National Park   12
Mongoose   330
Mother–infant relationship
    age changes in   46
    caretaking and   145–51
    conflict in   66–8, 253–4, 338–9
    death of mother   73–81
    dominance of mother   56, 132–4,
        171–5
    dynamics of   66–8, 70–3
    feedback in   70–3
    'ideal' mothering   254

matriline differences   132–4
and others   89–93
parturition and   134–9
and peer relationships   157–9, 171–5
separation   70–3
sex of infant   53–7, 143–5
siblings and   139–46
social situation and   130–1
Mountain Zebra National Park   12
Musth   318, 322–3

Norms   69, 336–7
Notifying   302

Orang-utan   308–9, 313
Orphans   73–81, 260

Parent–offspring conflict   66–8, 253–4, 338–9
Parity and behaviour   47–53, 142–5
Patas monkey   146, 165, 176
Personality (*see* Individual characteristics)
Play   22–7, 81–9
   sex differences in   84–9
Population control   251
Prairie dog   329, 330
Predation   258–9
Principles of explanation   3
   causal   3–5
   developmental   5, 65–70
   evolutionary   6–7
   functional   6–7, 250–5
Protection   260, 313

Rank (*see* Dominance)
Reciprocity versus complementarity   19
Reciprocity of preference   21–7, 116–20
Red deer   328
Resource holding potential   254–5, 267
Relationships (*see also* Alliances,
   Mother–infant relationship, Triadic
   interactions)
   consort (*see* Consortship)
   description   2, 18–19, 20–7
   development of   65–120
   differentiation of   89–93
   dynamics of   4–5, 65–120, 303–4
   female–female   89–116, 157–9,
      239–40, 266, 275–97
   male–female (*see also* Consortship)
      129–30, 200–8, 288–9
   male–infant   79–80
   male–male   166, 243–8
   peer   20–7, 57–64, 82–9, 90–3

sibling   20–7, 75–9, 134–42, 179
social influences on   130–51
'special' male–female   112–16, 262–6

Relatives and rank acquisition   157–9
Rhesus monkey (*see* Macaque)
Roles   336
Ruaha National Park   12
Ruff   326

Saimiri (squirrel monkey)   45–6, 128–30, 225
Schedule feeding   338
Seasonality   82–3, 181, 225–9
Situation versus individual variables   28–30
Social approval   68–9, 179, 336
Social insects   252, 328
Social networks   89–93
Social structure (*see also* Dominance)
   2–3, 89–91, 176–229
   block-model approach to   221–5
   demography and   181, 225, 303–4
   description of   176–8
   ecology and   181, 225–9
   of elephants   315–25
   and grooming   182–200
   and oestrus   305–6
   and parturition   305
   principles   178–82, 300–6, 336–9
   species differences   177–82, 300–6 308–14
   subgroups   220–1, 300–2
Status (*see* Dominance)

Techniques   13–16
   assessment of individual characteristics   29–30
Territory   231–49
Titi monkey   232
Triadic interactions (*see also* Alliances)
   103–12, 152–75, 206–8, 214–19
Tripartite interactions   152–3, 206–8

Ultimate factors   250–97

Vervet monkey   9–10, 13, 35–44, 58, 73–9, 82–9, 134–9, 145–51, 153, 162, 185–8, 189, 225–9, 232, 234–49, 251, 272, 278–86

Weaning conflict   7
Wild dogs   329, 330–1
Woodpecker, acorn   326

Zebra   329